C0-ALP-137

Biological Magnetic Resonance
Volume 15

In Vivo Carbon-13 NMR

Biological Magnetic Resonance
Volume 15

In Vivo Carbon-13 NMR

Edited by

Lawrence J. Berliner and
Pierre-Marie Robitaille

Ohio State University
Columbus, Ohio

KLUWER ACADEMIC / PLENUM PUBLISHERS
NEW YORK, BOSTON, DORDRECHT, LONDON, MOSCOW

The Library of Congress has catalogued the first volume of this series as follows:

Main entry under title:

Biological magnetic resonance:

Includes bibliographies and indexes.
1. Magnetic resonance. 2. Biology–Technique. I. Berliner, Lawrence, J. II. Reuben, Jacques.
QH324.9.M28B56 574.19'285 78-16035
AACR1

ISBN 0-306-45886-1

233 Spring Street, New York, N.Y. 10013

10 9 8 7 6 5 4 3 2 1

A C.I.P. record for this book is available from the Library of Congress.

Printed in the United States of America

To Noël Antoine Robitaille

Contributors

Edwin M. Chance • Department of Radiology and Radiological Sciences, Division of NMR Research, The Johns Hopkins University School of Medicine, Baltimore, Maryland 21205-2195; and Department of Biochemistry and Molecular Biology, University College London, Gower Street, London WC1E 6BT, England

John C. Chatham • Department of Radiology and Radiological Sciences, Division of NMR Research, The Johns Hopkins University School of Medicine, Baltimore, Maryland 21205-2195

Gary Gerstenblith • Carnegie 584, The Johns Hopkins Hospital, 600 N. Wolfe Street, Baltimore, Maryland 21287-6568

Joanne K. Kelleher • Department of Physiology, Ross Hall 450, The George Washington University Medical Center, 2300 Eye Street, N.W., Washington, D.C. 20037

Maren R. Laughlin • NIDDK, Natcher, 451 5AN-24J, 45 Center Drive, MSC 600, Bethesda, Maryland 20892-6600

E. Douglas Lewandowski • NMR Center, Massachusetts General Hospital, Harvard Medical School, Bldg. 149, 13th Street, Charlestown, Massachusetts 02129

Craig R. Malloy • Departments of Radiology and Internal Medicine, University of Texas Southwestern Medical Center at Dallas and Dallas Veterans Affairs Medical Center, The Mary Nell and Ralph B. Rogers Magnetic Resonance Center, 5801 Forest Park Road, Dallas, Texas 75235-9085

Graeme F. Mason • Magnetic Resonance Center, Yale University, School of Medicine, 333 Cedar Street, P.O. Box 208043, New Haven, Connecticut 06520-8043

Pierre-Marie Luc Robitaille • Departments of Medical Biochemistry and Radiology, The Ohio State University, Columbus, Ohio 43210

A. Dean Sherry • Department of Chemistry, University of Texas at Dallas, P.O. Box 830688, Richardson, Texas 75083-0688; and the Department of Radiology, University of Texas Southwestern Medical Center, The Mary Nell and Ralph B. Rogers Magnetic Resonance Center, 5801 Forest Park Road, Dallas, Texas 75235-9085

Robert G. Weiss • Carnegie 584, The Johns Hopkins Hospital, 600 N. Wolfe Street, Baltimore, Maryland 21287-6568

Preface

This volume is dedicated to a rapidly evolving and maturing field in MRI/MRS, namely, the use of the stable isotope ^{13}C to probe the chemistry, mechanism, and function in living sytems. We are honored to have brought together contributions from some of the world's foremost experts in this field who have provided the broad leadership to bring this technique to where it is today. Maren Laughlin and Joanne Kelleher introduce tracer theory and the suitability of ^{13}C NMR to this end. Dean Sherry and Craig Malloy address metabolism in their coverage of ^{13}C isotopomer analysis of glutamate in the citric acid cycle. This is followed by an introduction to dynamic methods by John Chatham and Edwin Chance discussing the determination of metabolic fluxes by mathematical analysis of ^{13}C labeling kinetics. Dynamic methods are continued by Doug Lewandowski, who has contributed a chapter on metabolic flux and subcellular transport of metabolites. Bob Weiss presents an incisive chapter on assessing cardiac metabolic rates during pathologic conditions with dynamic ^{13}C NMR. Graeme Mason describes applications of ^{13}C labeling to studies of human brain metabolism. The volume is concluded with a chapter by one of the editors, Pierre-Marie Robitaille, on the dynamic analysis of flux and substrate selection in the tricarboxylic acid cycle.

We are extremely proud of this excellent compilation. As always we are open to suggestions, comments, and criticism from the reader for future volumes on this subject.

Lawrence J. Berliner
Pierre-Marie L. Robitaille

Contents

Chapter 1

Tracer Theory and ^{13}C NMR

Maren R. Laughlin and Joanne K. Kelleher

Chapter 3

Determination of Metabolic Fluxes by Mathematical Analysis of ^{13}C-Labeling Kinetics

John C. Chatham and Edwin M. Chance

Chapter 4

Metabolic Flux and Subcellular Transport of Metabolites

E. Douglas Lewandowski

Chapter 5

Assessing Cardiac Metabolic Rates during Pathologic Conditions with Dynamic ^{13}C NMR Spectra

Robert G. Weiss and Gary Gerstenblith

Chapter 6

Applications of ^{13}C Labeling to Studies of Human Brain Metabolism *in Vivo*

Graeme F. Mason

Chapter 7

In Vivo ^{13}C NMR Spectroscopy: A Unique Approach in the Dynamic Analysis of Tricarboxylic Acid Cycle Flux and Substrate Selection

Pierre-Marie Luc Robitaille

1

Tracer Theory and ^{13}C NMR

Maren R. Laughlin and Joanne K. Kelleher

1. INTRODUCTION

1.1. Overview

Researchers who use the stable isotope ^{13}C and ^{13}C NMR or GC-MS for the study of metabolism have drawn on the theory and models developed for radioactive ^{14}C tracer studies. In many ways, the strengths of these two carbon isotopes complement each other beautifully. ^{14}C has been used extensively to measure whole body uptake and production of metabolites such as glucose, fatty acids, or amino acids from the specific activity of these compounds in sampled blood, while liberated $^{14}CO_2$ in expired air is a good measure for ^{14}C-labeled substrate oxidation. ^{13}C is usually used for experiments in which the label is monitored in tissue, whether in situ or in biopsy. ^{13}C NMR has provided two very important advances in isotope studies. First, it is often trivial to determine the specific site of the labeled carbon in a given molecule, since each carbon has a unique chemical shift. In addition, labeled carbons that share a covalent bond have unique spectral patterns, and can

Maren R. Laughlin • NIDDK, Natcher, 451 5AN-24J, 45 Center Drive, MSC 600, Bethesda, Maryland 20892-6600. **Joanne K. Kelleher** • Department of Physiology, Ross Hall 450, The George Washington University Medical Center, 2300 Eye Street, N.W., Washington, D.C. 20037

Biological Magnetic Resonance, Volume 15: In Vivo Carbon-13 NMR, edited by L. J. Berliner and P.-M. L. Robitaille. Kluwer Academic / Plenum Publishers, New York, 1998.

thus be distinguished from a population of molecules with ^{13}C in the same sites, but different molecules. This makes it possible to detect multiple pathways with a common product, as long as the carbon is scrambled to unique positions in the molecule. Second, it is possible to follow the interconversion of labeled metabolites nondestructively in living tissue in real time. Direct measurements of ^{13}C compounds can be made in a variety of biological systems, including animals and people. These properties of biological ^{13}C NMR are beautifully illustrated in Cohen's studies of the perfused rat liver (Cohen, 1987a, b, c).

Physiological parameters are derived from tracer data through the use of mathematical models. There are a wide variety of models which are tailored to specific types of experiments, but in general they are all used to describe or predict the behavior of a biological system. Mathematical models have been applied to the analysis of biological systems since early in the 20th century. In the 1920s, Briggs and Haldane (1925) and Michaelis and Menten (1913) developed models of the behavior of isolated enzymes. For the generations since, science students have been introduced to the process of modeling by the rederivation of the classic Michaelis–Menten equation. With a simple spectrophotometer, pencil, and graph paper, one can estimate the two unknown parameters of this model, K_M and V_{max}. By the 1940s, however, it was clear that the complexity of biological systems would soon require scientists to obtain much more sophisticated tools for performing calculations and, in 1943, Britton Chance reported the first biochemical simulation of the Michaelis–Menten equation using a "differential analyzer," a forerunner of the stored program computer.

In parallel with the tools of enzyme kinetics and computing machines, the development of the use of isotopes as tracers opened new possibilities for understanding biochemistry, and presented new demands for quantitative tools. Stable isotopes were the first biological tracers. This new technique was pioneered by workers such as Shoenheimer and Rittenberg, who prepared deuterated triglycerides and analyzed their distribution in mice using mass spectrometry (1935a, b). In the 1940s radioisotopes became widely available, and detection of isotopes with liquid scintillation counting (LSC) was then more practical than any available method for detection of stable isotopes. The mathematical theory utilizing radiotracer data to estimate the biological turnover of a compound was advanced by the work of Zilversmit *et al.* (1943) and Sheppard (1948). By the 1960s, Berman *et al.* (1962), Garfinkel (1963), and others had begun to use digital computers and multicompartmental modeling to analyze radiotracer data in biologically complex systems.

In the 1970s, stable isotopes began to re-emerge as widely used tracers, and advances in computer, radiofrequency, and superconducting technology spurred the development of lower-cost mass and NMR spectrometers. Today, the availability of compounds labeled with stable isotopes rivals that of radioisotopes. Commercially available, highly sophisticated mathematical programs are found on the desktop computers of most scientists. This confluence of technologies now allows us to apply the time-dependent models developed for isolated enzyme systems to the study of living organisms. A seminal paper by Edwin Chance *et al.* (1983) uses ^{13}C NMR time-dependent measurement of glutamate fractional enrichment and about 200 simultaneous differential equations to model the citric acid cycle in hearts perfused with ^{13}C-labeled substrates.

In the following chapter, we present the vocabulary and principles used in stable isotope tracer studies of biological systems. The following books were drawn on extensively throughout: Wolfe (1992), *Radioactive and Stable Isotope Tracers in Biomedicine*; Jacquez (1985), *Compartmental Analysis in Biology and Medicine*; Sheppard (1962), *Basic Principles of the Tracer Method*; and Carson, Cobelli, and Finkelstein (1983), *The Mathematical Modeling of Metabolic and Endocrine Systems*.

1.2. Definitions

A lexicon of terms has been developed for tracer theory, but has not been uniformly adopted for routine use in ^{13}C experiments, and many terms are in fact misleading when applied to ^{13}C NMR studies.

1.2.1. Modeling Terms

Tracer. The isotope-labeled, detectable molecule is known as the *tracer*, and is used to "trace" the fate of the unlabeled species. Tracer as used here does not refer to a vanishingly small amount of isotope. Typically NMR experiments "trace" a pathway using substantial amounts of isotope.

Tracee. The *tracee* is the unlabeled species, which denotes the entire pool of interest in a radioactive tracer experiment. The term has little meaning in the ^{13}C NMR experiment where a significant proportion of a metabolic pool is likely to be labeled.

Natural Abundance. The *natural abundance* is the naturally occurring fractional enrichment of the isotope of interest. Radioactive species usually have negligible natural abundance, but it can be appreciable with stable isotopes (1.1% for ^{13}C).

Fractional Enrichment. In an NMR experiment, the *fractional enrichment* of a molecular species A (F_A) is the fraction of the total pool that is labeled (denoted by *), both from specifically labeled substrates, and because of its natural abundance ^{13}C. The fractional enrichment can be different at different carbon sites in the same molecule (denoted by j) (Malloy *et al.*, 1988, 1990). Often, fractional enrichment is multiplied by 100 and reported as a % of total species:

$$F_{A,j} = A_j^* / (A_j^* + A_j) = A_j^* / \text{total} A_j \tag{1}$$

Readers should note that the above definition of enrichment, widely used in NMR research, is not identical to the commonly held definition of enrichment as used by the GC/MS community. In GC/MS literature, enrichment indicates the presence of ^{13}C *above* the natural ^{13}C abundance. If ^{13}C is present at the natural abundance of 1.1%, the carbon is said to have an enrichment of 0. *Abundance* is used by the GC/MS community to indicate the absolute amount of ^{13}C atoms as a fraction of the total as indicated by Eq. (1).

Atom Percent. Atom percent quantifies the isotopic composition of a specific atom in a molecule, and is usually used by commercial suppliers of stable isotopes. Atom percent is identical to fractional enrichment as defined above.

Specific Activity. The *specific activity* is similar to the *fractional enrichment*, but is commonly applied to radiolabeled compounds where the molarity of the tracer is negligible. It is expressed as a fraction (mol label/mol total) or directly in the measured units of disintegrations per minute (dpm):

$$\text{specific activity (dpm/mol)} = {}^{14}C/{}^{12}C$$

Models. Mathematical *models* are used to plan tracer experiments, and to interpret or predict the resultant data. A model is a formal quantitative relationship specifying the relationship among the variables and parameters of the system.

Parameters and Variables. In contrast to variables, parameters are unknown entities in the equations describing a model that cannot be directly measured. By way of review, the classic Michaelis–Menten equation consists of an independent variable $[S]$, a dependent variable v (velocity), and the two unknown parameters, V_{max} and K_M. If parameters are functions of time or other independent variables, this should be designated. In compartmental modeling, the accepted practice is to use uppercase letters for parameters that are constant and lowercase letters for time-varying parameters. For example, $g(t)$ indicates a parameter, g, that varies with time.

Compartment. A *compartment* is an idealized store of a substance. It is a theoretical construct used to describe molecules that exhibit the same behavior (extent or rate of fractional enrichment, for instance) in a tracer experiment. Its *size* Q is the amount of the compound of interest. A compartment can comprise several different chemical species, or it can represent only part of a larger metabolite pool.

Pool. Like a compartment, a *pool* is completely defined by the model describing the system. A pool can contain all the members of a particular chemical species, for instance, even if they are in several different compartments. A pool can also describe those molecules in a particular physical environment, such as pyruvate in plasma, mitochondria, or cytosol. It can also be a purely theoretical physical space; when a tracer is infused into the bloodstream of an intact animal, the volume of its initial distribution (in plasma, intracellular space, etc.) is thought to describe its pool of entry.

Assumptions. Those often unmeasurable aspects of an experiment required to fit the data into a mathematical model. Common assumptions include that of metabolic steady state, or of a well-mixed metabolite pool.

Steady State. The term *steady state* is applied to both the isotopic concentration and the metabolic state of the tissue. *Metabolic steady state* implies that the size of metabolic compartments and the fluxes between them in the pathway of interest are constant. Compartments representing end products will increase in size linearly with time. Metabolic steady state requires a constant physiologic state (work, hormone levels, substrate and oxygen availability, temperature, pH, level of anesthesia, etc.).

The system is at *isotopic steady state* when the fractional enrichment of all metabolites are constant in time. End products may continue to increase in enrichment. The time to reach isotopic steady state depends on the volumes of the metabolic compartments and the fluxes between them. The final fractional enrichment depends on the relative influx into the pathway of labeled and unlabeled substrates. An *apparent isotopic steady state* of tracer fractional enrichment in the plasma can be achieved much faster than a true steady state by suitable tailoring of the infusion.

Equilibrium. The term *equilibrium* applies to reactions in which neither substrate or product concentrations are changing, and where the ratio of products to substrates is a function of the reaction equilibrium constant:

$$A + B \underset{k_{-1}}{\overset{k_1}{\rightleftharpoons}} C + D \qquad \text{where } k_1 = k_{-1}$$

$$k_1 / k_{-1} = K_{eq} = [C][D]/[A][B]$$

When the entry or exit of label from two or more compartments is slow relative to the exchange between them, a state of *near-equilibrium* exists. There are few true equilibria in living cells, but near-equilibrium reactions can sometimes be treated as true equilibria in tracer experiments. Figure 1 shows that two molecular compartments participating in a near-equilibrium reaction are often modeled as a single compartment.

Volume of Distribution. The *volume of distribution* (VOD) is the total apparent volume (expressed in ml, cm^3, or g) in which a metabolite pool is distributed. The concentration of the metabolite is its size divided by its volume of distribution. The calculated apparent VOD can be difficult to interpret, as it may not correspond to any known physiologic compartment.

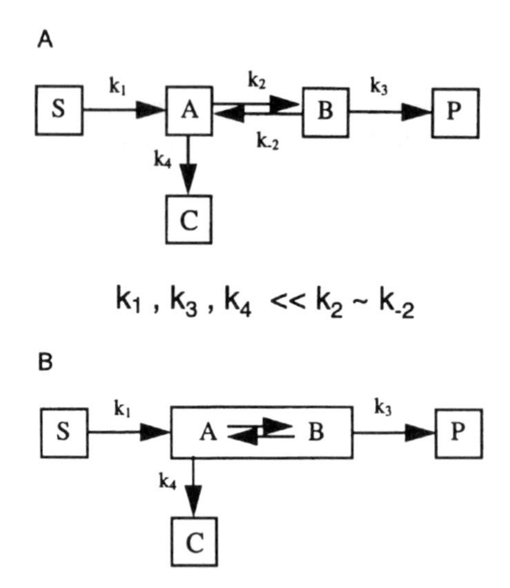

Figure 1. (A) A multicompartmental model, where the fluxes between A and B are fast relative to fluxes in and out of these pools. (B) Compartments A and B can be well represented by a single compartment.

Rate of Appearance and Rate of Disappearance. Ra (rate of appearance) and *Rd* (rate of disappearance) are terms used in radioactive tracer experiments. As shown in Fig. 2A, they describe the rate of entry of a metabolite into or exit from the sampled pool, such as the plasma.

Flux. Flux (designated as *J*) is the rate of movement of material from one metabolic compartment to another, usually through a chemical reaction, expressed in mass/unit time (Fig. 2B).

Fractional Transfer Coefficient or *Rate Constant.* The *rate constant k* is flux divided by concentration, and usually has units of time^{-1}. It is useful when describing flux rates that are dependent on the concentration of a substrate (Fig. 2B):

$$-dA/dt = k_A A$$

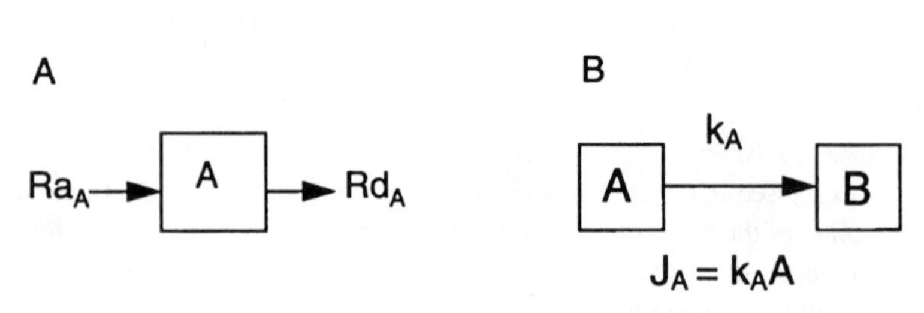

Figure 2. (A) Single-compartment model. (B) Two-compartment model.

Turnover. When a pathway is at metabolic steady state, concentrations of the compounds in the pathway do not change in time, and the rates of production and destruction of each compound are equal. This flux through a metabolite pool is called its *turnover.*

Clearance. Clearance is defined as the volume of plasma (or any pool) which is cleared of a compound in a given time. Clearance is used to relate the concentration of a substrate to its rate of disappearance. It is not typically used in ^{13}C NMR experiments.

$$\text{Clearance (ml/min)} = Rd/[\text{A}]$$

Rate-Determining Step. The slowest step in a metabolic pathway, often at a point of regulation. It can be transport across a membrane, diffusion through solution, or a chemical reaction, either spontaneous or catalyzed by an enzyme. Often, the control of a pathway is shared among several enzymes, and a true single rate-determining step does not exist (Newsholme and Start, 1974).

V_{max}, K_M. V_{max} (maximal velocity) and K_{M} (the Michaelis constant) are parameters gleaned from the kinetic treatment of the initial rates of isolated enzymes studied as a function of substrate, activator, and inhibitor concentrations. The simplest equation, describing a single enzyme with a single substrate, is the Michaelis–Menten equation, shown below and plotted in Fig. 3. The quantity V_{max} is the maximal rate achievable at infinitely high substrate levels, and K_{M} is the substrate concentration resulting in a flux that is half of maximal. The reaction velocity will depend strongly on substrate concentration at or below K_{M}, and

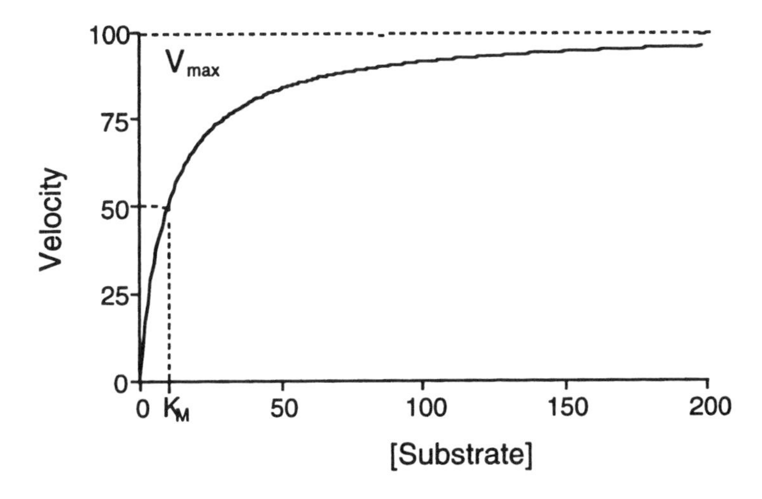

Figure 3. The Michaelis–Menten curve.

becomes independent of substrate concentration at values well above K_M. Although these terms do not apply rigorously to kinetics measured in living systems, they can sometimes be useful, since many biological phenomena are saturable and can be described by a similar curve. Like *clearance*, K_M is useful to relate the velocity of flux to the concentration of a substrate.

Michaelis–Menten Equation:

$$\text{velocity} = V_{max}\,[S]/(K_M + [S]) \tag{2}$$

1.2.2. NMR Specific Terms

Region of Interest. The *region of interest* (ROI) is the specific volume of tissue or solution that is interrogated by the NMR coil. If the sample is macroscopically inhomogeneous, like a perfused heart or the brain of an intact animal, spatial selection is achieved either by choice of coil (surface, volume coil), by tailored pulses (e.g., "depth" pulses), or by use of pulsed magnetic field gradients combined with selective excitation pulses. In a time-dependent NMR tracer experiment, care must be taken so that the ROI remains constant in space, and that changes in the volume of tracer distribution (i.e., plasma volume, extracellular space due to edema, etc.) are known.

Visibility. Visibility in the NMR experiment refers to the differential ability to detect a nuclide in different chemical or physical environments. It is somewhat analogous to quenching in a fluorimeter or scintillation counter. NMR signal intensity can be altered by several mechanisms. *Relaxation broadening* happens when a small molecule is bound to a large, immobile structure like a membrane or protein, or is in close proximity to a paramagnetic compound (heavy metals, deoxyhemoglobin). The line width is inversely proportional to the transverse relaxation time (T_2^*). When the relaxation time is decreased, the signals become broad and low, and can be difficult to observe. These molecules can comprise so-called "invisible pools" which may be metabolically active, but are undetectable by *in vivo* NMR.

Other mechanisms can influence signal intensity. *Saturation* occurs when the longitudinal relaxation time (T_1) is very long relative to the experimental time allowed for relaxation between excitation pulses (RD). The Nuclear Overhauser Effect (NOE) is an enhancement of the intensity of those carbons bound to protons that occurs when the proton frequency is irradiated during signal acquisition. Since ^{13}C in different sites may have different NOE and T_1, the uncorrected relative signal intensities may not accurately reflect relative pool sizes. Intensity can also change with the introduction of covalent bonds between labeled carbons, which causes multiplicity in the spectrum. If resolution is poor, the transition from singlet to multiplet can look like a reduction in peak height and an increase in peak width.

Multiplet Analysis. A ^{13}C NMR resonance can exhibit a multiplet structure if the detected ^{13}C is covalently bound to other ^{13}C carbons. Analysis of the splitting patterns can yield unique information about a metabolic pathway. Most often, this analysis is made in extracts of tissue which has been freeze-clamped at isotopic and metabolic steady state (Malloy *et al.*, 1988, 1990). Occasionally, the resolution is sufficient in studies of intact tissues to observe fine structure (Gruetter *et al.*, 1994).

Direct Detection, Inverse Detection. ^{13}C in tissue or extracts can be directly detected using ^{13}C NMR, with or without proton decoupling (which eliminates peak splitting due to adjacent protons). There are also a variety of ^{1}H editing techniques that can be used to detect only those protons that are bound to ^{13}C carbons. The predominant reason for using inverse detection is that NMR is much more sensitive for the proton than the ^{13}C nucleus.

2. CHARACTERISTICS OF A PERFECT TRACER

The description of a perfect tracer is an attempt to minimize the assumptions made in a tracer experiment. In many texts about tracer methodology, the attributes of a perfect tracer are those resembling a radioisotope. We will describe a perfect (albeit nonexistent!) tracer that combines the useful characteristics of both radioactive and stable tracers. The characteristics of different isotopes of carbon are compared in Table 1.

Chemical Equivalence. A tracer should behave exactly the same as the unlabeled molecule. It should participate in the same reactions, visit the same compartments, and react at the same rates.

Uniform Mixing. As it enters a pool or compartment, the tracer should mix uniformly and instantly throughout.

Control of Substrate Fractional Enrichment. The fractional enrichment of the primary substrate in a system (such as plasma glucose, if [1-^{13}C]glucose is being infused) should be completely under the control of the investigator at all times.

No Metabolic Perturbation. A perfect tracer would not alter the system that it is measuring. This is possible either when the isotope is present in vanishingly small amounts (10^{-3}–1 ppm), or when labeled substance replaces unlabeled substance that is being provided exogenously.

Stability of Labeling. The tracer should be stable for longer than the period of the experiment, in that it should not decay into another isotope. The tracer must also stay with the molecule being traced; it should not exchange with solvent or be lost in chemical reactions, unless this loss constitutes the desired information. If so, the label should not show up in other molecules through nonspecific reactions or exchanges.

Table 1
Properties of Carbon Isotopes Used in Tracer Experiments

	^{14}C	^{13}C GC/MS	^{13}C NMR	^{11}C PET
Real time *in vivo* studies	no	no	yes	yes
Regional distribution in tissue	no	no	yes	yes
Chemical specificity *in vivo*	no	no	yes	no
Yields intramolecular label distribution	no	yes	yes	no
Requires separation of metabolites prior to analysis	yes	yes	no	NA
Specific activity readily determined	no	yes	sometimes	no
"Massless"	yes	no	no	yes
Low natural abundance	yes	no	no	yes
Radioactive	yes	no	no	yes
Short half-life	no	NA	NA	yes

Low Background. The natural abundance of the tracer should be very low so that small quantities can be detected with insignificant error.

Nondestructive Detection. A perfect tracer should be detectable in situ without disturbing the biological system.

Site-Specific Detection. The researcher would be able to detect or sample a perfect tracer in any physical space (brain, muscle, liver, plasma, interstitial fluid, intracellular) uniquely and at will.

Time-Independent Detection. A perfect tracer would be detected with very high sensitivity in a time that is negligible relative to the reaction it is tracing.

Chemical Specificity. The detection system should be able to distinguish among labeled molecules, and determine where in the molecule of interest the label resides. It should be able to uniquely determine the labeling patterns in a given molecule. Ideally, each atom in a tracer molecule could be followed throughout a metabolic pathway.

Quantitation and Enrichment. A perfect tracer could be directly measured either as tracer pool size (mol tracer), concentration (mol tracer/liter), or as fractional enrichment (mol tracer/mol metabolite). The ability to measure each of these independently would reduce the error associated with calculation, since all yield unique information about the biological system.

Safety. A perfect tracer would be easy to handle and present no danger to the researcher or test subject.

3. COMPARTMENTAL MODELS

3.1. Introduction

A compartmental model is a quantitative description of the fluxes among compartments. As indicated above, a compartment is defined as an idealized store of a substance. Compartments may refer to different chemical species or to different populations of the same species. Compartments need not necessarily correspond to physically bounded spaces. The important point is that a compartment is defined functionally. Any collection of molecules where each molecule has an identical probability of leaving the compartment by any of the allowed exit paths comprises a compartment. The mathematical basis of compartmental models and their uses in tracer experiments is beyond the scope of this presentation and can be found elsewhere (Jacquez, 1985; Carson *et al.*, 1983). The objective of this simplified presentation is to describe how models may be used in tracer metabolic studies and to point out some of the advantages as well as some of the pitfalls of their use.

3.2. Objectives and Identifiability

Before examining specific models it is important to have a clear view of the purposes of modeling. Models are simplifications of the real world and need not and should not contain all possible compartments. The number and types of compartments in a model are chosen based on the purpose of the model and the type of data to be collected. In general, compartmental models are used to accomplish one or more of the following objectives: (1) to identify the structure of the system, (2) to estimate internal parameters, and (3) to predict the response of the model to external factors. Classically, a compartmental model is presented as a diagram with arrows to indicate permitted fluxes. Frequently the arrows are labeled with the *fractional transfer coefficient*, that is, the fraction of the compartment transferred per unit time, indicated as k. Alternatively, in some cases here the arrow will be designated with a J to indicate *flux* in mass per unit time.

Once a specific model has been chosen, an important issue is to determine whether the experimental measurements to be performed are adequate to estimate the unknown parameters of the model. The modeling term applicable here is "Identifiability." The concept of Model Identification is derived from systems theory and the theory of statistical estimation and has been described in detail for a physiologically oriented reader by others (DiStefano *et al.*, 1990; Jacquez, 1985). In essence, "Identifiability" simply means whether or not it is possible to use the collected data and the chosen model to estimate the desired parameters. Identification thus relates to both the structure of the model and the experimental techniques for estimating the internal parameters. Identifiability has two components. *A priori* Identifiability is established with a model in hand but before the experiment is performed. To determine whether a model is *a priori* Identifiable, it is assumed that the data collected are ideal, i.e., without error. A model is said to be *a priori* Identifiable if it can be determined from a proposed model and an ideal data set that it is possible to obtain unique estimates for all the unknown parameters of the model. For complicated experiments it may be necessary to perform simulations to generate the ideal data before the issue of *a priori* Identifiability can be determined. Alternatively, one may consider *a priori* Identifiability as part of the experimental design. Here the relevant question is: Given a specific model structure, what experiments can be used to determine uniquely each of the parameters? In summary, *a priori* Identifiability provides a way of distinguishing those experiments that cannot succeed (unidentifiable) from those that might succeed (Carson *et al.*, 1983).

Once an appropriate model and experimental design have been selected and data are collected, the issue of *practical* or *a posteriori* Identifiability is raised. *A posteriori* Identifiability concerns the model and the data actually collected. A model is said to be *a posteriori* Identifiable if, from a model and the experimental data, unique estimates are obtained for all the unknown parameters of the model. The difference in the two types of identifiability results from measurement errors and other disturbances that will inevitably appear in experimental measurements. *A posteriori* Identifiability analysis may indicate that the model is not uniquely identifiable for one or more parameters or that the confidence limits for the estimation of a specific parameter are unacceptably large. Often a model will fail to yield parameter values because it is too complex relative to the amount and the kinds of data that can be collected. If so, the model can often be reduced in complexity. One such process is that of lumping parameters or compartments. Lumped parameter representations will be adequate if the combined parameters behave approximately as if they are a single parameter. A common example of a lumped compartment is the representation of the concentration of a metabolite in plasma as a single number. In actuality, most metabolites in plasma are present in gradients. A strength of compartmental modeling is that simplifications such as lumping compartments together are clearly visible and can be evaluated. However, a weakness is that compartmental models do not easily accommodate a distributive process such as a gradient. In summary, identifiability is a protocol for determining if a model is capable of providing estimates for the parameters of interest. A model must be sufficiently complex to accomplish its purpose but not so complex as to render it unidentifiable. Perhaps this is best expressed as "Everything should be made as simple as possible, but not simpler."—Albert Einstein.

3.3. Parameter Estimation and Goodness of Fit

All modeling involves some data manipulation to extract the parameters from the data and model. In some instances parameter estimation is a simple process, such as when the number of equations are few and equal the number of parameters to be estimated. However, stable isotope tracer experiments are information-rich, especially using NMR, and may provide a number of possibilities for estimating each parameter. When multiple relationships can be used to estimate a parameter, good modeling practice requires that it be used, and that the data are weighted properly. The "best fit" solution should be obtained by a process such as weighted linear or nonlinear least-squares fitting. Once parameters have been estimated a model should be evaluated for goodness of fit. For a single model, standard statistical procedures including the sum-of-squares error and the covariance matrix will indicate the accuracy of the parameter estimates. If two competing models are under consideration for the same data, an F test can be used to determine whether a model with fewer parameters is better than a model with one or more additional parameters. The procedure involves determining whether adding additional parameters to a model reduces the sum-of-squares error sufficiently to justify the more complex model with the additional parameters. Glantz and Slinker (1990) can be consulted for a simple presentation of this process, and Carson *et al.* (1983) for a more sophisticated analysis.

3.4. Linearity and Tracer Models

Perhaps the most significant distinction in modeling is that between linear and nonlinear models. An example of a linear system commonly used in NMR and other tracer kinetic studies is the labeling of a single compartment at metabolic steady state (Fig. 2A). An isotope is introduced into the compartment beginning at $t = 0$ and the labeling of the metabolite, A, is observed until a steady state is reached, where the fractional enrichment $F_A = F_T$, the fractional enrichment of the tracer; J is the constant rate of flux through the compartment, k is the fractional transfer coefficient from the compartment, and Q is the size of the compartment in mol. The time dependence of tracer (A^*, in mol) and the fractional enrichment of A satisfy

$$dA^*/dt = F_T J - kA^* \tag{3a}$$

$$dF_A/dt = (F_T J - F_A J)/Q \tag{3b}$$

Linearity is important because it determines both the kinds of behavior available to a model and the mathematical approaches which can be used to investigate the model. Linear systems are identified by their response to several stimuli imposed simultaneously. A linear system responds to multiple simultaneous stimuli with a response that is the sum of the responses to each stimulus when applied individually. This is the "Principle of Superposition," a well established concept in controls engineering (DiStefano *et al.*, 1990). Linear systems can often be represented by linear differential equations. Although models in general may be either linear or nonlinear, an important feature of compartmental modeling and tracer studies is that of linearity as described by Jacquez (1985): "For all compartmental systems, linear and nonlinear, if the system is in a steady state and one adds a tracer to any compartment of the system, the distribution of the tracer labeled material follows the kinetics of a linear compartmental system with constant coefficients." This is an important reason for rigorously using compartmental models in tracer studies.

Compartmental models are not without their drawbacks, both biological and conceptual. Although a compartment may be described and studied with tracer kinetics, it can be difficult to identify the physical entity that corresponds to the specific compartment. Strictly speaking, compartmental models are always described by equations such as (3a) and (3b), where the flux from a compartment is simply proportional to the size of the compartment (kA) or its equivalent. This does not allow for well recognized effects such as saturability or the analysis of effectors of flux such as hormones, that can alter the value of k. Another difficulty is that a compartmental representation for some simple, common types of nonlinear systems, such as the formation of a product as the condensation of two molecules, can be unexpectedly complicated.

$$A + B \leftrightarrow A\text{–}B$$

A true representation of this process by compartmental modeling actually requires the addition of a fictitious input into one of the compartments (Jacquez, 1985). Such a requirement may be difficult to convey to the physiologically oriented reader. Investigators must therefore decide whether a compartmental model is appropriate to meet a specific quantitative or qualitative objective. In the end it is the task of the model maker to distinguish between the superficial and the essential.

4. THE BASIC TRACER EXPERIMENT

4.1. General Considerations

Isotopic carbon tracers are used to interrogate metabolic phenomena by introducing into the living system some compound that can be uniquely identified, and monitoring the fate of that compound in the compartment of interest. An extensive body of theory has been developed to allow the calculation of physiologically important variables from observables in the tracer experiment (Jacquez, 1985; Sheppard, 1962). Much of this theory was developed for the specific case of *in vivo* studies with radioactive tracers, in which the tracer is infused into the blood compartment, and all the information about tissue metabolism is derived from measurements made in blood sampled from the same compartment. This type of experiment can be thought of as *indirect sampling* of the pool of interest. Another large body of literature addresses the situation in which tissues are removed from animals or people after infusion of metabolic tracers, or cells and isolated organs are extracted after incubation with labeled substrates, and information is derived from the concentration and fractional enrichment of tracer in specific molecules in the tissue. Many NMR ^{13}C experiments fall into this second category, which we will call *direct sampling* of the pool of interest. A third category contains those experiments made possible by the nondestructive nature of NMR, in which living tissue is observed, whether *in vivo* or in intact organ or cell models, in real time during active metabolism of the supplied ^{13}C-labeled substrate. This will be called *in situ sampling* or detection.

The tracer experiment can utilize either time-dependent or time-independent detection. Time-independent detection implies that the system is sampled only once, either at a steady state, or at a defined point in the experiment. Many ^{14}C experiments that employ direct sampling of tissues use time-independent tracer detection. Such an experiment may involve sacrifice of an animal after injection of ^{14}C-alanine so that liver tissue can be extracted for the measurement of ^{14}C-glycogen specific activity, or the Folsh extraction of a dish of cultured cells 60 min after adding [2-^{14}C]acetate, in order to measure the ^{14}C content of newly manufactured lipid. Both of these use direct sampling, time-independent detection, and are in an isotopic and metabolic steady state with detection of an end product.

Time-dependent studies are those in which a series of samples are used to define kinetic phenomena, such as serial blood samples taken from a person to describe the curve of fractional enrichment in the plasma space during ^{14}C tracer infusion (Farrace and Rossetti, 1992; Rossetti and Giaccari, 1990). With *in vivo* NMR and PET, it has become possible to combine time-dependent detection and direct or in situ sampling. As an example, *in vivo* ^{13}C NMR has been used to monitor the kinetic curves describing the appearance of tracer from plasma [1-^{13}C]glucose in human brain tissue metabolites (glutamate, GABA, lactate) (Gruetter *et al.*, 1994; van Zijl *et al.*, 1993).

Finally, all tracer experiments are either sampled at a steady state, or under nonsteady-state conditions of some sort. As noted above, metabolic steady state implies that all metabolite intermediate pools and fluxes are constant. At isotopic steady state, the fractional enrichments of all metabolite pools are constant. When at metabolic steady state, data can be acquired at either isotopic steady state or nonsteady state or used to detect an end product. If the system is at isotopic steady state, measurements can be taken at metabolic nonsteady state. If both the isotopic fractional enrichments and the metabolism are far from steady state, the experimental data become very difficult to interpret.

All mathematical treatment of tracer data is model-dependent, and a few basic types of models will be discussed below. The following simple treatment of tracer data is taken from the *single pool model* (Wolfe, 1992). The equations are taken from radioactive tracer theory, and therefore are altered to reflect the fact that the

^{13}C experiment often uses large concentrations of labeled molecules. The following definitions will be used:

A^* is the infused ^{13}C-labeled compound
A, B, C are pool sizes (mol) in the compartments
B^*, C^*, etc. are the pool sizes (mol) of ^{13}C-labeled metabolites
A_T, B_T, C_T, etc. are the total pool sizes (mol) in the compartments
F_A, F_B, F_C, etc. are the fractional enrichments of each pool
J_A, J_B, J_C, etc. are the fluxes between pools, with subscripts identifying the exiting compartment

4.2. Single Pool Model

Description:

> A^* is infused into the pool of interest (plasma) at a constant known rate. In time, the system will reach metabolic and isotopic steady state. The blood is sampled at this time, and F_A is measured. The experiment yields the rates of appearance (Ra) and disappearance (Rd) of A in the pool of interest.

In a dynamic system, a compound A is constantly being produced from one set of substrates, and then used as a substrate in other chemical reactions (Fig. 2A). In metabolic steady state, the absolute amount, or pool size of A does not change, but material is constantly entering and leaving at the same rate. Even when the pool size of A is changing, a measurement of this change in time yields only the net change, the difference between its absolute rates of production and disposal:

$$dA/dt = \text{Appearance of A} - \text{Disappearance of A} = Ra - Rd \tag{4a}$$

If $dA/dt = 0$ (metabolic steady state)

$$Ra = Rd \tag{4b}$$

These two absolute rates can be measured independently with a tracer. The most important basic tool used to achieve this is that of the *fractional enrichment* (F_A) of the molecule of interest in the metabolic compartment of interest; F_A equals the fraction of the total metabolite which is labeled (denoted A^*). As noted above, different detection methods lend themselves to slightly different definitions of this quantity, but for NMR the most common definition is the following:

$$\text{Fractional Enrichment of A} = F_A = A^*/(A + A^*) = A^*/A_T \tag{5a}$$

The fractional enrichment of the compound A shown in the simple, single-compartment model of Fig. 4A is a function only of the relative rates of entry of A (Ra) and A^* (tracer appearance rate, Ra^*):

$$F_A = Ra^*_A/(Ra_A + Ra^*_A) \tag{5b}$$

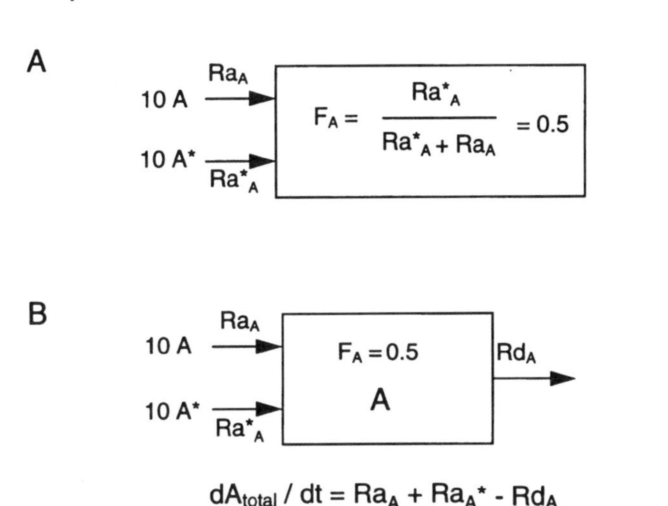

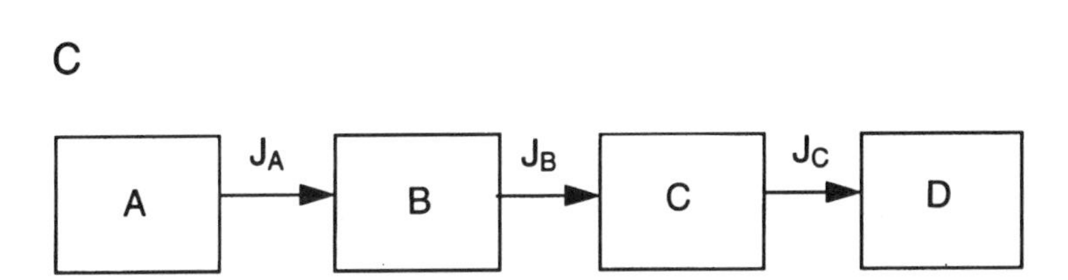

Figure 4. (A) The single-compartment model with input flux, or rate of appearance of tracer (Ra^*_A) and of endogenously produced substance A (Ra_A). The fractional enrichment of A (F_A) is completely determined by the inputs to the compartment. (B) The single-compartment model with the addition of a rate of disappearance, *Rd*. (C) A catenary multicompartment model, where fluxes between compartments are represented by *J*.

If there is flux of A out of the compartment of interest, as in Fig. 4B, the molecules leaving will always have the same fractional enrichment as those staying behind. This means that F_A yields information only about the rate of influx of A into a compartment, never efflux.

If the system is at steady state, the pool size of A will not change in time, and entry and exit from the pool are equal (Fig. 4B):

$$\text{if } dA_T/dt = 0 \text{ (metabolic steady state)}$$

$$Rd_A = Ra_A + Ra^*_A$$

$$\text{and if } dF_A/dt = 0 \text{ (isotopic steady state)}$$

$$Rd_A = Ra^*_A / F_A \tag{6}$$

Therefore, if the rate of label entry Ra^* is known, one can measure F_A and calculate *Ra* and *Rd*. If the pool size is changing in time, then *Rd* can be calculated only if dA/dt is known:

$$\text{if } dA_T/dt \neq 0 \text{ and } dF_A/dt = 0$$

$$Rd_A = Ra_A + Ra^*_A - dA_{total}/dt = Ra^*_A / F_A - dA_{total}/dt$$

At this point, we will abandon the terminology "*Ra*" and "*Rd*" in favor of "flux," denoted by *J*.

4.3. Multicompartmental Catenary Model

Description:

The general model is shown in Fig. 4C, and consists of 4 pools. Tracer A^* is found in the first pool at a constant known concentration and F_A. Label flows from pool A to pool B to pool C to pool D. The label in each pool, A*, B*, C* and D*, can be sampled over time. The ideal curves for A^*, B^*, etc., for A, B, etc., and for F_A, F_B, etc., can be constructed for the general case, and a variety of special cases. A fit of the experimental data to these curves yields the rates of flux between the pools.

The time-dependent, in situ ^{13}C NMR experiment is usually used to interrogate a metabolic pool in the tissue of interest, and directly yields the content of the ^{13}C-labeled metabolite, for instance, B^* (Fig. 4C). If several sequential spectra are taken, the slope of the intensity of the NMR signal plotted vs. time will be dB^*/dt. The actual flux rate J_A must be calculated, and the correct equations to use depend on the choice of model. A common model to choose is a multicompartmental catenary model, where several pools exist in such a way that label flows sequentially from one to another. We have chosen to discuss the simplified, special case in which there is only one input into each pool, and no backward fluxes. The following models describe the special cases of metabolic steady state and pre- or post-isotopic steady state, or isotopic steady state with a metabolic perturbation. For the simple model pictured in Fig. 4C, if we assume that all fluxes are a linear function of the size of the substrate pool ($J_A = k_A A$, where k_A is a rate constant), then the general descriptions for $F_B(t)$, $B(t)$, and $B^*(t)$ are:

$$F_B = B^*/(B + B^*) = B^*/B_T \tag{7}$$

$$dB_T/dt = J_A - J_B = k_A A_T - k_B B_T \tag{8}$$

and

$$dB^*/dt = J^*_A - J^*_B = k_A A^* - k_B B^* \tag{9}$$

Analogous equations exist for A, C, and D. The solution of these equations for any specific case will yield the behavior of label in pool B. For the following discussion, we will assume that pool A is the pool of tracer entry, and that its concentration and fractional enrichment are constant and known ($F_A = 0.5$). The first case that we will describe is when $k_B = 0$, i.e., there is no exit of material from pool B. B is then an end product that will build up in time (Fig. 5A). This example could be described as a special case of isotopic and metabolic steady state. dB^*/dt is measured in the ^{13}C NMR experiment, and used to calculate the rate of production of B (dB/dt).

4.3.1. Case 1: $k_B = 0, F_A = F_B$. End Product of System in Isotopic and Metabolic Steady State

With reference to Eqs. (7)–(9),

$$dB_T/dt = k_A A_T$$

$$dB^*/dt = J_A = k_A A^* = k_A A_T F_A = F_A dB_T/dt$$

$$.dB_T/dt = d(B^*/F_A)/dt$$

If we define B_0 as the concentration of B at time 0 and solve for B_T, we get a line whose slope is the rate of flux into pool B (Fig. 5A):

$$B_T = B^*/F_A + B_0 = k_A A_T t + B_0$$

$$B^*/F_A = J_A t \tag{10}$$

The next case to consider is a metabolic perturbation applied during isotopic steady state. Here, $k_B > 0$. The system has reached isotopic steady state ($F_A = F_B$), followed by a metabolic perturbation (indicated by an arrow in Fig. 5B) which alters J_A. We will describe the approach of the system to a new metabolic steady state.

4.3.2. Case 2: $k_B \neq 0, F_A = F_B =$ constant. Isotopic Steady State, Metabolic Perturbation

$$dB_T/dt = k_A A_T - k_B B_T$$

$$dB^*/dt = k_A A^* - k_B B^*$$

$$dB^*/dt = k_A A_T F_A - k_B B_T F_B = F_A (k_A A_T - k_B B_T) = F_A dB_T/dt$$

$$dB_T/dt = d(B^*/F_A)/dt$$

Again, we reached the conclusion that the change in B can be calculated simply by dividing the behavior of B^* by the fractional enrichment of the precursor F_A, which

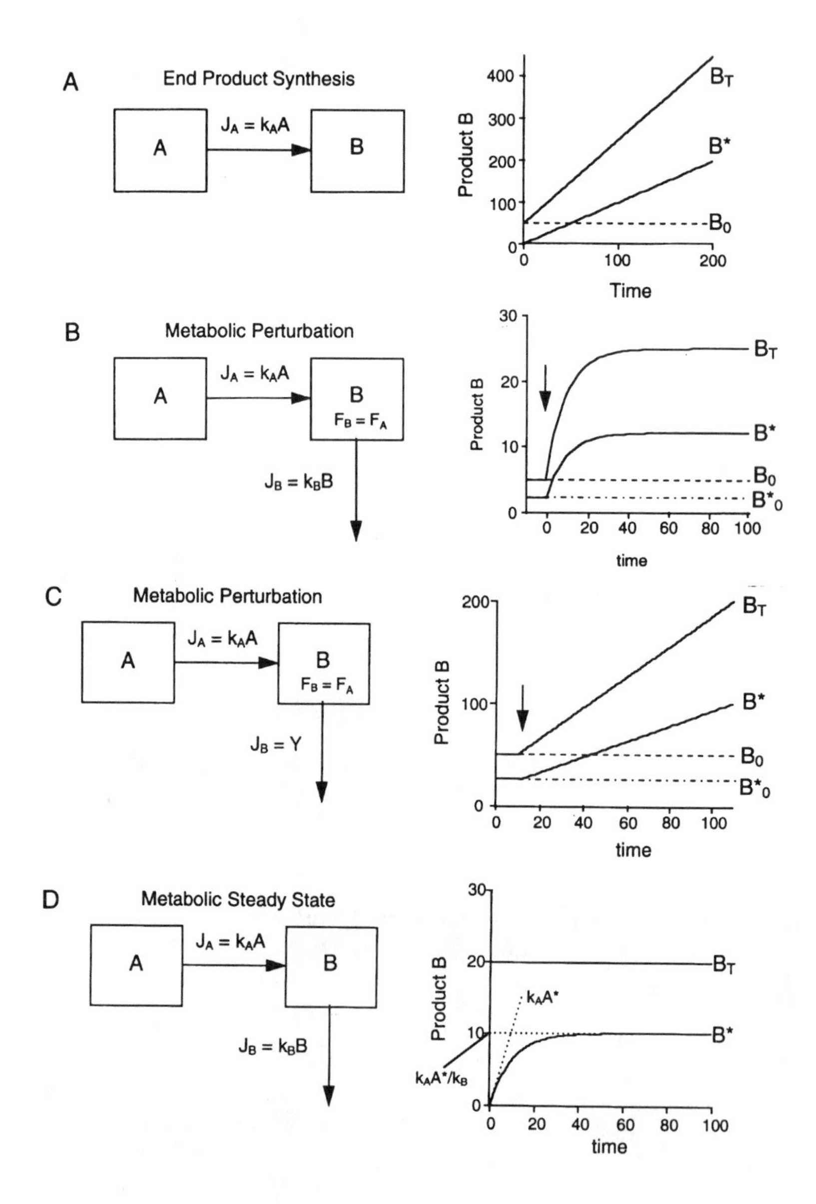
A
End Product Synthesis
A
$J_A = k_A A$
B
Product B
Time
B_T
B^*
B_0
B
Metabolic Perturbation
A
$J_A = k_A A$
B
$F_B = F_A$
$J_B = k_B B$
Product B
time
B_T
B^*
B_0
B^*_0
C
Metabolic Perturbation
A
$J_A = k_A A$
B
$F_B = F_A$
$J_B = Y$
Product B
time
B_T
B^*
B_0
B^*_0
D
Metabolic Steady State
A
$J_A = k_A A$
B
$J_B = k_B B$
Product B
time
B_T
$k_A A^*$
B^*
$k_A A^*/k_B$

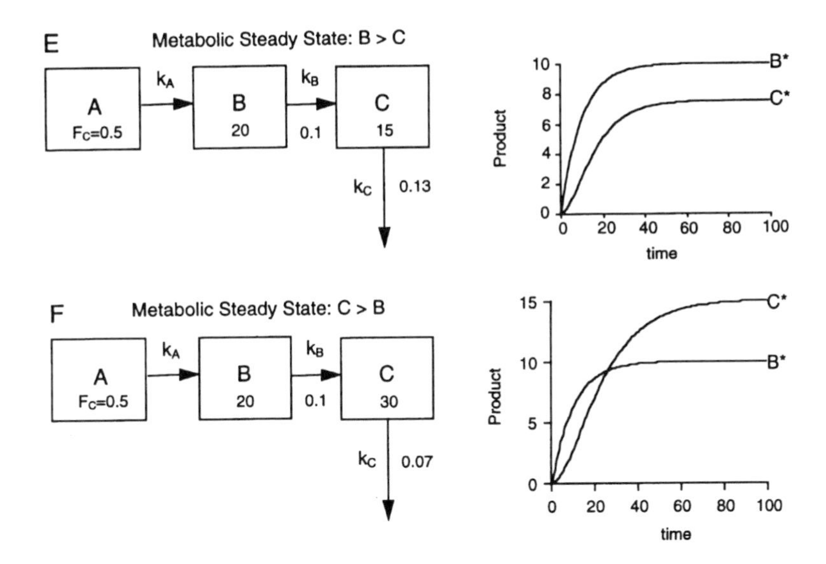

Figure 5. (A) A two-compartment model of end product synthesis in isotopic and metabolic steady state (Case 1). The pool size and fractional enrichment F_A of compartment A are constant. J_A is the flux from compartment A to B, and k_A is the rate constant for this flux. B_0 is product existing before label introduction at time 0. Labeled product (B^*) and total product (B_T) are both produced at a constant rate. (B) A two-compartment model showing flow from B (J_B) in isotopic steady state. At the arrow in the graph, a metabolic perturbation is applied that increases the flux J_A. B and B^* approach their new steady-state values via the curves shown (Case 2). (C) Similar to the model in Fig. 5B, except that J_B is now a constant ($J_B < k_A A$), and independent of the size of B (Case 2′). (D) A two-compartment model at metabolic steady state, where $J_A = J_B$. Tracer is introduced from A starting at time 0. The initial rate of tracer entry ($k_A A^*$) and the steady-state value of B^* ($k_A A^*/k_B$) can be read directly from the graph. (E) A three-compartment model at metabolic steady state, where the pool size of B is greater than the pool size of C. Tracer is introduced from compartment A at time 0, and flows into B and C. The time-dependent behavior of B^* and C^* are shown. (F) Same as for (E), except the pool size of C is greater than B.

is constant and known. If we solve for B [Eq. (11)], we find that both B and B^* will reach a point where they will no longer change, i.e., they will reach a new metabolic steady state. This is the correct model to use when a system at isotopic steady state is subjected to a metabolic perturbation which increases pool size B, and the data are collected during its movement toward a new metabolic steady state (Fig. 5B). The subscript "new" denotes B or B^* made following the perturbation at the arrow:

$$B_T = (B^*_{\text{new}} + B^*_0)/F_A = \{(k_A A_T/k_B) - B_0\}\{1 - \exp(-k_B t)\} + B_0 \qquad (11)$$

A special case of this exists where the flux from pool B is nonzero and constant (i.e., $J_B = Y$, and does not depend on the pool size of B). For the diagram in Fig. 5C, Y is less than the J_A into pool B. For this case, we will again assume that the system is in isotopic steady state, and a metabolic perturbation is applied at the arrow in Fig. 5C which changes J_A. In this case, the solution B is a line, where the slope is the difference between the synthetic flux and the breakdown flux [Fig. 5C, Eq. (12)].

Case 2′: $J_B = Y$, $F_A = F_B$ = constant. Isotopic Steady State, Metabolic Perturbation

$$dB_T/dt = k_A A_T - Y$$

$$dB^*/dt = k_A A^* - Y^* = k_A A_T F_A - Y F_B = F_A (k_A A_T - Y) = F_A\, dB_T/dt$$

$$B_T = (B^*_{new} + B^*_0)/F_A = (k_A A_T - Y)t + B_0 \tag{12}$$

For the third case, we will assume that $F_B \neq F_A$, i.e., it is the general case describing time points before an isotopic steady state has been achieved. Although we can describe the system without the imposition of metabolic steady state, we will assume metabolic steady state in order to solve for B^* and B. It is important when designing a model to have either the isotopic or the metabolic state of the system in steady state (Fig. 5D).

4.3.3. Case 3: $F_B \neq F_A$. General Case of Metabolic Steady State, Isotopic Nonsteady State

$$dB_T/dt = k_A A_T - k_B B$$

$$dB^*/dt = k_A A^* - k_B B^*$$

Sometimes it is easy and reasonable to determine the initial rate of flux from the slope of the B^* curve at very early time points, before appreciable product B^* has a chance to appear. This is a way to get some kinetic information from a curve without fitting it to a model. The initial rate, at time 0 and $B^* = 0$, is given by

$$(dB^*/dt)_{t=0} = k_A A^*$$

Solving the general equation for B^* (Fig. 5D), we see that B^* approaches its steady-state value exponentially. The reciprocal of the time constant of this exponential is the rate constant for flux leaving pool B:

$$B^* = (k_A A^*/k_B)\{1 - \exp(-k_B t)\} \tag{13}$$

We can use this equation to solve for F_B as well:

$$F_B = B^*/B_T = (k_A A^*/B_T k_B)\{1 - \exp(-k_B t)\} \tag{14}$$

In this form the equation is not very useful, since the ^{13}C NMR experiment doesn't report B_T. If we invoke the fact that the experiment is done at metabolic steady state, where $dB_T/dt = 0$ and B_T is simply a constant, we see that when B^* reaches its isotopic steady state, F_B will equal F_A. The parameter k_B from a fit of experimental data to this equation can be used to calculate $k_A A = J_A$.

$$dB_T/dt = 0, \text{ so}$$

$$k_A A_T = k_B B_T$$

$$F_B = B^*/B_T = (A^*/A_T)\{1 - \exp(-k_B t)\} = F_A\{1 - \exp(-k_B t)\} \quad (15)$$

Now suppose that we actually want to observe C^*, the next pool in our model. We know from the above analysis that the flux rate of label into pool C, $k_B B^*$, is a time-dependent term. If we solve the general equation assuming metabolic steady state (shown in Fig. 5E, F):

4.3.4. Case 4: $dC/dt = 0$, $F_A \neq F_B \neq F_C$. Metabolic Steady State, Isotopic Nonsteady State

$$dC_T/dt = k_B B_T - k_C C_T = 0$$

$$dC^*/dt = k_B B^* - k_C C^*$$

$$C^* = (k_B B^*/k_C)[1 - \{(1/k_B)\exp(-k_B t) - (1/k_C)\exp(-k_C t)\}/\{1/k_B - 1/k_C\}] \quad (16)$$

Figure 5E shows a case in which $B > C$, and Fig. 5F shows a case in which $C > B$. Note the sigmoidal shape of the C^* curve, and the lag between the time to steady state for B and C. By analogy to our work above, we can also solve for $F_C = C^*/C_T$ at metabolic steady state where $dC_T/dt = 0$:

$$F_C = (B^*/B_T)[1 - \{(1/k_B)\exp(-k_B t) - (1/k_C)\exp(-k_C t)\}/\{1/k_B - 1/k_C\}] \quad (17)$$

which, at very long times, reduces to isotopic steady state:

$$F_C = A^*/A_T$$

The general solution for D^* in metabolic steady state is presented below. For ease of presentation, we will define the time constant T as the inverse of the rate constant k, i.e., $T_B = 1/k_B$. Examples are shown in Fig. 9.

$$D^* = (T_D C^*/T_C)\left[1 + \frac{\{T_B^2 \exp(-t/T_B)\}}{(T_B - T_C)(T_D - T_B)} + \frac{\{T_C^2 \exp(-t/T_C)\}}{(T_B - T_C)(T_C - T_D)} + \frac{\{T_D^2 \exp(-t/T_D)\}}{(T_D - T_B)(T_C - T_D)}\right] \quad (18)$$

Most of the samples shown above are for metabolic steady state, where the compartment sizes do not change, or for very specific nonsteady-state cases. A general solution for a two-compartment model when not in steady state can be found in Jacquez (1985), and is beyond the scope of this chapter. Except for a few, multicompartment models do not have simple solutions, and it may be necessary to perform numeric rather than symbolic integration for more complex models. For a beautiful example of a complete numerical solution of the equations describing TCA cycle flux, see Chance *et al.* (1983).

5. SATURABLE KINETIC PROCESSES

Thus far, we have represented fluxes as the product of a constant k and the pool size of the substrate. This formalism is in general correct for a single measurement of a biological system at metabolic steady state. Most processes that occur in tissues are not linear functions of a rate constant, however, but instead reach a constant rate at very high substrate levels. These are saturable processes. Most reactions or transport phenomena that are dependent on a catalyst, such as an enzyme or a transporter protein, become saturated at substrate levels high enough to employ all the available catalyst. The Michaelis–Menten equation is the simplest relationship between a unidirectional, saturable flux and the concentration of the rate-controlling substrate [Fig. 3, Eq. (2)]. It is often used in its linearized form:

$$1/\text{velocity} = 1/V_{max} + K_M/(V_{max}\,[S]) \qquad (2a)$$

It was developed to describe a first-order reaction (one substrate) catalyzed by a single, isolated enzyme, and describes the general behavior of the flux from one compartment to the next as a function of substrate, not a single experiment conducted under specific conditions. It contains two parameters, V_{max} which is the maximal flux possible, and K_M, the Michaelis constant, which is the substrate concentration at half-maximal velocity. At very low substrate concentrations, the apparent first-order rate constant k is equal to V_{max}/K_M, and at very high substrate levels, $k = V_{max}/[\text{substrate}]$. Special forms of this equation exist for higher-order reactions (more than one substrate) and those in which the catalyst responds to either activators or inhibitors of flux (Segel, 1975).

The parameters V_{max} and K_M can be extremely useful to describe the general nature of a saturable flux *in vivo*, even if the experimental data do not fit the actual Michaelis–Menten equation very well. If the maximal velocity, the substrate concentration at half-maximal velocity, and the physiologic range of the substrate are known, we can determine the nature of the regulation of that metabolic pathway under normal physiologic conditions. Sometimes it is tempting to make the assumption that the reaction of interest can be perfectly described by the Michaelis–Menten equation, because V_{max} and K_M can then be calculated by only two experimental points. Unfortunately, this assumption may not be valid under *in vivo* conditions.

6. CONDENSATION REACTIONS

^{13}C NMR provides a unique tool for the study of reactions which produce a single product from two labeled substrates because of the property of multiplicity. Multiplicity arises in a ^{13}C NMR spectrum due to splitting of a resonance of one nucleus by another covalently bound, NMR-visible nucleus. These are usually protons, other ^{13}C carbons, and occasionally a phosphate moiety. Splittings that arise from neighboring ^{13}C-labeled carbons ($J = 34–55$ Hz in glutamate and lactate) can result in a messy spectrum where the peaks appear either as multiplets (well-resolved) or as broad, irregularly shaped peaks (often the case *in vivo*). It is usually important to take as a measure of spectral intensity the area of the multiplet, rather than peak height.

The multiplet structure due to ^{13}C–^{13}C J-coupling yields unique information about pathways in which bonds are made or broken between two labeled compounds. Consider the reaction shown in Fig. 6. A 4-carbon molecule labeled with ^{13}C at C1, and a 2-carbon, C2-labeled compound react to form a 6-carbon molecule with label in C2 and C3. Each molecule in the figure is accompanied by its fractional enrichment. The enrichment of the molecule labeled at both C2 and C3 is 30%,

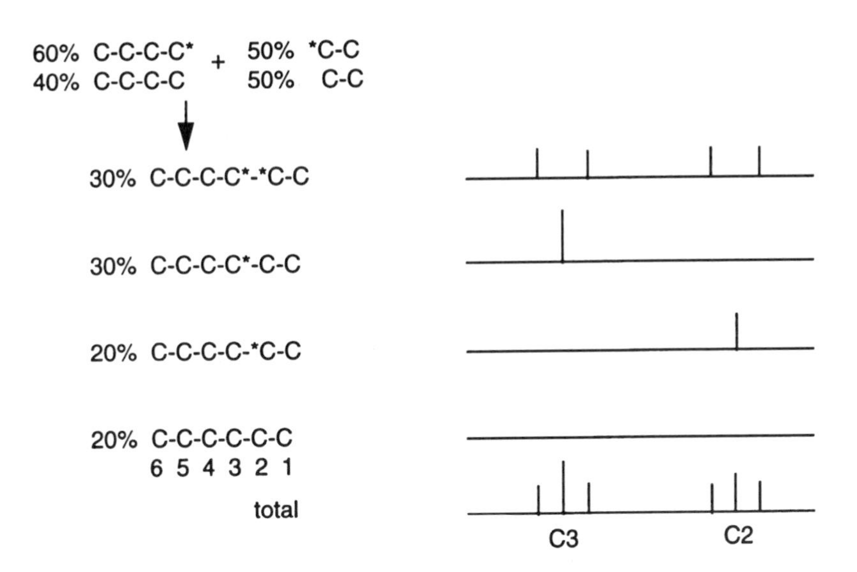

Figure 6. A condensation reaction of two ^{13}C-labeled compounds, one which is 60% labeled and the other 50% labeled. The products are shown with their relative sizes. On the right is the ^{13}C spectrum for each, and a composite of all spectra is shown at the bottom. Products with only one ^{13}C will have singlets at the appropriate frequency, while that with two ^{13}C sharing a covalent bond will have two doublets in the spectrum due to J coupling.

equal to the product of the enrichments of their precursors. Likewise, the enrichment of the unlabeled molecule is $0.4 \times 0.5 = 0.2$ or 20%. The ^{13}C NMR spectrum of each molecule is shown, as well as the composite spectrum. When either C2 or C3 is labeled, a singlet with an intensity proportional to its fractional enrichment is found in the spectrum. When C2 and C3 are labeled in the same molecule, there is a doublet at both the C2 and C3 frequencies due to *J*-coupling. The ratio of the areas of the doublet at C2 due to C2–C3 splitting and the singlet at C2 yields the ratio of C–C–C–C*–C*–C to C–C–C–C–C*–C (30%/30% = 1). The doublet/total ratio at C2 is the fractional enrichment of the substrate which labels the other peak (C3) in the reaction. Likewise, the ratio of doublet to total signal in C3 yields the fractional enrichment at C2, i.e., the fractional enrichment of the 2-carbon molecule [$1/(1+1) = 50\%$]. This simple property of ^{13}C NMR has proven very useful to study condensation reactions in biological tissues. ^{13}C–^{13}C splitting could even be seen in spectra of human brain during [1-^{13}C]glucose infusion, allowing quantitation of ^{13}C fractional enrichment at glutamate C3 and C4 (Gruetter *et al.*, 1994).

The TCA cycle "starts" with a condensation between oxaloacetate and acetyl-CoA, allowing this pathway to be studied very successfully with multiplet analysis (Fig. 7). A detailed and beautifully presented treatment is found in Malloy *et al.* (1988, 1990). If C2 of oxaloacetate and C2 of acetyl-CoA are both labeled with ^{13}C, they will combine to form [3,4-^{13}C]citrate, which will in turn become [3,4-^{13}C]α-ketoglutarate, then [3,4-^{13}C]glutamate, in which the C3 and C4 resonances are split into doublets similar to the general example in Fig. 6.

The splitting in the C3 and C4 resonances that come from ^{13}C carbons originally found in C2 of oxaloacetate and C2 of acetyl-CoA are propagated to the other carbons of the glutamate molecule as the TCA cycle turns. The labeled pair, glutamate C3–C4, becomes C2–C3 in fumarate, and eventually, C2–C3 in oxaloacetate. If this in turn condenses with [2-^{13}C]acetyl-CoA, [2,3,4-^{13}C]glutamate

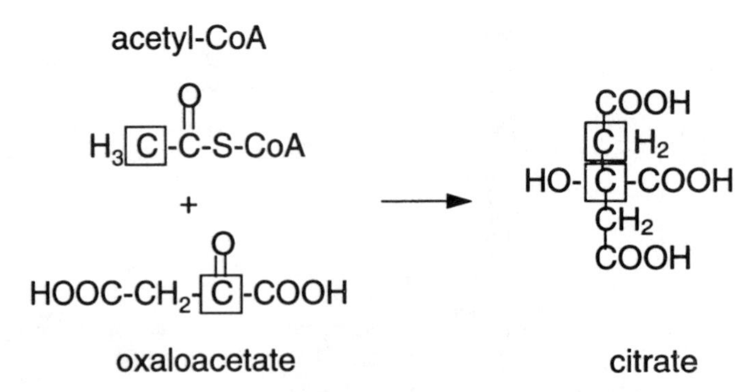

Figure 7. The condensation reaction catalyzed by citrate synthase in the TCA cycle between [2-^{13}C]acetyl-CoA and [2-^{13}C]oxaloacetate results in a ^{13}C–^{13}C bond.

results. The resonances of C2 and C4 glutamate will be doublets, of course, but the resonance of C3 will be split twice into a triplet. In the next turn of the TCA cycle, a new phenomenon crops up; label in C2 of glutamate flows to C1 of fumarate. Fumarate is a symmetric molecule, so C1 and C4 are identical; the generated oxaloacetate will either have a labeled triplet at C2–C3–C4 or at C1–C2–C3. Adding the next labeled acetyl-CoA yields [1,2,3,4-^{13}C]glutamate and more [2,3,4-^{13}C]glutamate. These two molecules have the same splitting patterns in C4 and C3, but in the first C2 will be a quartet, and in the second, a doublet. It is clear that even though the only new bond that is ever made is that between C2 of oxaloacetate and C2 of acetyl-CoA, the splitting patterns become increasingly more complex with more turns of the cycle. The patterns can therefore be used to determine the relative length of time (or number of turns) that each molecule remains in the TCA cycle. If the labeled substrate can enter the cycle through an anaplerotic pathway, such as pyruvate carboxylase, the isotopic pattern will be altered again in a predictable way. Malloy *et al.* (1988, 1990) have worked out the equations that relate the isotopic pattern found in glutamate at metabolic and isotopic steady state to the following kinetic parameters: y, the ratio of anaplerotic flux to citrate synthase flux; F_c, the fractional enrichment of acetyl-CoA; and F_a, the fractional enrichment of the anaplerotic substrate.

7. TISSUE HETEROGENEITY

Even if NMR spectra are successfully localized to a well-defined tissue volume, it is likely that the interrogated region of interest will contain more than one cell type. In heart, the predominant cell type is the cardiomyocyte and others—the vasculature and neural cells, etc.—are a small fraction of total volume. On the other hand, workload can vary dramatically across the heart wall, and between the two ventricles and the atria. Robitaille *et al.* (1993) demonstrated a large gradient in creatine kinase flux across the wall of the left ventricle. It has also been shown that the PCr/ATP ratio is elevated *in vivo* in the right ventricle muscle, but not the left ventricle, during catecholamine infusion. This is likely due to similar increases in fatty acid oxidation, but differences in workload between the ventricles (Schwartz *et al.*, 1994; Katz *et al.*, 1989). Most other NMR accessible organs are even more heterogeneous. Brain spectra report on a mixture of neurons, glia, and astrocytes, and ^{13}C studies have shown metabolic heterogeneity (Brainard *et al.*, 1989). Exercising skeletal muscle is comprised of fast twitch and slow twitch fibers, which are very heterogeneous with regard to metabolism (Ball-Burnett *et al.*, 1991).

The model development process should therefore consider the question: How will the validity of the model be affected by heterogeneity in this tissue? While it is impossible to provide a general answer, a consideration of three specific types of model will illustrate some principles that apply when tissue heterogeneity is a possibility.

7.1. Metabolic and Isotopic Steady State, Time-Dependent Experiment

First consider a system in metabolic and isotopic steady state. The simple example shown in Fig. 8 is the production of labeled lactate (*lactate) as an end product from labeled glucose (*glucose) in anaerobic tissue. The system is in metabolic steady state and reaches isotopic steady state by time t1, so that thereafter *lactate will be produced at a constant rate [Eq. (10)]. The system may contain other precursors of lactate, but the experiment has been designed to measure the rate of production of *lactate from exogenous *glucose, J^*_{lactate}. The calculation of J^*_{lactate} shows that it is simply the amount of labeled lactate produced in the time interval t2 – t1 = Δt:

$$J^*_{\text{lactate}} = (^*\text{lactate moles at t2} - {}^*\text{lactate moles at t1})/\Delta\text{t} \quad (19)$$

This experiment could be conducted with ^{14}C or ^{13}C labeled glucose using liquid scintillation counting, GC/MS, or NMR for data collection. If ^{14}C-glucose, or ^{13}C-glucose and GC/MS are to be used, two identical biological systems are required. The first experiment is terminated at time t1, while the other is sampled at t2. Labeled lactate is then separated from labeled glucose and other components by a method such as ion exchange chromatography. The eluted lactate is either counted by liquid scintillation, or derivatized for GC/MS. In the latter case, the choice of [U-^{13}C]glucose as substrate would allow *lactate to be monitored as the M + 3 peak. If ^{13}C NMR is used, data may be collected as the experiment proceeds, and J^*_{lactate} is then estimated from the change in the *lactate signal over time. The data collected by any of these methods can be used in the current model to illustrate a general feature of tissue heterogeneity in metabolic and isotopic steady state.

If tissue is homogeneous as shown in Fig. 8A, glycolysis may be considered to occur in a single compartment of arbitrary size X, and *lactate will accumulate at steady state with a constant rate of 6 per unit time. This is shown as line A on the accompanying graph. In contrast, the model in Fig. 8B shows a heterogeneous tissue comprised of two equal cell populations of size X/2, which have different glycolytic rates. Here, one collection of cells (gly_{fast}) produces lactate at a rate of 5 per unit time (line B1), which is five times faster than the rate of the cell population gly_{slow}, 1 per unit time (line B2). Under conditions of relative hypoxia, it would not be surprising if a population of cells corresponding to gly_{fast} extracts more glucose and exports more lactate than the neighboring more highly oxygenated cells represented by gly_{slow}. The *lactate produced by the heterogeneous system in Fig. 8B can be summed to yield a line which is indistinguishable from line A for the homogeneous model. Therefore, the calculated rate of *lactate production will be 6/time with either model.

A. Homogeneous Tissue

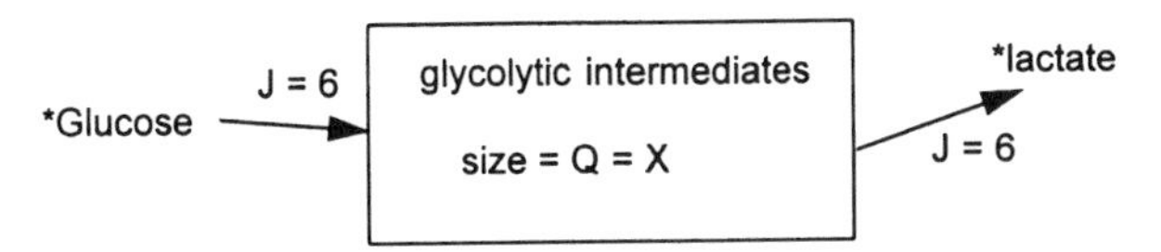

B. Heterogenous Tissue

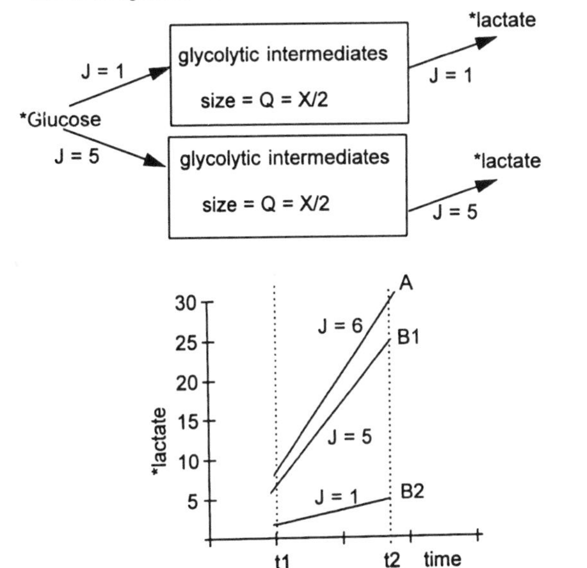

Figure 8. Metabolic and isotopic steady-state simulation of labeled lactate (*lactate) production. Model A depicts homogeneous tissue producing lactate at rate 6 per time. Model B depicts heterogeneous tissue comprised of two compartments of equal size (in moles of glycolytic intermediates) but different fluxes. In both cases it is assumed that all glycolytic intermediates can effectively be lumped into a single compartment. The production of labeled lactate versus time is graphed in steady state. The graph demonstrates that rate of labeled lactate production measure for the single compartment will equal that measured for the sum of the two compartments provided all systems are in metabolic and isotopic steady state.

Under these circumstances, heterogeneity of the biological sample will not be apparent to the investigator. Moreover, this result is independent of the choice of tracer and detection system. In general, tracer flux through systems in metabolic and isotopic steady state do not reveal any information about possible tissue heterogeneity. The positive aspect of this result is that no errors are produced due to heterogeneity. The negative aspect is that no information is gained about the organization of the metabolic system from the measurement of tracer fluxes.

7.2. Tissue Heterogeneity Measured in Pre-Isotopic Steady State

The results obtained from comparing homogeneous and heterogeneous tissue in metabolic and isotopic steady state contrast with those from a system sampled prior to isotopic steady state. Consider a system in metabolic steady state which is examined by NMR from the moment tracer is first added at $t = 0$ until it reaches isotopic steady state. Again, the glycolytic pathway will be used to illustrate the general features of this model. In this case, however, *lactate is an intermediate during $^*CO_2$ production from *glucose. First, examine a homogeneous system where the tissue lactate pool is well represented by a single compartment of size $Q = 30$ (Fig. 9). The flux through the lactate compartment is $J = 2$, and *lactate production and fractional enrichment F_{lac} are both described by the classic exponential equation [Eqs. (20) and (21)]. The quantity J/Q is flux over pool size, and is equal to the fractional rate constant k used in previous examples. The time constant of the exponential is the time required for the variable of interest to reach 63% (= 1 − 1/e) of its final, equilibrium value.

The example shown here represents ideal, error-free data, and any given data point along the plotted curve will yield a correct estimate of J/Q as described by Eq. (22). In real NMR experiments, appropriately weighted nonlinear regression would be used to estimate the value of J/Q from the entire data set plotted as a F_{lac} or *lactate time-dependent profile. Once J/Q is determined experimentally by fitting, the size of the pool (Q) is measured with other techniques and J or flux rate is estimated.

Now consider the consequences of employing this experiment in a heterogeneous system as shown in Fig. 10. The total flux through this system is the same as for the homogeneous case, $J = 2$, but it is divided equally between two compartments of $J = 1$ each. One path flows through a lactate compartment of size 25, Q_{big}, while the other flows through Q_{small} of size 5. Note that the *total* flux ($J = 2$) and the *total* lactate pool size (30) are equivalent to the example of Fig. 9. This type of heterogeneity could easily result from enzyme regulation at a rate-limiting step; consider the possibility that pyruvate dehydrogenase is partially inactivated in Q_{big} so that its rate constant is smaller than that of Q_{small}. The pyruvate and lactate pool (considered to be near-equilibrium) would have to increase in the Q_{big} fraction in order to maintain flux.

The labeling of each lactate compartment will occur over time as shown in panel A. However, the investigator is examining the tissue as a whole and will not observe these two distinct profiles; he will see a single profile corresponding to the weighted sum of the two traces. This curve (labeled $Q_{big} + Q_{small}$) looks superficially like a very good approximation to an equation for a single compartment [Eq. (20)]. However, as illustrated by Eq. (23), this labeling pattern is not equivalent to a simple exponential expression.

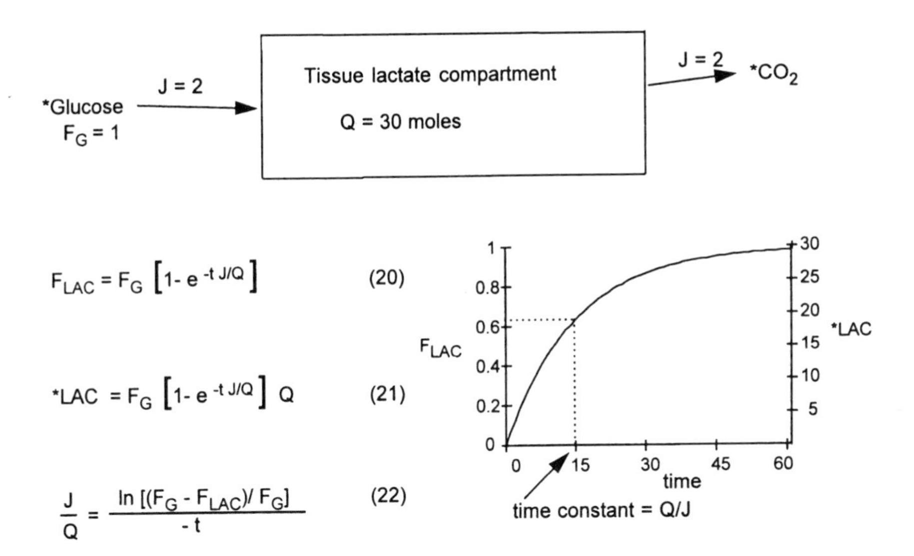

Figure 9. A single compartment representing tissue lactate at metabolic steady state but pre-isotopic steady state. Equations (20) and (21) described the labeling profile for the enrichment of lactate, F_{LAC}, and the total amount of labeled lactate, *LAC, in the tissue with time as shown in the graph. It is assumed that the enrichment of glucose, F_G, is 1. Q represents compartment size in moles and J is flux. Equation (22) demonstrates the determination of J/Q from the labeling profile.

Unless the data are very precise, it is not likely that this type of heterogeneity will be detected. The plots in panel B illustrate how pre-steady-state tracer data from a heterogeneous tissue might be misinterpreted. An investigator observing the solid line might assume that this profile represents a single compartment and use Eq. (20) (Fig. 9) to find the best fit estimate of J/Q. Fitting the data, shown as the solid line in panel B, to the single exponential curve would yield the dashed line, which has $J/Q = 1/22$. To examine the consequences of heterogeneity in the estimation of the glycolytic rate, assume that the investigator again measures the total lactate pool and finds, as for the homogeneous case of Fig. 9, that it is 30. The glycolytic flux is then estimated by the relationship

$$J = Q/22 = 30/22 = 1.36$$

This estimate is only 68% of the true total flux value of 2 for this tissue bed. This indicates that a significant error may be generated when heterogeneous tissues are treated as if they were homogeneous in pre-steady-state tracer experiments. Because the error in the case shown in Fig. 10 results from the fact that the two pools reach their isotopic steady state at different times, a better estimation of the true

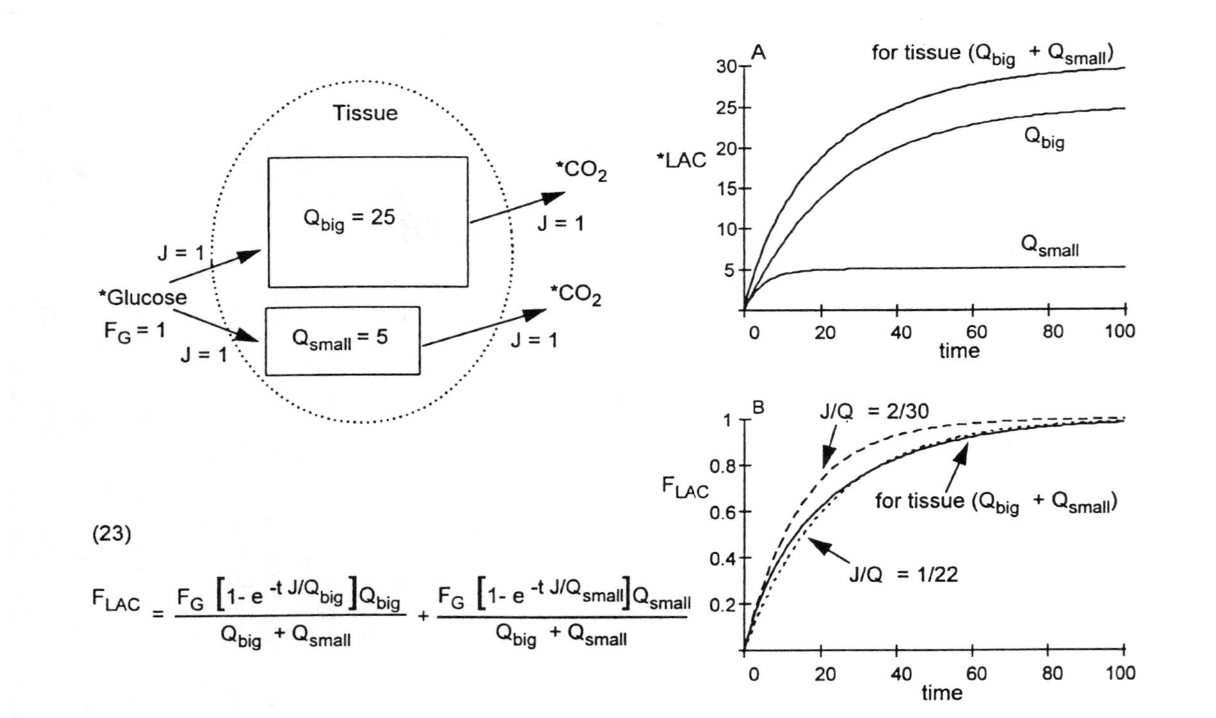

$$F_{LAC} = \frac{F_G \left[1 - e^{-t\,J/Q_{big}}\right] Q_{big}}{Q_{big} + Q_{small}} + \frac{F_G \left[1 - e^{-t\,J/Q_{small}}\right] Q_{small}}{Q_{big} + Q_{small}} \qquad (23)$$

Figure 10. Heterogeneous tissue comprised of two compartments of lactate at metabolic steady state but pre-isotopic steady state. The size of each compartment is represented by Q_{big} and Q_{small} ; other symbols are described in Fig. 9. Equation (23) describes the enrichment of lactate, F_{LAC}, with time in this heterogeneous tissue. Panel A graphs the amount of labeled lactate in each compartment with time as well as the sum of the two compartments representing the tissue. Panel B compares the labeling profile for the tissue as described by Eq. (23), solid line, with that for the single compartment from Fig. 9 ($J/Q = 2/30$) and with that for a single compartment with J/Q of 1/22.

flux through this particular system is the initial rate, estimated from the slope of data taken very early in the experiment. If very good data are available, the heterogeneity shown here would be most obvious by performing a "runs" test, which would indicate that the observed values systematically fall above the model solution at $t < 30$, and below the fit at $t > 30$ (Glantz and Slinker, 1990). In summary, this example illustrates a general property of pre-isotopic steady state experiments: the kinetic parameters estimated from a fit of time data are not independent of tissue heterogeneity, and may be significantly different from their true values.

7.3. Fractional Enrichment in the Metabolic and Isotopic Steady-State Experiment

Occasionally, a special situation will occur in which pool fractional enrichments will report on metabolic heterogeneity in a tissue. If a product B is derived directly from a single precursor A in a cell population, its measured ^{13}C fractional enrichment F_B will be equal to the measured F_A at all carbons (Fig. 11). Metabolic compartmentation is a possibility if the measured F_A and F_B are different from predictions based on a single well-mixed intracellular pool model, or on the presumption of a near-equilibrium between two compartments. In the second example shown in Fig. 11, A exists in two cell populations, one in which it gives rise to B (population 1), and one in which it is metabolically inactive (population 2). When F_A is measured, it is the average of F_{A1} and F_{A2} weighted by their respective pool sizes. When F_B is measured, it is also an average weighted by the fact that B_2 is at its natural abundance of 1.1%, and $F_B < F_A$. If the model for the metabolic process under consideration is known to be that of Fig. 5B, and $F_B \neq F_A$, the hypothesis that a single metabolic compartment exists must be rejected.

Brainard *et al.* (1989) used this unique ^{13}C NMR ability to detect differential labeling in C2 and C4 of glutamate and GABA (made from glutamate) to demonstrate the presence of multiple metabolic compartments in rat brain after a [1-^{13}C]glucose bolus. Total brain glutamate showed 30% fractional enrichment at C2 and 40% at C4. C2 is labeled through both pyruvate carboxylase and pyruvate dehydrogenase (PDH), while C4 is labeled only via PDH activity. The corresponding pools of GABA showed 60% and 0% labeling. The bulk of glutamate must have been made in a compartment with high pyruvate dehydrogenase activity, while the GABA was made in a compartment where oxaloacetate was highly labeled through pyruvate carboxylase, but where there was little pyruvate dehydrogenase.

Another example of compartmentation is provided by the following radioactive tracer study of glycogen and glucose utilization in perfused heart. Glycogen and exogenous glucose both enter the glycolytic pathway to produce pyruvate, which can either enter the TCA cycle and be oxidized to CO_2, or be converted directly to lactate. Henning *et al.* labeled the glycogen pool using either [5-^{3}H] or [U-^{14}C]glu-

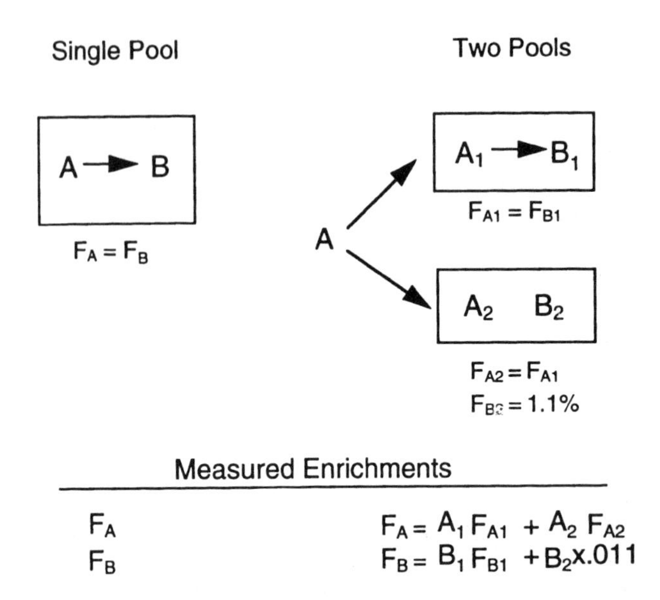

Figure 11. Metabolic heterogeneity can appear as altered fractional enrichments measured in whole tissues. If B is produced only from A in a single metabolic pool, $F_A = F_B$ as shown on the left. If A reacts to give B in only one pool, and other pools containing both A and B also exist in the tissue as shown on the right, the resultant fractional enrichments of A and B are weighted averages of the two pools, and will not be the same.

cose, washed away the original tracer, then presented the hearts with glucose labeled with the other. They measured effluent ^3H_2O (glycolysis) and $^{14}CO_2$ (oxidation) and found that more than half of the glycogen utilized by these hearts was oxidized, while less than 20% of the exogenous glucose entering glycolysis was oxidized (Henning *et al.*, 1996). This indicates that there may be at least two different pyruvate compartments in the beating heart. ^{13}C NMR has also provided evidence that this is the case. The fractional enrichments of glutamate, lactate, and alanine were measured in extracts from both perfused hearts (Lewandowski, 1992) and *in vivo* dog hearts given [3-^{13}C]pyruvate (Laughlin *et al.*, 1993). The lactate fractional enrichment was less than that of alanine and glutamate, even though all compounds are direct products of pyruvate metabolism. Taken together, these studies imply that one pool of pyruvate equilibrates with lactate, while another more rapidly equilibrates with amino acids and enters the TCA cycle. Therefore, ^{13}C NMR can be invaluable for the study of metabolic compartmentation because it lends itself well to the determination of the fractional enrichment of a variety of compounds in a single pathway. It is becoming clear that in a tracer study, it may be unwise to assume that the cell behaves like a well-mixed solution of enzymes and substrates, or that all sampled cells are identical.

8. THE ^{13}C NMR EXPERIMENT

The following section will discuss a few of the important assumptions, requirements, and technical aspects of ^{13}C NMR tracer experiments. Some considerations apply only to the *in vivo* experiment, while others also apply to isolated tissues.

8.1. Chemical Shift and Spectral Resolution

A radioactive tracer is quantitated as disintegrations per minute (dpm). It is not possible to tell where the tracer is located: which compound, or the location of the tracer atom in that molecule. Experimenters must be very clever and choose the specific labeled substrate and sampling method very carefully, and in addition must isolate potentially contaminating labeled chemical species before counting dpms in the species of interest.

In ^{13}C NMR, each carbon that experiences a unique chemical and physical environment will have a unique chemical shift. This means that it is usually possible to determine both the molecule and the position within the molecule where the ^{13}C label is located. Resolution is the ability to distinguish two peaks from one another. Peaks can be resolved if the chemical shift difference is larger than the average width of the two peaks, where peak width is an inverse function of magnetic field homogeneity and transverse relaxation time (T_2^*). The carbon spectrum is also reasonably large—slightly more than 200 ppm—which means that compounds of interest are widely dispersed and can usually be resolved quite well. In addition, the ^{13}C NMR carbon chemical shift and peak width tend to be much less sensitive to ion concentration and pH than either ^{1}H or ^{31}P NMR.

Many biological molecules have very similar functional groups, which resonate at similar frequencies. This makes it difficult to tell the internal carbons in long fatty acid acyl chains apart, for instance. The problem is most severe in *in vivo* studies where peaks tend to be very broad. Usually, the problem is solved by going to higher field strength, or to more homogeneous systems (like isolated organs, cells, or solutions of extracted metabolites). A very powerful technique is to use $\{^{13}C{-}^{1}H\}$ 2-dimensional shift-correlated spectroscopy to enhance resolution (van Zijl *et al.*, 1993). This was successfully used in the brain of intact cats to separate resonances from ^{13}C-labeled glutamate, glutamine, lactate, and glucose.

8.2. Metabolic Perturbation

In the *in vivo* experiment, it is important to choose methods that minimally perturb the system under study. From one standpoint, this is achieved beautifully by *in vivo* ^{13}C NMR in that multiple "samples" can be taken over time in a nondestructive way. From another point of view, however, the large concentration of label that must be present for adequate signal-to-noise in active, biological systems may constitute in itself an overwhelming metabolic perturbation. It is possible to detect molecular pools containing > 0.5 mM ^{13}C in a few minutes. To study *in vivo* lactate metabolism in the heart, therefore, it may be necessary to raise blood [3-^{13}C]lactate to 5–10 mM in order to detect enrichment in intracellular compounds (Laughlin *et al.*, 1993). However, plasma lactate is normally about 0.5 mM, and heart uptake of lactate is proportional to its plasma concentration. When lactate uptake is high, the concentration of NADH and pyruvate, products of lactate metabolism, are increased in both the cytosol and the mitochondrial compartments. This creates a somewhat artificial metabolic state in the heart. For NMR studies of brain metabolism, it has become common to raise blood glucose with [1-^{13}C]glucose to as high as 13 mM–16 mM (van Zijl *et al.*, 1993, 1994; Fitzpatrick *et al.*, 1990). However, glucose uptake and intracellular glucose is a function of plasma concentration, with a plasma glucose concentration at half-maximal uptake of between 4 and 6 mM (van Zijl *et al.*, 1997; Mason *et al.*, 1992; Gruetter *et al.*, 1996). An example of large metabolic perturbation due to plasma substrate concentration is glycogen synthesis in response to plasma glucose in rat skeletal muscle. Glucose uptake and glycogen synthesis are sensitive to plasma glucose and insulin concentration (Farace and Rossetti, 1992; Rossetti and Giaccari, 1990). At both basal (180 pM) and elevated insulin (2500–3000 pM vs. normal fed insulin of 450 pM) rat muscle glycogen synthesis is increased by raising plasma glucose from 5 mM to 15 mM; in the hyperinsulinemic rats, this amounted to an increase from 77 to 210 μmol/kg·min. Muscle glucose-6-phosphate (G6P), a regulator of glycogen synthesis, is increased more than 50% by glucose elevation at the lower insulin concentration, but not at the higher. If a perturbation is imposed during a ^{13}C NMR experiment with [1-^{13}C]glucose that is expected to raise skeletal muscle G6P and therefore the glycogen synthesis rate, the effect is likely to be most apparent at basal glucose and insulin, and may be masked by the experimental conditions under which muscle ^{13}C-glycogen can be observed in rats. The researcher must be aware of the metabolic perturbations, which are sometimes extreme, imposed by his experimental design.

8.3. Detection Limits

The natural abundance of ^{13}C in all carbon is 1.1%, and the enrichment of a typical substrate in a ^{13}C NMR experiment is often between 50 and 99.9%. In contrast, the plasma tracer specific activity of radionuclides used in a typical experiment is 0.01% (10^4 dpm/μmol). It is easy to see that ^{13}C meets neither the criteria for low background signal or that for a "massless" tracer. However, it is not always necessary to provide very large fractional enrichments for the sake of detection. If the tissue can be excised and extracted, the labeled molecules are stable and ^{13}C NMR spectra can be taken over many hours at much higher fields (yielding higher resolution and S/N) than are generally available for in situ experiments. A lower limit for the fractional enrichment of the substrate in this instance is set not by the ability to detect label, but by the error in detecting ^{13}C over the natural abundance. With GC-MS, this limit is about 0.1% atom percent above natural abundance (Bier, 1982). In the NMR experiment, the intensity of the carbon of interest can sometimes be compared with the intensities of the unlabeled carbons in the same molecule in a fully relaxed spectrum, and excess fractions of less than 1% can be measured with acceptable errors.

8.4. Correction for Natural Abundance Fractional Enrichment

In tracer theory established with radionucleides, it is customary to assume that the tracer is essentially massless, i.e., that the mass of the infused tracer can be neglected when calculating specific activity, which equals mol tracer/mol tracee. In the ^{13}C experiment, this is almost never appropriate, and if we define A^* as the labeled and A as the unlabeled molecule, the fractional enrichment is defined as $A^*/(A+A^*)$ [Eq. (1)]. In addition, the endogenously produced molecule has a natural abundance ^{13}C of 1.1%, which will contribute to the total measured A^*. This means that measured A^* = infused A^* + $0.011A$. Often, the natural abundance ^{13}C is subtracted out of the NMR data as a baseline, natural abundance spectrum, and poses no problem for calculations. If substrate fractional enrichment is specifically measured, the natural abundance ^{13}C is included in this measurement, and can pose few problems, especially if the substrate fractional enrichment is very high. Occasionally, a very high rate of labeled substrate infusion, or the presence of appreciable natural abundance ^{13}C in a metabolite pool, can make calculations more difficult.

Consider the imaginary experiment in which ^{13}C glucose is infused into a 350 g rat, and we would like to measure *Ra* by the single pool model (Fig. 4B). This method requires only that we infuse tracer at a constant, known rate, and measure plasma fractional enrichment of the molecule of interest at steady state. In our experiment, the infusion rate of 55 μmol/min of 99.9% enriched [1-^{13}C]glucose results in a steady-state plasma glucose fractional enrichment of 40%. Let us define F^*_{A} as fractional enrichment of the infusate, F^{N}_{A} as fractional enrichment of the glucose produced naturally by the animal, and F_{A} as the measured plasma fractional enrichment of glucose; J^*_{A} is the infusion rate of labeled glucose (mol glucose/min), Ra_{A} is the production rate of natural glucose (mol glucose/min), and Rd_{A} is the total rate of disappearance of glucose (mol glucose/min). If we were to calculate Ra_{A}, ignoring both the mass of the infused [1-^{13}C]glucose and the naturally occurring [1-^{13}C]glucose, we would choose Eq. (24), and make the assumption that calculated Ra_A and Rd_A were equal:

$$Ra_A = F^*_A\ J^*_A\ /F_A = Rd_A \tag{24}$$

The correction for mass of infused tracer requires a correction of F_A, and Rd_A no longer equals calculated Ra_A:

$$F_A = \text{infused } A^*/\text{total } A = F^*_A\ J^*_A\ /(J^*_A + Ra_A)$$

$$Ra_A = J^*_A\ (F^*_A - F_A)/F_A = Rd_A - J^*_A \tag{25}$$

Finally, the correction for naturally occurring ^{13}C:

$$F_A = (\text{infused } A^* + \text{naturally occurring } A^*)/\text{total } A = (F^*_A\ J^*_A + F^N_A\ \ Ra_A)/(J^*_A + Ra_A)$$

$$Ra_A = J^*_A\ (F^*_A - F_A)/(F_A - F^N_A\ ) = Rd_A - J^*_A \tag{26}$$

For the above example, Ra_A was calculated by Eqs. (24)–(26) and the results are shown in Table 2. When F_A is varied, the error associated with neglecting the natural abundance fractional enrichment is low as long as F_A is high. However, as F_A drops, the error becomes large.

A second example (Table 3) demonstrates that fractional enrichment of a large pool like glycogen must be corrected for the natural abundance of glycogen that was there before the experiment (Laughlin *et al.*, 1988). Glycogen is synthesized in heart during a 100 min perfusion with ^{13}C glucose containing different fractional enrichments, the hearts are freeze-clamped, and the glycogen ^{13}C content and fractional enrichment is measured. The aim of the experiment is to measure the net rate of glycogen synthesis. Heart glycogen is made from exogenous glucose, and

Table 2
Uncorrected *Ra* (I), or *Ra* with Correction for Mass of Infused A* (II) and for Naturally Occurring ^{13}C (III)

		I No Correction		II Correct for Infused Mass		III Correct for F^N_A		Error in *Ra*
$F^*_A J^*_A$	F_A	*Ra*	*Rd*	*Ra*	*Rd*	*Ra*	*Rd*	(II vs. III)
55	40%	138	138	82	137	85	140	3.5%
3.5	5%	70	70	66	70	85	89	22%

Table 3
Net Glycogen Synthesis Calculated With and Without Correction for Natural Abundance ^{13}C

F_g	T (total glycogen)	F_{gly}	^{13}C glycogen	Glycogen synthesis rate (uncorrected)	Glycogen synthesis rate (corrected)	Error
100%	50 mM	36%	18 mM	0.180 mM/min	0.176 mM/min	2%
50%	50 mM	18%	9 mM	0.180 mM/min	0.173 mM/min	4%
5%	50 mM	1.8%	0.9 mM	0.180 mM/min	0.090 mM/min	100%

the precursor, UDPG, is expected to have the same enrichment as perfusate glucose. Glycogen is very difficult to deplete in the heart, so the extracted glycogen will consist of an old pool with its natural abundance fractional enrichment, and a new pool with a fractional enrichment equal to that of perfusate glucose. The fractional enrichment of the total pool is a weighted average of the two; F_g = plasma glucose fractional enrichment, F_{gly} = total glycogen fractional enrichment, F^N is the natural abundance fractional enrichment, O is the size of the old glycogen pool (mmol), N is mmol of new glycogen, T is the total glycogen pool, and t is time.

We can easily calculate glycogen synthesis from measured T and F_{gly} without correction for the natural abundance enrichment:

$$\text{glycogen synthesis} = N/t = {}^{13}\text{C glycogen}/(F_g t) = (T\,F_{gly})/(F_g\,t) \qquad (27)$$

Equation (27) will overestimate the rate of glycogen synthesis if there is a large fraction of naturally occurring glycogen. To correct for F^N, we can solve the two following equations;

$$\text{total glycogen} = T = O + N$$

$$T\,F_{gly} = F^N\,O + F_g N = F^N\,(T - N) + F_g N$$

Solving for N we obtain

$$N = T\,(F_{gly} - F^N)/(F_g - F^N)$$

$$\text{glycogen synthesis} = \{T\,(F_{gly} - F^N)/(F_g - F^N)\}/t \qquad (28)$$

Table 3 lists the rates of glycogen synthesis calculated without [Eq. (27)] and with [Eq. (28)] correction for the natural abundance of pre-existing glycogen. Again, the error in calculating the rate becomes negligible at very high substrate and product fractional enrichments, but can be appreciable if the substrate fractional enrichment approaches the natural abundance level of 1.1%.

The high natural abundance of ^{13}C can be exploited, since some very large metabolite pools, such as fatty acid acyl chains and glycogen, are sometimes visible in the *in vivo* NMR experiment without being labeled. If so, the changes in ^{13}C NMR signals are directly related to changes in total pool sizes. Shulman *et al.* were able to measure the changes in human leg and liver glycogen in time with exercise and fasting using natural abundance ^{13}C NMR (Price *et al.*, 1996).

8.5. Subtraction of Natural Abundance Spectra in *In Vivo* Experiments

An *in vivo* ^{13}C NMR spectrum of heart, skeletal muscle, or liver will show large, broad, irregular peaks centered at 30, 130, and 180 ppm due to the natural abundance ^{13}C carbon of lipids (Laughlin *et al.*, 1993). Any large reservoirs of carbohydrate, such as glycogen, will be represented by a broad collection of peaks between 60 and 100 ppm. These background peaks (especially those of lipids) are often larger than the peaks that will arise later on from metabolism of an infused tracer. It is usually necessary, then, to subtract an appropriate natural abundance control spectrum from each experimental spectrum in order to best detect peak intensities. In most *in vivo* studies, the control spectrum must be acquired before the exposure of the experimental subject to tracer, so the two spectra which must be combined are separated in time, sometimes considerably. Care must be taken that the control spectrum was taken under exactly the same physical conditions as those from which it will be subtracted. If a change is made in the workload imposed on skeletal muscle or heart, a new control spectrum (taken in the absence of tracer) is required. Movement of or in the sample, especially with respect to the detection coil, will often make subtraction impossible. Changes in volume, such as tissue edema or cellular swelling, can cause large subtraction errors. Changes in line widths, caused by altered magnetic susceptibility in the sample compartment during the experiment, will also make subtraction difficult.

This problem is eliminated if the compound of interest resonates in a clean region of the spectrum that has little natural abundance signal. Flux of [1-^{13}C]glucose (92.6 and 96.7 ppm) into [1-^{13}C]glycogen (100.6 ppm) can be easily observed without subtraction of a control spectrum.

8.6. Sites of Label Entry and Sampling, and Substrate Enrichment

The blood vessels for infusion and sampling site, or the sampling point in a perfusion circuit, should be chosen so as to best control and measure the intracellular concentrations of the infused molecule in the organ of interest. The circulation can be considered unidirectional, starting from the right ventricle of the heart, picking up oxygen in the lungs, leaving the left ventricle, and delivering labeled substrate to the body tissues. Blood picks up metabolic products along its way back through the venous circulation of skeletal muscle, brain, liver, etc., and finally back

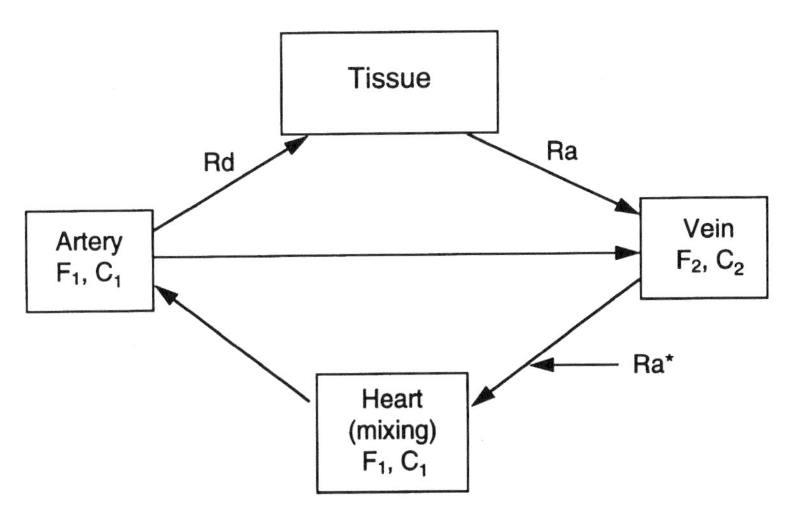

Figure 12. The unidirectional blood circuit from heart to tissues through the arterial vessels, returning through the venous system. In the ^{13}C NMR experiment, label is most commonly introduced into the vein (Ra^*) so it can be mixed to produce a homogeneous concentration (C_1) and F_1 in heart. Material disappears into tissues (*Rd*) and is produced endogenously (*Ra*), which yields a new C_2 and F_2 in the vein.

to the heart (Fig. 12). The problem is that substrates are simultaneously drawn from the arterial circulation into tissues (*Rd*), and produced from tissue into the venous circulation (*Ra*). At steady state, both the substrate concentration and fractional enrichment would be reduced due to *Rd* and *Ra* in venous blood relative to arterial blood. For the case where uptake of the plasma substrate is the only entry path into the cell, substrate is infused into the venous circulation to ensure good mixing in the heart before entering the arteries and being delivered to the organ of interest. The sampling site should be after the points of entry of labeled and unlabeled substrate (Ra^*), but before considerable substrate has been removed from the circulation (*Rd*). The fractional enrichment of substrate being presented to the target organ is usually best represented by, and should be sampled in, the arterial blood ("V–A mode") (Wolfe, 1992).

A reasonably straightforward example is that of [2-^{13}C]acetate metabolism in heart. Because acetate is a minor endogenous substrate in nonruminant mammals, but is avidly consumed, the major point of entry is the infusion. The intracellular acetate ^{13}C fractional enrichment is likely to be similar to that in the blood which is presented to the heart, i.e., the arterial blood. For this experiment, the infusion point should be in a vein, so that mixing can occur in the left ventricle before blood again reaches the coronary arteries.

Consider an *in vivo* ^{13}C NMR experiment in an anesthetized rat in which the labeled molecule is lactate, and the organ of interest is again the heart. Since the heart simultaneously extracts blood lactate and produces lactate from glucose, the intracellular lactate fractional enrichment will be less than the fractional enrichment of the lactate in blood. However, we would like to measure the fractional enrichment of the lactate taken up by the heart. If [3-^{13}C]lactate is infused into the jugular vein, it will be taken directly to the heart to be mixed with other venous blood and oxygenated in lung. Unlabeled lactate is produced predominantly in skeletal muscle and blood cells, and lactate is cleared by the liver and heart. In addition, [3–^{13}C]lactate is exchanged with unlabeled pyruvate by muscle and erythrocyte lactate dehydrogenase, and the resultant [3-^{13}C]pyruvate is also avidly taken up by the heart. The heart is perfused directly with blood from the left ventricle, and therefore the measurement of exogenous lactate and pyruvate fractional enrichment should be made in the arterial circulation as close to the heart as possible (i.e., the carotid artery). If the fractional enrichment measurement is made in the femoral vein, the [3-^{13}C]lactate fractional enrichment will be lower than arterial blood due to lactate production in skeletal muscle, and the [3-^{13}C]pyruvate/[3-^{13}C]lactate will be higher.

8.6.1. Bolus, Constant or Primed Infusion

When designing a tracer experiment, it is necessary to use a tracer input function that is appropriate for the desired information. This is relatively trivial if the study subject is an isolated organ or cell system, and the label is mixed homogeneously in the perfusate. The following is therefore concerned with the *in vivo* experiment. In general, if the experiment is to be done in isotopic and metabolic steady state, and if it is convenient to wait for isotopic steady state, a constant tracer infusion is warranted, where the time to steady-state fractional enrichment is described by one or more exponentials. Figure 4B shows a pool A into which label is being infused at a rate Ra^*. In this case, if $Rd = kA$, and V is the volume of distribution, the fractional enrichment in pool A is F_A:

$$F_A = (Ra^*/V\,[A_{total}]\,k)\,\{1 - \exp(-kt)\} \tag{29}$$

This function may become complex if $[A_{total}]$ and Rd are changing with time, and is discussed in detail in Patlad and Pettigrew (1976).

If the experiment requires metabolic steady state, but yields information from changes in substrate fractional enrichment, a bolus infusion may be best. After injection of a perfect tracer bolus (introduction of entire dose in infinitely short time) the plasma or perfusate fractional enrichment F_A rises instantaneously to its maximal value F_0, then falls exponentially over time, dependent only on k, the rate constant for flux from the pool of interest:

$$F_A = F_0 \exp(-kt) \tag{30}$$

The maximum fractional enrichment, found at time 0, is determined by

$$F_0 = \text{dose (mol)}/\{V[\text{A}] + \text{dose (mol)}\} \tag{31}$$

Major problems are the estimation of the pool size, the strong dependency of the model on the assumption of instantaneous mixing in the infusion pool, and the ability of the researcher to deliver the entire dose in a very short time relative to the loss of tracer determined by *Rd*. This is especially difficult if a very large total amount of tracer is used. Because of the difficulty of knowing the fractional enrichment at any time, a bolus infusion is usually used only in a ^{13}C NMR experiment where the researcher is looking for differential fractional enrichments (Brainard *et al.*, 1989).

If it is desirable to measure the flux of label from infused substrate into a metabolite pool at metabolic steady state and isotopic nonsteady state, the plasma substrate fractional enrichment should be raised quickly to its steady-state value and kept constant thereafter. If the total plasma pool volume and concentration remain constant during the infusion, this can be accomplished with a primed continuous infusion. Tracer is delivered first as a bolus to raise the fractional enrichment of tracer to the desired level, followed by a continuous infusion to keep it there. This results in an apparent steady-state fractional enrichment at times much earlier than steady state is achievable by a continuous infusion. In an ideal, well-mixed single-pool system, the label from the bolus will fall by the same function that the label from the continuous infusion rises, and the fractional enrichment at each time point is constant if the bolus dose and infusion rate are chosen properly (Wolfe, 1992). Combining Eqs. (29) and (30) we obtain

$$F_{\text{A}}(t) = F_{\text{bolus}}(t) + F_{\text{continuous}}(t) = F_{0,\text{B}} \exp(-kt) + F_{\infty,\text{C}} \{1 - \exp(-kt)\}$$

$$F_{\text{A}}(t) = F_{\infty,\text{C}} \qquad \text{when } F_{0,\text{B}} = F_{\infty,\text{C}} \tag{32}$$

The primed continuous infusion can be used for studies with radioisotopes, where the pool size is unlikely to change significantly during the tracer infusion. This is rarely the case for studies using stable isotopes. When raising the total glucose concentration in humans, the procedure of the hyperglycemic glucose clamp is used (DeFronzo *et al.*, 1979; Patlad and Pettigrew, 1976). [1-^{13}C]glucose is infused as a priming dose at a rate which is decreased every minute following an exponential curve for the first 15 min of the clamp. The priming dose to raise blood glucose by 125 mg/dl in an adult human is 9,622 mg/m^2 (DeFronzo *et al.*, 1979). At the end of this period, blood glucose is sampled every 5 min, and the glucose infusion rate is adjusted accordingly. A similar infusion protocol has been devised for animals, whereby a bolus of glucose is given which is sufficient to increase total plasma glucose to the desired concentration, followed by a continuous infusion that is decreased every 30–60 sec following an exponential curve, until the infusion rate and *Rd* are matched (Patlad and Pettigrew, 1976; Fitzpatrick *et al.*, 1990).

8.6.2. Local Infusion of Tracer

In some limited cases, it may be advantageous to infuse tracer locally, near the site of detection. Because large concentrations of label are often desired for ease of detection (sometimes as much as 10 mM), the impact on whole body metabolism can be huge. For a large animal, the cost of such an experiment can also be prohibitive. If the tissue to be interrogated is fed predominantly by a single arterial vessel, a catheter placed directly into that vessel can deliver a large concentration of trace directly to the area under the coil, which is then diluted as it passes through the rest of the body. This method has been used successfully in dog heart with a variety of infused ^{13}C substrates (Laughlin *et al.*, 1992, 1993; Robitaille *et al.*, 1993). The delivery catheter was placed into the lumen of the LAD, and the detection coil was over the anterior apex of the heart. LAD blood flow was measured downstream of the catheter tip. Enrichment of delivered substrate in the LAD was calculated as follows for the example of infused glucose, where $[A^*]_{inf}$ = tracer concentration in infusate, F_{inf} = fractional enrichment of glucose in infusate, $flow_{Inf}$ = infusate flow (ml/min), J^*_{ing} = rate of appearance of tracer (mol/min), F_1 = fractional enrichment of glucose in arterial blood entering LAD, $flow_1$ = flow of blood into LAD, J_1 = rate of appearance of endogenous substrate, F_{LAD} = fractional enrichment of glucose in LAD, $flow_{LAD}$ = total flow in LAD (ml/min) (Fig. 13):

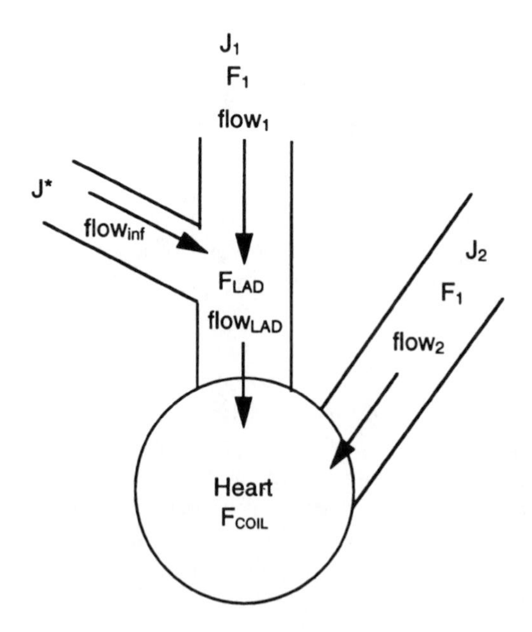

Figure 13. Local infusion of tracer into the LAD of dog heart for detection by a surface coil located over the anterior apex. Tracer is infused at rate J^*_{inf} (mol/min) with $flow_{inf}$ (ml/min) into a catheter positioned in the LAD. Blood moving into the LAD has $flow_1$, appearance of endogenous substrate J_1, with a fractional enrichment of F_1. Below the point of the catheter, we measure $flow_{LAD}$ and F_{LAD} due to both infused tracer and normal blood delivery. A second vessel labeled 2 perfuses the region of interest under the coil. The fractional enrichment of detected substrate is F_{COIL}.

$$\text{flow}_{\text{inf}} = J^*_{\text{ing}} / [\text{A}^*]_{\text{inf}}$$

$$\text{flow}_{\text{LAD}} = \text{flow}_{\text{inf}} + \text{flow}_1$$

$$F_{\text{LAD}} = \frac{J^*_{\text{ing}} + F_1 J_1}{J^*_{\text{ing}} + J_1} = \frac{\text{flow}_{\text{inf}}[\text{A}^*]_{\text{inf}} + F_1(\text{flow}_{\text{LAD}} - \text{flow}_{\text{inf}})[\text{A}]_{\text{B}}}{\text{flow}_{\text{inf}}[\text{A}^*]_{\text{inf}} + (\text{flow}_{\text{LAD}} - \text{flow}_{\text{inf}})[\text{A}]_{\text{B}}} \tag{33}$$

The fractional enrichment of glucose that reaches the tissue at the apex underneath the coil (F_{COIL}) will be equal to or less than the fractional enrichment of glucose calculated here for the LAD, and is a function of the amount of blood that reaches the same tissue from other vessels (Fig. 13). In order to keep the F_{LAD} constant, the infusion rate must be increased by the same factor as flow of blood into the LAD. Also, any label that stays in the circulation and reflows back to the heart in arterial blood will appear as F_1, and must be accounted for.

This model will work well as long as the blood flow from the vessel of label entry to the heart apex is a constant fraction of all blood flow reaching the apex. It will fail in the case where blood flow from other vessels is differentially altered. In Table 4, F_{LAD} and F_{COIL} are compared for an imaginary case of changing heart work loads where the ratio of flow from the LAD and other vessels is constant or altered. The tracer infusion rate is always matched to the flow in vessel 1. Thus, any metabolic rates obtained from tracer data in workload III cannot be directly compared to rates measured during workload I, and these rates will be in error if calculated assuming a constant precursor fractional enrichment of 50%.

8.7. Fractional Enrichment

The following several sections will be concerned with fractional enrichment. In any tracer experiment, substrate fractional enrichment must be known in order to calculate fluxes from the experimental data. In the ^{13}C NMR experiment, the

Table 4
Calculated Substrate Fractional Enrichments During a Local Infusion

	Workload I $\text{flow}_1 = 5\ \text{flow}_2$	Workload II $\text{flow}_1 = 5\ \text{flow}_2$	Workload III $\text{flow}_1 = \text{flow}_2$
J^*	10	20	20
J_1	10	20	20
J_2	2	4	20
F_{LAD}	50%	50%	50%
F_{COIL}	45%	45%	33%

directly measured peak intensities report on the total pool size of a labeled metabolite A^*, and its behavior with time dA^*/dt. The total pool size $A_T = A + A^*$ is determined from Eq. (1) by dividing A^* by F_A. Therefore, the ^{13}C peak intensities of two compounds yield only their relative labeled pool sizes, not the actual pool sizes. This must be kept in mind whenever changes in A^* are measured, since they could be due to either changing A_T or F_A.

8.7.1. Measuring Fractional Enrichment of the Substrate

The ^{13}C fractional enrichment of the substrate in plasma or perfusate samples can be measured with GC-MS (van Zijl *et al.*, 1997), or by coinfusion of a radioactive tracer (Shulman *et al.*, 1985). If the ^{13}C-labeled carbon of interest is covalently bound to protons, a ^1H NMR spectrum may be useful for directly obtaining the ^{13}C fractional enrichment of that compound (Lewandowski, 1992; Laughlin *et al.*, 1993; Mason *et al.*, 1992). This technique works well for methyl groups ([3-^{13}C]lactate, [3-^{13}C]alanine) and other moieties that have very simple proton spectral features ([1-^{13}C]glucose). Because both ^1H and ^{13}C have a spin quantum number $S = \pm 1/2$, each resonance will be split if they are covalently bound to each other. The proton resonance will be split into a doublet with an interpeak distance J of 20–200 Hz, and the carbon into $N + 1$ peaks, where N is the number of identical bound protons. All molecules that contain ^{12}C instead of ^{13}C will have a singlet proton signal. The ratio of the doublet to the total signal at the proton resonance of interest is equal to the fractional enrichment of the bound carbon with ^{13}C. This phenomenon can be used *in vivo* to monitor the fractional enrichment of a metabolite in time. Rothman *et al.* (1991) were able to monitor the fractional enrichment of [3-^{13}C]lactate at the site of the infarct in the brain of a stroke patient during an infusion of [1-^{13}C]glucose. They showed that the elevated tissue lactate pool was labeled in time to a fractional enrichment that was 50% of that of the plasma glucose, indicating that the lactate was metabolically active, and did not represent ischemic tissue. Zhao *et al.* (1992) used proton NMR to monitor the fractional enrichment of alanine and lactate in KCl-arrested perfused rat hearts that had been supplied with [3-^{13}C]pyruvate. They showed that the fractional enrichment of alanine was similar to that in extracts of the same hearts (about 60%), while the lactate fractional enrichment was much higher in the intact hearts (85% vs. 67% in the extracts).

8.7.2. Time Dependence of Pool Fractional Enrichment

A label usually flows through several pools, from the blood or perfusate to the interstitial space, to the intracellular space (denoted by ic), and through chemical reactions to other molecules (Fig. 14). If there is no endogenous production of the substrate in any of the pools shown, at steady state, the fractional enrichment of each pool should be equal to that of the plasma. Even if uniform mixing of label

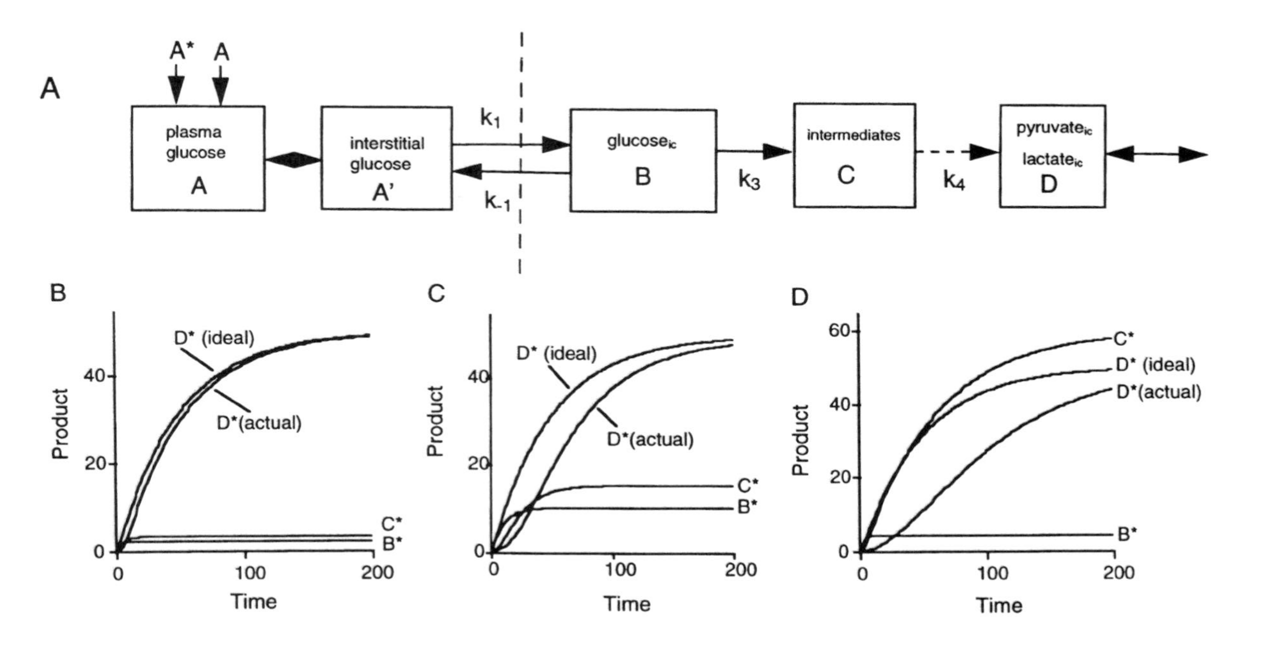

Figure 14. (A) A sequential multi-pool system representing the uptake and metabolism of glucose in a tissue. (B) Labeled metabolite in pools B, C, and D are plotted for the case where B and C are small relative to D (Table 5). D^*(actual) is the calculated curve under these conditions. D^*(ideal) is calculated assuming that the pool sizes are all 0 between D and plasma, where label is found at a constant concentration and fractional enrichment. (C) B and C are 20 and 30% of D, respectively. (D) C is 120% of D. The difference between D^*(actual) and D^*(ideal) is a function of relative pool sizes.

with each entire pool is instantaneous, however, the enrichment with label in each of these pools follows an exponential behavior in time, determined by pool size and by the rate of flux between pools [Eqs. (14), (17), and (18)]. Transport of glucose into the brain is a very good example of this phenomenon, because transport is relatively slow, only about 2.4 times the rate of glucose utilization at normal glucose concentrations, and the intracellular glucose pool can be quite large (brain glucose/plasma glucose is from 0.2–0.45 μmol/g/μmol/ml). Therefore, [1-^{13}C]glucose appearance in brain from plasma [1-^{13}C]glucose has been measured in time with *in vivo* NMR, and used to calculate brain glucose transport in animals and people (Mason *et al.*, 1992; Gruetter *et al.*, 1992, 1996; van Zijl *et al.*, 1997). The apparent first-order rate constant for glucose transport in cat brain was found to be about 0.054/min at a plasma glucose concentration of 25 mM (van Zijl *et al.*, 1997).

Often, the researcher would like to measure a flux later in a metabolic pathway, and would like to know the fractional enrichment in the immediate precursor pool for the reaction of interest. It is tempting to assume that this is the same as the substrate fractional enrichment in plasma, since this quantity can be easily measured. This is most likely to be the case when uptake of substrate from the plasma pool is the single route of entry, and there is no local production of unlabeled substrate to dilute the intracellular fraction enrichment relative to the blood. It is also a good estimate if the size of pools between the plasma and the products of interest are small and rapidly replenished relative to the total flux through the pathway. If there is a large metabolic pool like brain glucose that turns over relatively slowly between the point of tracer entry (plasma) and the observed compartment of interest (lactate, for instance), the correct substrate fractional enrichment will differ significantly from that of plasma at early time points.

The fractional enrichment of each pool follows an exponential time course, which means that the input function into the next pool in the pathway is also a time-dependent exponential. A multiple-pool catenary model describes the situation in Fig. 14A. If we assume that the fractional enrichment of plasma glucose (pool A) is constant, i.e., it was raised to its steady-state value instantaneously at time 0, then Eq. (18) is a general solution for the behavior of tracer flow into pool D. Dividing Eq. (18) by D_T yields F_D.

Figure 14 shows the calculated D^*(actual) from Eq. (18) in metabolic steady state for three cases where the relative pool sizes of B, C, and D are varied, but the flux between the pools is always the same at 2 mol/min. D^*(actual) is compared to D^*(ideal), which is calculated with the same pool size and flux, but this time we assume that the immediate precursor pool (C) has a constant steady-state fractional enrichment, i.e., F_C equals the plasma substrate F_A of 0.50. In Fig. 14B, the pool sizes B and C are much smaller than D, and the assumption that $F_C = F_A$ is not unreasonable. In Fig. 14C, pools B and C are 20% and 30% of D, respectively, and D^* has a clearly sigmoidal shape with a substantial lag in labeling. In Fig. 14D, pool C is 20% larger than D, and the resultant lag is also very large. Table 5 shows the

Table 5
Error in Fluxes from Assumption that Plasma F_A = Substrate F_C

	Flux (real)	Flux (from fit)	B/D	C/D	Error
A	2	1.81	0.04	0.06	10%
B	2	1.28	0.20	0.30	36%
C	2	0.85	0.04	1.20	57%

real fluxes for these cases calculated with the true $F_C(t)$, or with the assumption that $F_C = F_A = 0.50$ (Flux from fit). In the latter case, the curves of D^*(actual) in Fig. 14 are fit to Eq. (13), which is the exponential equation that the researcher would choose if he assumed that $F_C = F_A$. The error in this calculation quickly becomes unacceptable as the intermediate pools approach the size of D. Therefore, it is easiest to calculate fluxes from pre-isotopic steady-state data if all pools between the site of entry and the measured product are small and the fluxes between them are fast. For instance, Yu *et al.*.(1995) calculated that acetyl-CoA in perfused heart (0.2 μmol/g dry wt) is labeled from exogenous [2-^{13}C]acetate to 78% of its steady-state value in 10 sec. When intermediate pool sizes are not negligible, flux rates can be estimated from time data of D^* with an appropriate choice of model.

8.8. Sensitivity and Time

The most important instrumentation advance for application of NMR to biological systems was FT-NMR, which allows many acquisitions to be collected as FIDs (free induction decay) in the time domain, summed, and filtered before Fourier transformation to the frequency domain. Coherent signal is added together, while randomly distributed noise is not. This means that the sensitivity of detection, or spectral signal-to-noise, is dependent on the number of acquisitions as well as the field strength and sample size. The signal-to-noise is increased as a function of the square root of the number of scans. Since each scan is acquired in a defined length of time, the sensitivity is proportional to the square root of the time over which the spectrum is accumulated.

For tracer experiments done at isotopic and metabolic steady state, the time available for the acquisition of data with adequate signal-to-noise is limited only by the stability of the biological preparation (or the patience of the person in the magnet!). For dynamic phenomena, several time points with adequate signal-to-noise must be acquired within the time before steady state is achieved in order to define the system. If the data are to be fit to a simple model, such as a single exponential or a straight line, fewer time points are required than for a multiexponential curve. Therefore, the acceptable period of time available for data acquisition must be a small fraction of the time to achieve steady state, and is dependent on the expected shape of the dynamic curve. If the maximum signal-to-noise achievable in the available time is still too low, substrate fractional enrichment, sample size, field strength, or detection coil efficiency must be increased. Figure 15A shows an example, where the actual biological system is described by the D^*(actual) curve of Fig. 14C. The sigmoidal shape is apparent when data are acquired with a time resolution of 5 min (small open circles). The true shape cannot be ascertained from data collected every 100 min, as the filled triangles indicate. A fit of the data acquired with poor time resolution to Eq. (13), which might appear to be a reasonable modeling choice, yields a very poor fit to the true data. In Fig. 15A, the 100 min data were plotted at the midpoint of the acquisition time, so that the intensity of the spectrum that was completed at 100 min is plotted at 50 min, that for the 200 min spectrum is shown at 150 min, etc. Figure 15B shows the situation when the data are plotted at the end of the acquisition time. The resultant curve bears little resemblance to the real data. It is important to keep in mind that while many physiologic or metabolic parameters are measured at a specific point in time, NMR data are usually acquired over a period of time.

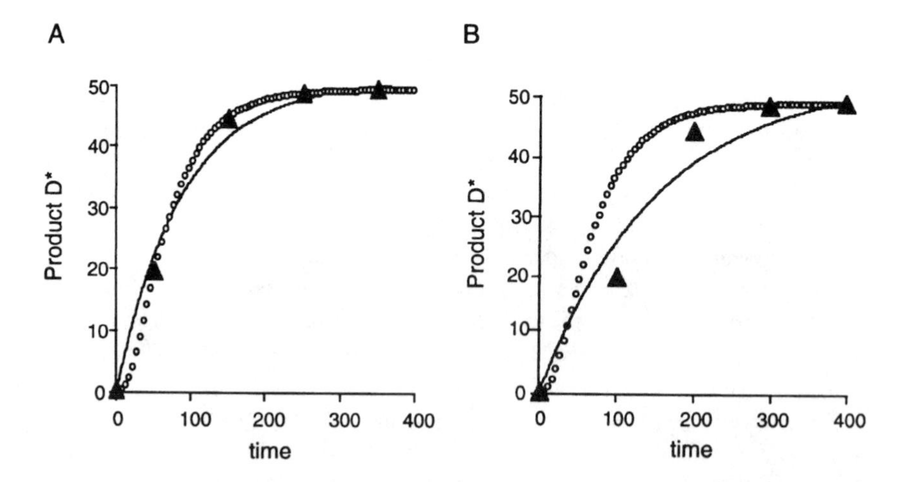

Figure 15. (A) Data acquired with 5 min time resolution (small open circles) describes a sigmoidal shape at early time points, which is not apparent when data are acquired with 100 min time resolution, and plotted at the midpoint time of spectral acquisition (filled triangles). The solid line is a best fit of the 100 min data to a single exponential, which is inadequate to describe the system. (B) The same data are shown, but plotted at the end of the acquisition period, rather than the midpoint. The resultant best fit curve bears little resemblance to the real data.

NMR is rather insensitive to ^{13}C for two reasons: low natural abundance (which is increased artificially in the tracer experiment) and a low nuclear magnetic moment μ (due to γ, the magnetogyric ratio). The same number of protons would give a signal that is 63 times stronger. Therefore, the sensitivity and time resolution of detection would be increased if we could detect the protons coupled to ^{13}C, rather than detect ^{13}C directly. There are many techniques that have been proven successful in biological ^{13}C-tracer experiments to do just that, and are lumped under the name of indirect detection. All exploit the fact that protons attached to ^{13}C carbons experience a perturbation (J-coupling) that is sensitive to the magnetic state of the attached carbon nucleus. The POCE, or proton-observe-carbon-edit experiment, yields excellent signal-to-noise. For this technique, two scans are taken, one that detects all protons, and one that detects all protons, but inverts those coupled to ^{13}C with a selected value of j. The difference of the two yields just those protons that are attached to ^{13}C (Fitzpatrick *et al.*, 1990). Another approach is to use coherence selection to directly detect in a single scan only those protons bound to ^{13}C. Gradient-enhanced heteronuclear multiple quantum coherence, or HMQC, was used successfully to detect ^{13}C glucose, glutamate, and lactate in cat brain (van Zijl *et al.*, 1993).

8.9. Use of Reporter Molecules

We will define a reporter molecule as an NMR-visible compound which is not directly involved in the pathway of interest, but which can be used to derive information when the pool sizes of the compounds of interest are too small to observe. Often, one molecule can be used to report on the fractional enrichment of another if the two pools are at near-equilibrium with each other through a highly active enzyme. For example, alanine and pyruvate are interconverted in heart by a transaminase whose activity is thought to be much higher than pyruvate turnover (Peuhkurinen *et al.*, 1983, Peuhkurinen and Hassinen, 1982; Zhao *et al.*, 1992; Lewandowski, 1992). If true, the fractional enrichment of alanine and pyruvate must be equal. Since the pyruvate pool is too small to be visible, measurement of the alanine fractional enrichment would yield pyruvate fractional enrichment. This idea can be explored experimentally in hearts receiving [3-^{13}C]pyruvate as substrate, which then constitutes the primary source of acetyl CoA for the TCA cycle. The fractional enrichment of this acetyl-CoA (F_C) can be estimated by isotopic analysis of the glutamate produced in the TCA cycle (Malloy *et al.*, 1988, 1990), which will then determine how much this pool is diluted with other unlabeled substrates. If pyruvate and alanine are equilibrated in the cytosol, the [3–^{13}C]alanine fractional enrichment (and therefore that of [3-^{13}C]pyruvate) must be larger than or equal to F_C. This idea is supported by measurements made in heart extracts after *in vivo* [3-^{13}C]pyruvate or [3-^{13}C]lactate infusion in dogs (Laughlin *et al.*, 1993), or with [3-^{13}C]pyruvate and unlabeled acetate or glucose as substrate in perfused hearts (Zhao *et al.*, 1992) which showed that alanine fractional enrichment was always equal to or greater than F_C. In another study, hearts were perfused with [3-^{13}C]pyruvate, and pyruvate dehydrogenase was activated with dichloroacetate (Lewandowski, 1992). In the control hearts of this study, [3-^{13}C]alanine fractional enrichment was only 60% relative to an F_C of 80%. When dichloroacetate was included, the [3-^{13}C]alanine fractional enrichment remained unaltered at 60%, but F_C rose to well over 90%. Flux of exogenous pyruvate through pyruvate dehydrogenase therefore appeared to be faster (and could be increased with dichloroacetate) than its equilibration with alanine. These results indicate that compartmentation of alanine or pyruvate could well exist within the heart cell, and that the relationship between the fractional enrichments of alanine and pyruvate may be a function of the physiologic and metabolic state of the heart.

This is almost sure to be the case for lactate and pyruvate as well, which are sometimes thought to be in near-equilibrium through lactate dehydrogenase. When ^{13}C-3-lactate is the substrate for heart *in vivo*, the ^{13}C lactate and ^{13}C alanine fractional enrichments are equal (Laughlin *et al.*, 1993). The common intermediate between these two pools is cytosolic pyruvate. Although blood lactate and pyruvate are also in constant exchange due to erythrocyte lactate dehydrogenase, tissue ^{13}C lactate and ^{13}C alanine fractional enrichment were also equal when unlabelled plasma pyruvate was artificially elevated, suggesting that the alanine, pyruvate, and lactate pools were all equilibrated. When [3-^{13}C]pyruvate was the substrate (in perfused heart or *in vivo*), however, the lactate pool had a fractional enrichment which was substantially below that of alanine in all cases (Zhao *et al.*, 1992; Laughlin *et al.*, 1993; Lewandowski, 1992). In this case, it is clear that lactate and pyruvate are not equilibrated in heart and lactate fractional enrichment is not a good measure of pyruvate fractional enrichment.

Glutamate is a large pool that is thought to be in fast exchange with α-ketoglutarate through fast transport across the mitochondrial membrane, and either a transaminase or a dehydrogenase. ^{13}C-labeling in glutamate is commonly used to report on flux rates and fractional enrichment of TCA intermediates in heart and brain (Malloy *et al.*, 1988, 1990; Lewandowski, 1992; Laughlin *et al.*, 1993; Chance *et al.*, 1983). When the system is at metabolic and isotopic steady state, the relative fractional enrichments of the 5 carbons of glutamate will report faithfully those of the corresponding carbons of α-ketoglutarate. If the system is not at steady state, the ability to use glutamate to detect α-ketoglutarate fractional enrichment is a function of the ratio of fluxes of α-ketoglutarate to glutamate vs. that of α-ketoglutarate to succinyl-CoA. In a recent study conducted in perfused rabbit heart, Yu *et al.* (1995) measured the time dependence of the ^{13}C signals in C2 and C4 of glutamate with [2-^{13}C]acetate or [2-^{13}C]butyrate as substrates. MVO_2 and glutamine transaminase enzyme kinetics were also measured. The data were fit to a model which allowed the estimation of TCA cycle flux rate (V_{TCA}) and the rate of

α-ketoglutarate flux to glutamate (F_1); F_1 was shown to be similar to V_{TCA} under normal workloads, but was about twice V_{TCA} in the KCl-arrested heart. The value of F_1 was 20-fold lower than transaminase activity, indicating that mitochondrial transport may in fact limit this flux. More importantly, the fact that F_1 is similar to V_{TCA} indicates that glutamate and α-ketoglutarate cannot be considered to be in a near-equilibrium state in the perfused heart.

A reporter molecule does not have to report on fractional enrichment, and therefore near-equilibrium with a molecular pool under study is not a requirement. In other words, glutamate can be used to interrogate α-ketoglutarate at steady state when the ratio of label in different carbons is the same for the two molecular pools. Likewise, the ratio of [3-^{13}C]– to [2-^{13}C]lactate made from [2-^{13}C]glucose in isolated cells can be used to interrogate the fraction of flux through the glycolytic pathway that proceeds via the pentose phosphate shunt. Figure 16 shows this

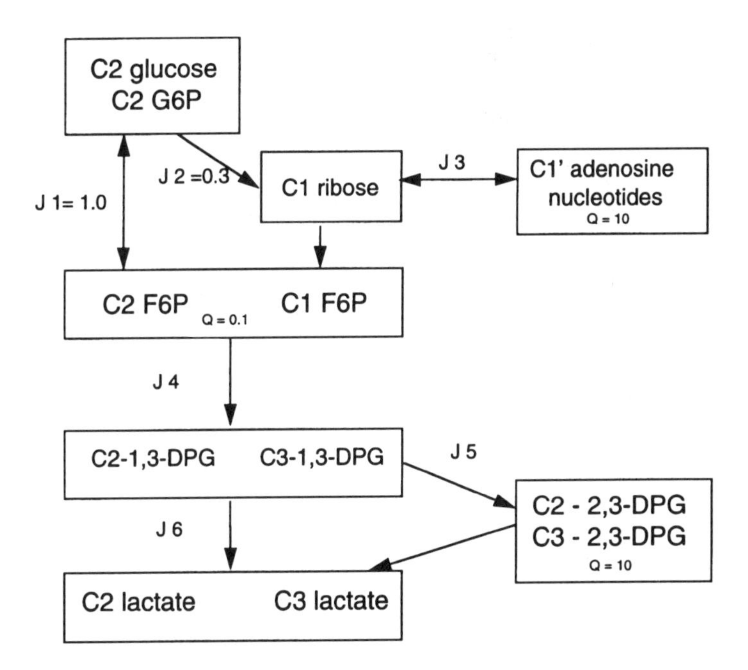

Figure 16. A compartmental model of glycolysis and pentose phosphate shunt in the erythrocyte, using [2-^{13}C]glucose as substrate, and showing pool sizes (Q) and fluxes (J) between compartments. Detected products are [1′-^{13}C]adenosine nucleotides, [2-^{13}C]-2,3-DPG, and [2-^{13}C]lactate produced via the pentose phosphate shunt, and [3-^{13}C]-2,3-DPG and [3-^{13}C]lactate produced via normal glycolysis. The ratio of C3 to C2 labeled products yields the fraction of glycolysis that proceeds via the pentose shunt.

experiment in the erythrocyte. [2-^{13}C]glucose yields one unlabeled and one [2-^{13}C]pyruvate (or [2-^{13}C]lactate) in the glycolytic pathway. The first carbon of glucose is liberated in the 6-phosphogluconate dehydrogenase step of the pentose phosphate shunt, and the resultant pentose can eventually return to the glycolytic pathway as [1-^{13}C]fructose-6-P (F6P). This, in turn, produces one unlabeled and one [3-^{13}C]pyruvate (or lactate). Care must be taken with this method, since the label in [2-^{13}C]pyruvate can be scrambled to [3-^{13}C]pyruvate through the TCA cycle and malic enzyme, as well as in the pentose phosphate shunt. In the absence of malic enzyme, as in the erythrocyte example shown in Fig. 16, the ratio of signal at the C3 and C2 positions of lactate can be used to calculate the flux through the pentose phosphate shunt as a fraction of total glycolysis (Shrader *et al.*, 1993; Laughlin and Thompson, 1996).

Lactate is an end product used in this example to report on an earlier event in the pathway. If the lactate pool has a constant concentration and the system is at isotopic and metabolic steady state, the labeling pattern in lactate at any time will faithfully report on the event (the pentose phosphate pathway) of interest. If lactate is a major product which rises throughout the experiment, as in packed erythrocytes, the label will arrive in lactate continuously, both before and during steady state, and the lactate fractional enrichment will report an average over time (the total label pattern reflects history as well as present). If this is the case, lactate will have the correct ratio of C2 and C3 only if no loss of label occurs via any other pathway prior to remixing of the molecules that have been relabeled in the two pathways (the fructose-6-phosphate pool). Loss of label after this point (into the 2,3-diphosphoglycerate pool) will not alter the ratio of [3-^{13}C]- to [2-^{13}C]pyruvate.

When the pentose phosphate shunt is stimulated with methylene blue, these cells produce adenine nucleotides from adenine and ^{13}C-1-ribose (Page *et al.*, 1998, Fig. 17A). They also have a large cycle through 2,3-DPG mutase and 2,3-DPG phosphatase which results in equilibration of label in the 2,3-DPG pool with newly formed glycolytic intermediates. At early time points, the adenine nucleotide pool becomes labeled with [1-^{13}C]ribose (liberating unlabeled ribose back into the pentose phosphate shunt) while [2-^{13}C]glucose-6-phosphate flows unimpeded to [2-^{13}C]lactate in the glycolytic pathway (Fig. 17B). During this time, both the 2,3-DPG and lactate labeling patterns will yield an underestimate of the true C3/C2 ^{13}C ratio (Fig. 17C). After isotopic steady state is reached, the 2,3-DPG pool will yield the correct ratio (since this pool has no "history"). The lactate pool C3/C2 will underestimate the true ratio due to label from the pentose phosphate shunt which is found in the [1-^{13}C]ribose of the adenine nucleotide pool instead of the [3-^{13}C]lactate pool (Fig. 17C). In this case, the ratio (R) of [3-^{13}C]/[2-^{13}C] in 2,3-DPG, which is the number needed to estimate the relative rates of J_2 and J_1 in Fig. 16, equals {[3-^{13}C]lactate + [1-^{13}C]ribose} / [2-^{13}C]lactate (Page *et al.*, 1998). The activity of the pentosephosphate shunt (PPS) can be estimated from

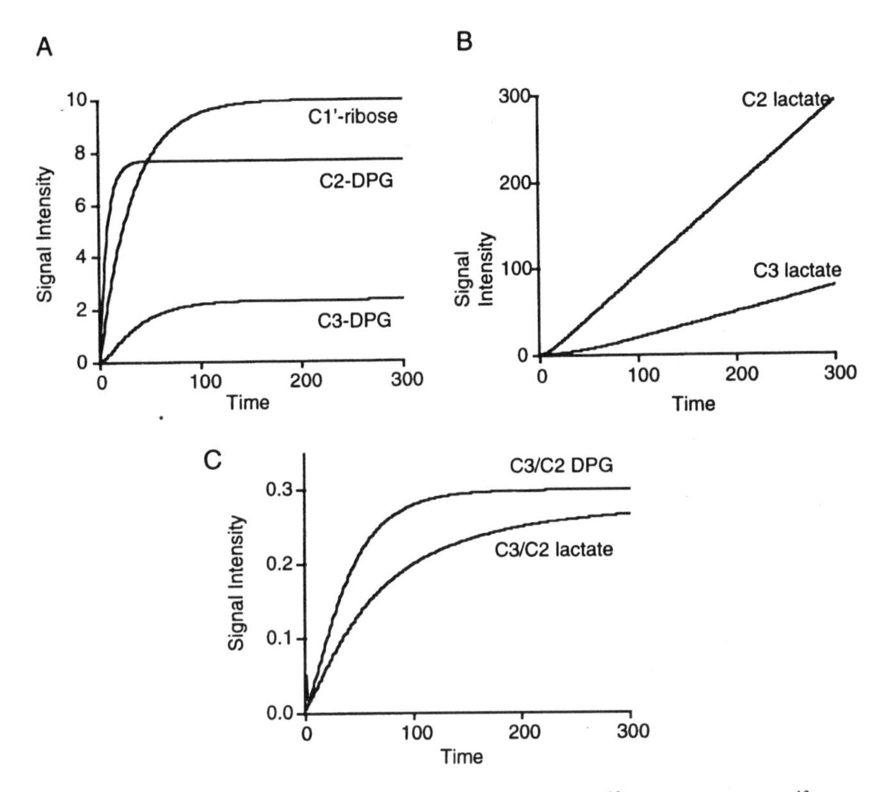

Figure 17. (A) The production of [1′-^{13}C]adenosine nucleotides, [2-^{13}C]-2,3-DPG, and [3-^{13}C]-2,3-DPG from the system of Fig. 16 plotted as a function of time. (B) End products [2-^{13}C]lactate and [3-^{13}C]lactate. (C) The ratio of C3/C2 of both 2,3-DPG and lactate. Note that once isotopic steady state is reached, the 2,3-DPG ratio is constant, and at the correct level ($R = J_2/J_3 = 0.3$). C2/C3 of lactate approaches, but does not reach, the correct ratio.

the true ratio (R) and the total rate of glucose consumption (G) (Shrader *et al.*, 1993):

$$R = \frac{2\ \text{PPS}/3\ \text{G}}{1 + (2\ \text{PPS}/3\ \text{G})} \tag{34}$$

9. CONCLUSIONS

The *in vivo* ^{13}C NMR technique opens up a rich new landscape for metabolic research, providing the ability to detect the flow of label in real time from one metabolite pool to the next in living tissue. However, the abundance of data that is gleaned from these experiments yields up useful results only via the process of mathematical modeling. The use of tracer kinetics in biological systems has a rich history dating back at least fifty years, and the body of theory expands as each new tracer, detection methodology, or biological system requires a new collection of models tailored to particular strengths and limitations. With ^{13}C NMR, we have been able to monitor multiple intracellular compounds in time, allowing the use of far more detailed and complex models than ever before (Chance *et al.*, 1983). The challenge presented to the investigator now is to use the particular strengths of ^{13}C NMR, those of in situ, time-dependent, and compound-specific detection, to investigate those aspects of biological systems that have constituted the black boxes of assumptions made for other tracer methodologies.

REFERENCES

Ball-Burnett, M., Green, H. J., and Houston, M. E., 1991, *J. Physiol.* **437**:257.

Berman, M., Shahn, E., and Weiss, M. F., 1962, *Biophys. J.* **2**:275.

Bier, D. M., 1982, *Nutrition Rev.* **40**:129.

Brainard, J. R., Kyner, E., and Rosenberg, G. A., 1989, *J. Neurochem.* **53**:1285.

Briggs, G. E., and Haldane, J. B. S., 1925, *Biochem. J.* **19**:338.

Carson, E. R., Cobelli C., and Finkelstein, L., 1983, The Mathematical Modeling of Metabolic and Endocrine Systems, John Wiley & Sons, New York.

Chance, B., 1943, *J. Biol. Chem.* **151:**553.

Chance, E. M., Seeholzer, S. M., Kobayashi, K., and Williamson, J. R., 1983, *J. Biol. Chem.* **258**:13785.

Cohen, S. M., 1987a, *Biochemistry* **26**:563.

Cohen, S. M., 1987b, *Biochemistry* **26**:573.

Cohen, S. M., 1987c, *Biochemistry* **26**:581.

DeFronzo, R. A., Tobin, J. D., and Andres, R., 1979, *Am. J. Physiol.* **237**: E214.

DiStefano, J. J., Stubberfeurd, A. L., and Williams, I. J., 1990, *Feedback and Control Systems*, II, McGraw-Hill, New York.

Farrace, S., and Rossetti, L., 1992, *Diabetes* **41**:1453.

Fitzpatrick, S. M., Hetherington, H. P., Behar, K. L., and Shulman, R.G., 1990, *J. Cereb. Blood Flow Metab.* **10**:170.

Garfinkel, D., 1963, *Ann. N.Y. Acad. Sci.* **108:**293.

Glantz, S. A., and Slinker, B. K., 1990, *Primer of Applied Regression and Analysis of Variance*, McGraw-Hill, New York.

Gruetter, R., Novotny, E. J., Boulware, S. D., Rothman, D. L., Mason, G. F., and Shulman, R. G., and Tamborlane, W. V., 1992, *Proc. Natl. Acad. Sci. U. S. A.* **89**:1109.

Gruetter, R., Novotny, E. J., Boulware, S. D., Mason, G. F., Rothman, D. L., Shulman, G. I., Prichard, J. W., and Shulman, R. G., 1994, *J. Neurochem.* **63**:1377.

Gruetter, R., Novotny, E. J., Boulware, S. D., Rothman, D. L., Shulman, R. G., 1996, *J. Cereb. Blood Flow Metab.* **16**:427.

Henning, S. L., Wambolt, R. B., Schonekess, B. O., Lopaschuk, G. D., and Allard, M. F., 1996, *Circulation* **93**:1549.

Jacquez, J. A., 1985, *Compartmental Analysis in Biology and Medicine*, The University of Michigan Press, Ann Arbor, MI.

Katz, L. A., Swain, J. A., Portman, M. A., and Balaban, R. S., 1989, *Am. J. Physiol.* **256**:H265.

Laughlin, M. R., Petit, W. A., Dizon, J. M., Shulman, R. G., and Barrett, E. J., 1988, *J. Biol. Chem.* **263**:2285.
Laughlin, M. R., Taylor, J. F., Chesnick, A. S., and Balaban, R. S. 1992, *Am. J. Physiol.* **262**:E875.
Laughlin, M. R., Taylor, J., Chesnick, A. S., DeGroot, M., and Balaban, R. S., 1993, *Am. J. Physiol.* **264**:H2068.
Laughlin, M. R., and Thompson, D., 1996, *J. Biol. Chem.* **271**:28977.
Lewandowski, E. D., 1992, *Biochemistry* **31**:8916.
Malloy, C. R., Sherry, A. D., and Jeffrey, F. M. H., 1988, *J. Biol. Chem.* **263**:6964.
Malloy, C. R., Sherry, A. D., and Jeffrey, F. M. H., 1990, *Am. J. Physiol.* **259:**H987.
Mason, G. F., Behar, K. L., Rothman, D. L., and Shulman, R. G., 1992, *J. Cereb. Blood Flow Metab.* **12**:448.
Michaelis, L., and Menten, M. L., 1913, *Biochem. Z.* **49**:333.
Newsholme, E. A., and Start, C., 1974, *Regulation in Metabolism*, John Wiley and Sons, London.
Page, S., Salem, M, and Laughlin, M. R., 1998, *Am. J. Physiol.* **274**:E920.
Patlad, C. S., and Pettigrew, K. D., 1976, *J. Appl. Physiol.* **40**:458.
Peuhkurinen, K. J., and Hassinen, I. E., 1982, *Biochem. J.* **202**:67.
Peuhkurinen, K. J., Hiltunen, J. K., and Hassinen, I. E.,1983, *Biochem. J.* **210**:193.
Price, T. B., Perseghin, G., Duleba, A., Cehn, W., Chase, J., Rothman, D. L., Shulman, R. G., and Shulman, G. I., 1996, *Proc. Natl. Acad. Sci. U. S. A.* **93**:5329.
Robitaille, P.-M. L., Abduljalil, A., Rath, D., Zhang, H., and Hamlin, R. L., 1993, *Magn. Res. Med.* **30**:4.
Robitaille, P.-M. L., Rath, D. P., Abduljalil, A. M., O'Donnell, J. M., Jiang, Z., Zhang, H., and Hamlin, R. L., 1993, *J. Biol. Chem.* **268**:26296.
Rossetti, L., and Giaccari, A., 1990, *J. Clin. Invest.* **85**:1785.
Rothman, D. L., Howseman, A. M., Graham, G. D., Petroff, O. A. C., Lantos, G., Fayad, P. B., Brass, L. M., Shulman G. I., Shulman, R. G., and Prichard, J. W., 1991, *Magn. Res. Med.* **21**:302.
Schwartz, G. G., Greyson, C., Wisneski, J. A., and Garcia, J., 1994, *Am. J. Physiol.* **267**:H224.
Segel, I., 1975, *Enzyme Kinetics*, John Wiley and Sons, New York.
Sheppard, C. W., 1948, *J. Appl. Physiol.* **19:**70.
Sheppard, C. W., 1962, *Basic Principles of the Tracer Method*, John Wiley and Sons, New York.
Shoenheimer, R., and Rittenberg, D., 1935a, *J. Biol. Chem.* **111:**163.
Shoenheimer, R., and Rittenberg, D., 1935b, *J. Biol. Chem.* **111:**175.
Shrader, M. C., Eskey, C. J., Simplaceanu, V., and Ho, C., 1993, *Biochim. Biophys. Acta* **1182**:162.
Shulman, G. I., Rothman, D. L., Smith, D., Johnson, C. M., Blair, J. B., Shulman, R. G., and DeFronzo, R. A., 1985, *J. Clin. Invest.* **76:**1229.
Shulman, G. I., Shulman, R. G., and Prichard, J. W., 1991, *Magn. Res. Med.* **21**:302.
van Zijl, P. C. M., Chesnick, A. S., DesPres, D., Moonen, C. T. W., Ruiz-Cabello, J., and van Gelderen, P., 1993, *Magn. Res. Med.* **30**:544.
van Zijl, P. C. M., Davis, D., Eleff, S., Moonen, C. T. W., Parker, R., and Strong, J., 1997, *Annu. Meeting Soc. Magn. Res. Med.*, submitted.
Wolfe, R. L., 1992, *Radioactive and Stable Isotope Tracers in Biomedicine*, Wiley-Liss, New York.
Yu, X., White, L. T., Doumen, C., Damico, L. A., LaNoue, K. F., Alpert, N. M., and Lewandowski, E. D., 1995, *Biophys. J.* **69**:2090.
Zhao, P., Sherry, A. D., Malloy, C. R., and Babcock, E. E., 1992, *FEBS Lett.* **303:**247.
Zilversmit, D. B., Entenman, C., and Fishler, M. C., 1943, *J. Gen. Physiol.* **26:**325.

2

^{13}C Isotopomer Analysis of Glutamate

A NMR Method to Probe Metabolic Pathways Intersecting in the Citric Acid Cycle

A. Dean Sherry and Craig R. Malloy

1. INTRODUCTION

Pathways intersecting in the citric acid cycle have been studied extensively using carbon tracers since the early 1950s (Strisower *et al.*, 1952; Weinman *et al.*, 1957). In those experiments, the distribution of radiolabeled carbons in cycle intermediates was fit to mathematical models of varying complexity to measure relative flux through critical pathways such as gluconeogenesis. ^{14}C tracers, however, have practical disadvantages related to radiation precautions and laborious sample handling procedures which restrict their utility in animal and clinical studies. The advantages of monitoring a stable isotope such as ^{13}C stimulated development of mass spectroscopy for metabolic studies and, most recently, ^{13}C

A. Dean Sherry • Department of Chemistry, University of Texas at Dallas, P.O. Box 830688, Richardson, Texas 75083-0688 and the Department of Radiology, University of Texas Southwestern Medical Center, The Mary Nell and Ralph B. Rogers Magnetic Resonance Center, 5801 Forest Park Road, Dallas, Texas 75235-9085. **Craig R. Malloy** • Departments of Radiology and Internal Medicine, University of Texas Southwestern Medical Center at Dallas and Dallas Veterans Affairs Medical Center, The Mary Nell and Ralph B. Rogers Magnetic Resonance Center, 5801 Forest Park Road, Dallas, Texas 75235-9085.

Biological Magnetic Resonance, Volume 15: In Vivo Carbon-13 NMR, edited by L. J. Berliner and P.-M. L. Robitaille. Kluwer Academic / Plenum Publishers, New York, 1998.

NMR. Although the sensitivity of ^{13}C NMR is poor compared to any of these techniques, its primary appeal lies in the information it provides about ^{13}C-labeling patterns that is encoded in cycle intermediates as a consequence of ^{13}C–^{13}C and ^{1}H–^{13}C spin–spin coupling.

^{13}C NMR was first used to examine intermediary metabolism in studies of *Saccharomyces cervesiae* almost 25 years ago (Eakin *et al.*, 1972). The first ^{13}C NMR kinetic analysis, reported a decade later, used the fractional enrichment in glutamate carbons (analogous to ^{14}C specific activity measurements) to calculate citric acid cycle flux in the isolated heart (Chance *et al.*, 1983). It was noted that each glutamate carbon resonance was composed of rather complex multiplets due to ^{13}C–^{13}C spin–spin coupling, and that these multiplets were a consequence of turnover of the citric acid cycle. At approximately that time it was realized (Cohen *et al.*, 1981; Walker *et al.*, 1982; Cohen, 1983; Walsh and Koshland, 1984) that this complexity reflected the activities of various pathways feeding the citric acid cycle, and the development of tools that relied on detection of these multiplets was initiated. The use of ^{13}C multiplet data to measure metabolic processes has become known as isotopomer analysis (London, 1988). The term "isotopomer," a contraction of the words isotope and isomer, indicates molecules that differ from one another solely by isotope distribution. ^{13}C isotopomers are described by listing each site of ^{13}C enrichment while the unnamed sites are assumed to be unenriched (natural abundance levels of ^{13}C). Thus, for example, [1,4,5-^{13}C]glutamate indicates a high level of ^{13}C enrichment (usually 99% or greater) in carbons 1, 4, and 5 while carbons 2 and 3 are essentially all ^{12}C. As we shall see, neither ^{13}C NMR nor any other carbon tracer technique directly measures the concentration of all isotopomers in glutamate (2^5 or 32 isotopomers) or glucose (2^6 or 64 isotopomers). However, the areas of the various multiplets that appear in a ^{13}C spectrum as a result of ^{13}C–^{13}C spin–spin coupling do reflect the relative concentrations of different groups of isotopomers and these, in turn, are determined by relative fluxes through various biochemical pathways. In this chapter, we will summarize those pathways briefly, discuss some general features of NMR detection of ^{13}C-enriched molecules, compare several basic metabolic models that relate NMR spectral information to metabolism, and finally review the tools which ^{13}C isotopomer analysis provides for exploration of the citric acid cycle and related reactions.

1.1. The Role of the Citric Acid Cycle in Substrate Oxidation

During his early investigations which lead to formulation of the citric acid cycle, Krebs (1970) searched for "any piece of information which might have a bearing on the intermediary stages of the combustion of foodstuffs." Even before its complete description, the role of the citric acid cycle as a common pathway for the oxidation of diverse substrates was well appreciated. Indeed, many tissues are equipped for the complete oxidation of ketones, fatty acids of various chain lengths and saturation, glucose, pyruvate, lactate, and other substrates (Fig. 1). The final steps in the stoichiometric oxidation of these compounds to CO_2 and water requires the citric acid cycle. Since the inflow of 2 carbon units from acetyl-CoA is exactly balanced by the production of CO_2, the concentration of citric acid cycle intermediates does not change. This role of the cycle in provision of energy by substrate oxidation is often emphasized in the analysis of metabolism in skeletal muscle and heart, but it is equally important in all respiring tissues.

Thus, a basic question in metabolism is: which compounds are oxidized in the citric acid cycle to produce energy used by the tissue? This question, often formulated as the "substrate preference" of a tissue, would seem relatively straightforward to address since the available substrates are well known and the pathways

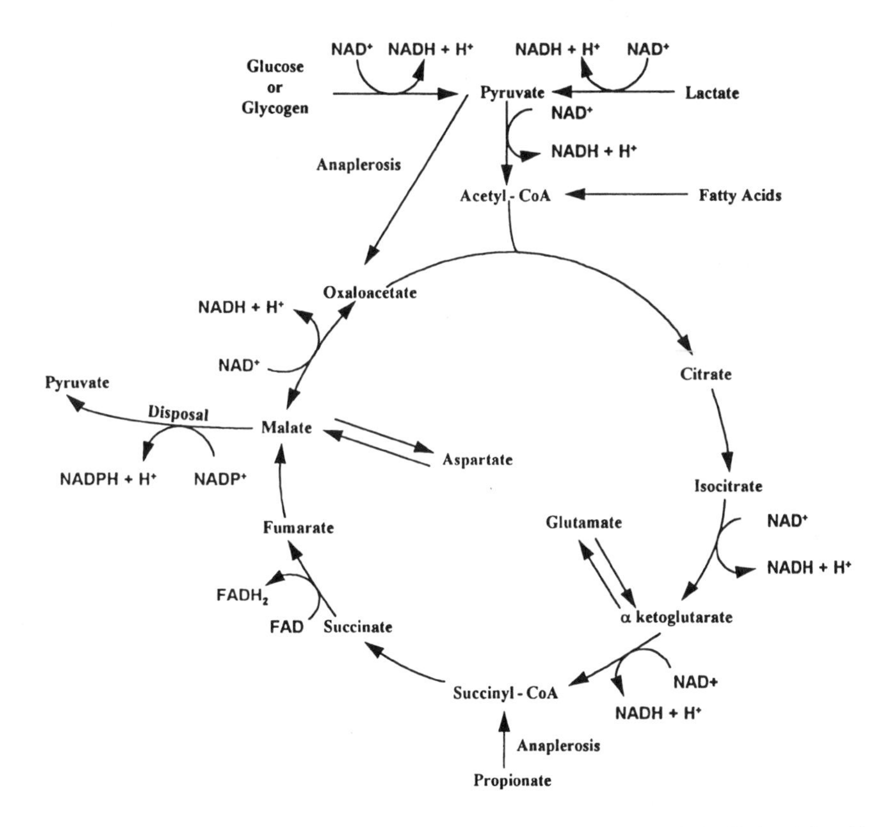

Figure 1. A schematic illustration of a few selected pathways associated with the Krebs citric acid cycle. The figure shows multiple pathways for substrates entering the acetyl-CoA pool (from fatty acids, ketone bodies, glucose, and lactate), three possible anaplerotic pathways (transamination of aspartate and glutamate, pyruvate carboxylation, propionate carboxylation), and one example of a disposal pathway (conversion of malate to pyruvate via the malic enzyme).

that oxidize the substrates available from the blood (ketones, glucose, fatty acids, etc.) or tissues stores (triglycerides or glycogen) to acetyl-CoA are well-established. The broad outlines of substrate selection (heart tissue prefers fatty acids, brain oxidizes glucose) were derived from arteriovenous concentration differences, respiratory quotient measurements, and $^{14}CO_2$ appearance from ^{14}C-enriched compounds. However, these techniques are not easily applied when tissues are metabolizing the mixtures of substrates commonly available *in vivo*.

This question could, in principle, be examined using standard analyses of the rates of $^{14}CO_2$ release from ^{14}C-labeled compounds, but practical difficulties arise. By definition, each source of acetyl-CoA yields $^{14}CO_2$, so to obtain a profile of sources of acetyl-CoA, complementary experiments must be performed in which every source of acetyl-CoA is separately examined. This becomes a tedious task if fatty acids, ketones, lactate, pyruvate, and glucose are all considered. $^{14}CO_2$ methods also assume that the tissue is in isotopic and metabolic steady state and this is difficult to confirm in some circumstances. For example, the activity of various pathways in exercising skeletal muscle, the liver from a septic animal, or the post-ischemic heart may be changing during the course of a study, and it may not be possible to establish or confirm steady-state conditions. ^{13}C NMR spectroscopy offers a simple solution to this problem by allowing a quantitative measure of the oxidation of multiple compounds under nonsteady-state conditions.

1.2. Anaplerotic Functions of the Citric Acid Cycle

In addition to energy production, the citric acid cycle plays an equally important role in degradative and synthetic pathways. This group of functions, the original target of carbon tracer studies, is often emphasized in studies of hepatic metabolism. Nevertheless, all respiring cells use intermediates of the cycle for synthetic purposes such as production of neurotransmitters, amino acids, or glucose. Clearly, carbon atoms must leave the cycle in a form other than CO_2 to serve this function, and mechanisms must be available for replenishing intermediates of the cycle. In mammalian tissues, several pathways are available, including pyruvate carboxylation to oxaloacetate or malate, and propionate carboxylation to succinyl-CoA. Carbon skeletons may flow into the cycle at high rates even when the concentration of citric acid cycle intermediates remains constant (for example, during hepatic gluconeogenesis from lactate). Thus, activation of a pathway that provides citric acid cycle intermediates under steady-state metabolic conditions also requires activation of a pathway(s) to remove citric acid cycle intermediates at an equal rate. In 1966, Kornberg suggested that any pathway that results in synthesis of a citric acid cycle intermediate be termed an "anaplerotic sequence" (Kornberg, 1966). Although he emphasized metabolism in nonmammalian systems, anaplerosis must be considered in the interpretation of ^{13}C NMR spectra of virtually all tissues.

The pathways that feed carbon skeletons into the cycle have historically been given the symbol *y*, defined as the ratio of all combined anaplerotic flux relative to citrate synthase flux (or citric acid cycle flux). In the original usage (Strisower *et al.*, 1952), only one anaplerotic reaction, carboxylation of pyruvate to form oxaloacetate, was considered. Since then, it has been appreciated that even skeletal muscle or the heart, often considered relatively simple from the standpoint of intermediary metabolism, may have multiple anaplerotic sources including the coupled transamination reactions of aspartate and alanine aminotransferase, glutamate dehydrogenase (Williamson *et al.*, 1967), the purine nucleotide cycle (Tornheim and Lowenstein, 1972), carboxylation of pyruvate via the malic enzyme (Sundqvist *et al.*, 1989), carboxylation of pyruvate via pyruvate carboxylase (Davis *et al.*, 1980), and carboxylation of propionate (Sherry *et al.*, 1988). Similarly, several pathways are theoretically available for disposal of citric acid cycle intermediates. The effects of anaplerosis on the appearance of a ^{13}C NMR spectrum can be quite dramatic, even when the anaplerotic substrate is not enriched in ^{13}C. This means that the ^{13}C NMR spectrum can be used to measure the activity of anaplerotic reactions in addition to measurement of the sources of acetyl-CoA.

2. QUANTITATION OF GLUTAMATE ISOTOPOMERS BY ^{13}C NMR

Metabolism of a ^{13}C-enriched compound in the citric acid cycle labels all of the citric acid cycle intermediates as well as the molecules in exchange with cycle intermediates (such as glutamate or aspartate) and anabolic products of the cycle (such as glucose in gluconeogenic tissues). Although the ^{13}C NMR spectrum of any of these compounds could be analyzed by an isotopomer analysis, glutamate is emphasized in this chapter for three reasons: (1) it is usually present in high concentration and hence most easily detected by ^{13}C NMR, (2) the multiplets in the ^{13}C NMR spectrum of glutamate are usually well resolved from other metabolites, and (3) glutamate and α-ketoglutarate are usually in rapid exchange relative to citric acid cycle flux. The latter assumption implies that the relative concentrations of ^{13}C isotopomers are identical in glutamate and α-ketoglutarate. Since this supposition may not be valid under all conditions, it is useful to define the exchanging pool of glutamate. Unless otherwise specified, this indicates the pool of glutamate whose relative concentrations of isotopomers are identical to the relative concentrations of equivalent isotopomers in α-ketoglutarate. Conceivably, a fraction of tissue glutamate may be completely sequestered and not in exchange with α-ketoglutarate. This nonexchanging pool of glutamate would not contain any ^{13}C above natural abundance. These concepts are important in the interpretation of both ^{13}C and ^{1}H NMR spectra of glutamate.

2.1. Relation of Glutamate Isotopomers to Multiplets in the ^{13}C NMR Spectrum

The maximum number of ^{13}C isotopomers in a molecule with n carbons is 2^n. Thus, glutamate with 5 carbons (referred to as C1, C2, C3, C4, and C5) can form 32 possible isotopomers. The relative concentrations of these isotopomers are sensitive to both fluxes into the cycle and the labeling pattern of compounds entering the cycle. The utility of ^{13}C NMR is that groups of these isotopomers (but not individual isotopomers) can be easily quantified in the ^{13}C NMR spectrum. For example, when the C4 of glutamate is the only site labeled with ^{13}C, a single resonance (a singlet) centered at 34.2 ppm is detected. For those glutamate molecules that also have a ^{13}C present in the adjacent C3, the C4 resonance is split into two signals of nearly equal intensity. Thus the C4 spectrum of [3,4-^{13}C]glutamate would appear as two resonances (a doublet) approximately centered around 34.2 ppm. The frequency difference between those two peaks is the coupling constant, abbreviated by the symbol J_{34}, which in this case just happens to be 34 Hz. Similarly, the C4 resonance may be split into two by an adjacent ^{13}C at C5 with a coupling constant J_{45} of different magnitude, about 52 Hz. Thus, the C4 spectrum of [3,4,5-^{13}C]glutamate would appear as four peaks nearly centered around 34.2 ppm (representing a doublet of doublets or a quartet) due to the nearest-neighbor effects of ^{13}C in both C3 and C5. The spectrum of C4 glutamate is not affected by ^{13}C present at C1 and/or C2 because the coupling constants, J_{14} and J_{24}, are too small for direct detection. A similar analysis applies to the other carbons of glutamate.

A convenient method to describe the relative areas of the multiplets in glutamate or other molecules is by a letter and number code. The first two characters identify the carbon atom (C1–C5). The third character identifies the type of multiplet, singlet (S), doublet (D), triplet (T), or quartet (doublet of doublets, Q), and the final 2 numbers (when necessary) identify the particular coupling (i.e., J_{12} versus J_{23} or J_{34} versus J_{45}) that may be identified with an observed doublet (D). In each of the isotopomer methods described below, we have chosen to define the multiplet areas as a fraction of the total resonance area. For example, C4S represents the area of the glutamate C4 singlet expressed as a fraction of total C4 resonance area, C4D34 refers to the area of the doublet due to J_{34} relative to the total C4 resonance area, C4D45 refers to the area of the doublet due to J_{45} relative to the total C4 resonance area, and C4Q represents the area of the C4 quartet expressed as a fraction of total C4 resonance area. Thus, by definition, C4S + C4D34 + C4D45 + C4Q = 1. Similarly, C5S + C5D = 1, C3S + C3D + C3T = 1, C2S + C2D23 + C2D12 + C2Q = 1, and C1S + C1D = 1. We have chosen this method because it describes the multiplets observed in the NMR spectrum without specifying a particular isotopomer. For example, C4D45 could have contributions from [4,5-

^{13}C]glutamate, [1,4,5-^{13}C]glutamate, [2,4,5-^{13}C]glutamate, and [1,2,4,5-^{13}C]glutamate, but not from [3,4,5-^{13}C]glutamate.

2.2. Acquisition of the ^{13}C NMR Spectrum

Proton decoupling is essential for the studies described in this chapter and is assumed in all cases. One consequence of the ^{1}H–^{13}C interaction, aside from spin–spin coupling, is through-space dipolar interactions between nuclei which provides a longitudinal relaxation (T_1) pathway for ^{13}C. A phenomenon related to this relaxation pathway, the nuclear Overhauser effect (NOE), can provide an increase in ^{13}C signal intensity if irradiation at the ^{1}H frequency is continuously applied between pulses. The NOE is maximal (3-fold) for small molecules when ^{13}C is covalently bonded to ^{1}H. However, carboxyl or carbonyl ^{13}C nuclei which are not covalently bonded to ^{1}H may also experience a small NOE due to interactions with protons 2 or more bonds away. ^{13}C–^{13}C dipolar coupling also provides a relaxation pathway for adjacent ^{13}C nuclei, but this interaction is quite weak and its effect on T_1 is ordinarily overwhelmed by the much stronger ^{1}H–^{13}C dipolar coupling, especially in carbons with directly bonded protons (Moreland and Carroll, 1974).

Quantitation of the ^{13}C spectrum is influenced by both T_1 and NOE effects. Since the ^{1}H–^{13}C dipolar interaction is approximately 100 times greater than the ^{13}C–^{13}C dipolar interaction, the T_1s of protonated carbons are not appreciably influenced by ^{13}C enrichments at adjacent carbons. Hence, relative multiplet intensities (see below) from a particular protonated carbon atom should not be influenced by pulsing rate. When comparing different carbon nuclei within the same molecule, and particularly when comparing protonated vs. carbonyl carbons, the relative resonance intensities will be sensitive to both pulsing rate and NOE.

2.3. Quantitation of ^{13}C Fractional Enrichment by ^{1}H NMR

Possibly the simplest quantitation of ^{13}C enrichment is to detect and measure the ^{13}C satellite resonances in each resolved proton resonance of glutamate by ^{1}H NMR. This approach takes advantage of ^{1}H sensitivity and provides a direct measure of the fraction of carbon in that site enriched in ^{13}C. It also provides a mechanism to determine whether the entire glutamate pool was in exchange with the citric acid cycle. For example, the fraction of glutamate carbons in the 4 position which are ^{13}C is denoted C4F. Since a ^{13}C NMR isotopomer analysis (described later) provides a measure of fractional enrichment in ^{13}C at carbon 4, this can be compared to a direct measurement of ^{13}C fractional enrichment in carbon 4 by ^{1}H NMR. Thus, the fraction of cell glutamate that was in exchange with the citric acid cycle under steady-state conditions (measured in tissue extracts where visibility is not an issue) is simply: C4F(by ^{1}H NMR)/C4F(by ^{13}C NMR). This is illustrated in

Fig. 2 by the ^{1}H spectrum of purified glutamate. The fraction of glutamate C4 enriched in ^{13}C is easily determined by direct inspection, in this case about 79%. In contrast, an isotopomer analysis of the ^{13}C spectrum found that 92% of the acetyl-CoA (and therefore 92% of glutamate C4) was enriched in ^{13}C. This indicates that 79/92 or 86% of the total glutamate found in this heart extract was in exchange with the citric acid cycle. Similar observations have been reported by Lewandowski (Lewandowski and Hulbert, 1991; Lewandowski, 1992). These observations could certainly reflect compartmentation of a pool of glutamate that was not exchanging with α-ketoglutarate in the citric acid cycle. Alternatively, it may simply demonstrate incomplete equilibration among glutamate pools that are in slow exchange. Although this distinction is important, we emphasize that strict compartmentation of a fraction of glutamate has no influence on the resulting ^{13}C isotopomer analysis of glutamate that was in exchange with α-ketoglutarate in the citric acid cycle but becomes an important parameter to know when addressing kinetic issues where the

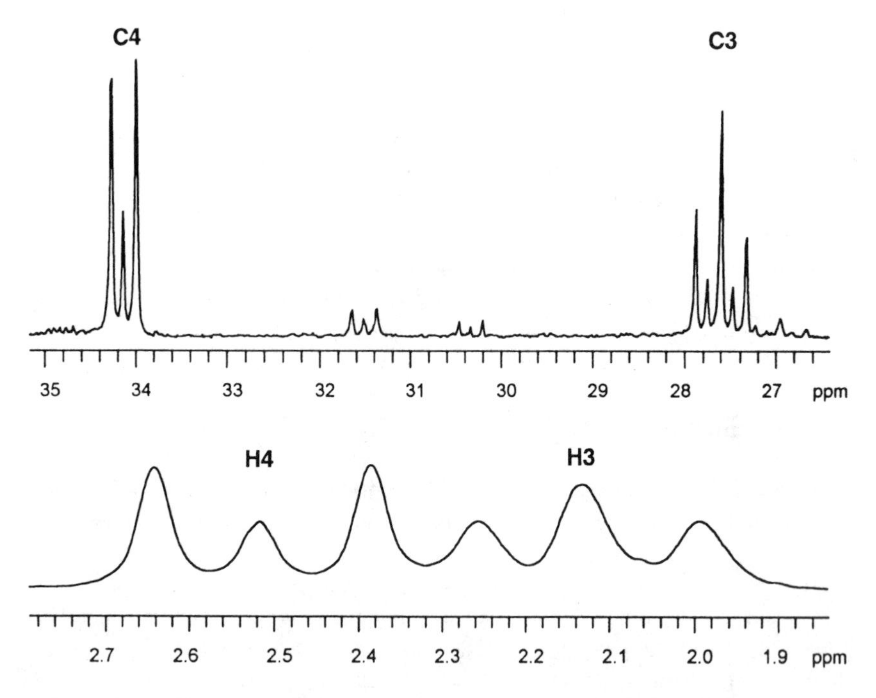

Figure 2. ^{1}H and ^{13}C NMR spectra of glutamate isolated and purified from a heart perfused to steady state with [2-^{13}C]acetate. An isotopomer analysis of the ^{13}C spectrum (top) indicated that F_{C2} was 0.92; hence, glutamate C4 of the exchanging glutamate pool should also be enriched with ^{13}C to a level of 92%. However, the ^{1}H spectrum (lower) shows that the actual level of ^{13}C enrichment in the C4 position of glutamate was about 79%. This result shows that only 86% (79/92) of the total tissue glutamate was in exchange with α-ketoglutarate in the critic acid cycle of this heart.

size of the glutamate pool detected by ^{13}C NMR (the glutamate exchanging with α-ketoglutarate) is often equated with total tissue glutamate as measured in tissue extracts.

2.4. Influence of Natural Abundance ^{13}C on an Isotopomer Analysis

Any singlet appearing in a ^{13}C NMR spectrum of glutamate could, of course, have a contribution from natural abundance levels (1.1%) of this isotope in addition to the singlets resulting from metabolic turnover of a ^{13}C-enriched molecule. These cannot be distinguished by NMR, so one must always be aware of a possible natural abundance contribution to any spectrum. Under most metabolic circumstances, corrections to singlet intensities from a natural abundance level of glutamate can be safely ignored. However, when the ^{13}C fractional enrichment of acetyl-CoA entering the citric acid cycle is low or when only a portion of total cell glutamate is in exchange with cycle pools, then corrections may become necessary. Prior experience with ^{13}C NMR spectra of extracts of perfused hearts suggests that acetyl-CoA must be enriched to a level of at least 20% for a reliable determination of glutamate multiplet areas. If one assumes for the moment that all glutamate within the cell is in exchange with α-ketoglutarate, then a 20% ^{13}C-enrichment in the methyl carbon of acetyl-CoA would translate into a 20% enrichment in glutamate C4, only part of which would appear as the singlet, C4S. Assuming that C4S was 0.80, then the singlet from natural abundance glutamate in the metabolically unenriched fraction ($80\% \times 1.1\%$ or 0.0088) would amount to about 5.5% of the singlet from the metabolically enriched fraction ($20\% \times 100\% \times 0.8 = 0.16$). That is significant enough to justify correcting the C4S for the natural abundance contribution. If a nonexchanging pool of glutamate was large relative to the exchanging pool, then the contribution of natural abundance ^{13}C to C4S would become even more significant.

3. MATHEMATICAL AND COMPUTER MODELS: APPLICATIONS FOR ISOTOPOMER ANALYSIS

For over 40 years carbon isotope tracer studies have been interpreted using mathematical models, and ^{13}C NMR isotopomer analysis is no exception. Indeed, mathematical and computer models are absolutely essential to fully exploit the power of ^{13}C isotopomer analysis. It is true that some valuable applications of ^{13}C NMR isotopomer analysis, such as the nonsteady-state analysis or measurement of anaplerosis, are mathematically simple and intuitively accessible. However, routine use of mathematical models and simulations is almost essential in two respects.

3.1. The Value of Mathematical Models

For the purpose of designing experiments, it is quite useful to be able to predict the ^{13}C NMR spectrum under virtually any metabolic condition with any labeling pattern in the metabolized substrates. In this application, the model assists by (1) requiring the experimentalist to set the relative activity of all pathways and labeling patterns, (2) calculating the relative concentrations of all isotopomers in all molecules of interest, and (3) displaying a ^{13}C NMR spectrum calculated from the isotopomer populations. Since the relative concentrations of isotopomers are calculated in every case, it is a simple step to also calculate predicted mass spectra or ^{14}C specific activity measurements. The consequences of systematically varying experimental parameters (site of label, relative activity of pathways, sizes of intermediate pools, etc.) on the relative concentration of isotopomers and the ^{13}C spectrum are easily explored. In doing so, the model may expose flaws in a planned experiment, or reveal results which are not intuitively obvious. Alternatively, "obvious" results may not be confirmed by the model.

Models are also required to solve the reverse problem, that is, to interpret the ^{13}C NMR spectra in terms of relative activities of various pathways. Simple questions, such as the sources of acetyl-CoA, are easily addressed with simple algebra, as shown below. Again, however, under realistic metabolic conditions, anaplerosis must be considered in every tissue. We have derived the relations among anaplerosis and the ^{13}C-labeling patterns in anaplerotic substrates and acetyl-CoA under steady-state conditions. Although it is beyond the scope of this chapter to derive these relations, we have shown that the relation between the ^{13}C NMR spectrum of glutamate and the relevant fluxes represents an overdetermined linear algebraic system which can be solved using least-squares methods. The essential programs are available in numerical analysis packages or in custom-designed programs such as *tcaCALC* (Jeffrey *et al.*, 1996).

3.2. Historical Background of Current Modeling Techniques

Somewhat arbitrarily, mathematical models of carbon distribution in the citric acid cycle may be grouped into four categories, depending on the mathematical foundation: geometric series, input–output analysis, computer simulations, and differential equations. The evolution of models has been driven primarily by technical advances which allow more comprehensive information about the fate of a tracer. Initially, ^{14}C specific activity was measured in CO_2 and the first applications of mathematical modeling to complex metabolic pathways were reported by Strisower *et al.* (1952) and Weinman *et al.* (1957). This work explicitly stated the requirements for isotopic and metabolic steady state: the labeling pattern and level of enrichment of the acetyl-CoA and other intermediate pools are constant. Later, Exton and Park (1967) extended this principle in their analysis of gluconeogenesis.

These early studies are relevant to ^{13}C NMR isotopomer analysis because the equations which relate ^{14}C specific activities to metabolic fluxes can be directly compared to those derived for NMR measurements. The geometric series method used by Strisower and later by Exton and Park was not easily extended to more complex metabolic problems, so the mathematical basis shifted away from a geometric series to what was termed an "input–output" analysis (Katz, 1985). Because of conservation of mass and the assumption of metabolic and isotopic steady state, the sum of enrichments in all possible products must equal the total input enrichment. The ratios of ^{14}C enrichment in each carbon of molecules such as citrate, glutamate, and glucose have subsequently been analyzed or described by Katz *et al.* (1989), Landau (Magnusson *et al.*, 1987), Goebel *et al.* (1982), Kelleher (1986), and others.

The third type of modeling, computer simulation, was first applied in the 60s (Haut and Basickes, 1967). Since the fate of label may be considered on each pass through the citric acid cycle, computer simulations are not conceptually different from the geometric series approach. Nevertheless, computer simulations are far more tractable for complex pathways and these simulations are increasingly attractive because of the wealth of information provided by simple display of isotopomers and simulated ^{13}C NMR spectra, mass spectra and ^{14}C fractional enrichment data. In each metabolic step, the distribution of carbon isotope in the reactants (for example, acetyl-CoA and oxaloacetate) is known. The distribution of isotope and the relative concentrations of isotopomers is then easily calculated for the product, citrate. The major advantages of computer simulations (aside from easy display of ^{13}C or ^{1}H spectra) is that each calculation can be set up rather easily and that portions of the program can be debugged separate from the entire citric acid cycle and its interacting pathways. We found that a computer simulation (Malloy *et al.*, 1990a; Jeffrey *et al.*, 1991) running on a personal computer serves both purposes quite well and has the additional benefit of confirming the results of the analytical methods described below. A simulation has also been reported by Cohen and Bergman (1995).

An example of a ^{13}C NMR spectrum from a simulated experiment of hepatic metabolism, generated using *tcaSIM*,* is shown in Fig. 3. The assumed metabolic conditions in this particular example were complex: exogenous lactate was 100% [3-^{13}C]lactate, exogenous fatty acids were 80% [U-^{13}C]fatty acids (with the remainder unlabeled), and 30% of the CO_2 was $^{13}CO_2$. Flux through pyruvate kinase and pyruvate carboxylase was set equal to citric acid cycle flux while flux through pyruvate dehydrogenase was 50% of cycle flux; 40% of the oxaloacetate formed via pyruvate carboxylation was scrambled "backwards" through the symmetric intermediate fumarate, and all succinyl-CoA yielded succinate that was free to

***tcaSIM* runs on a personal computer and is available from Dr. F. Mark Jeffrey at the Mary Nell and Ralph B. Rogers Magnetic Resonance Center, 5801 Forest Park Road, Dallas, Texas, 75235–9085.

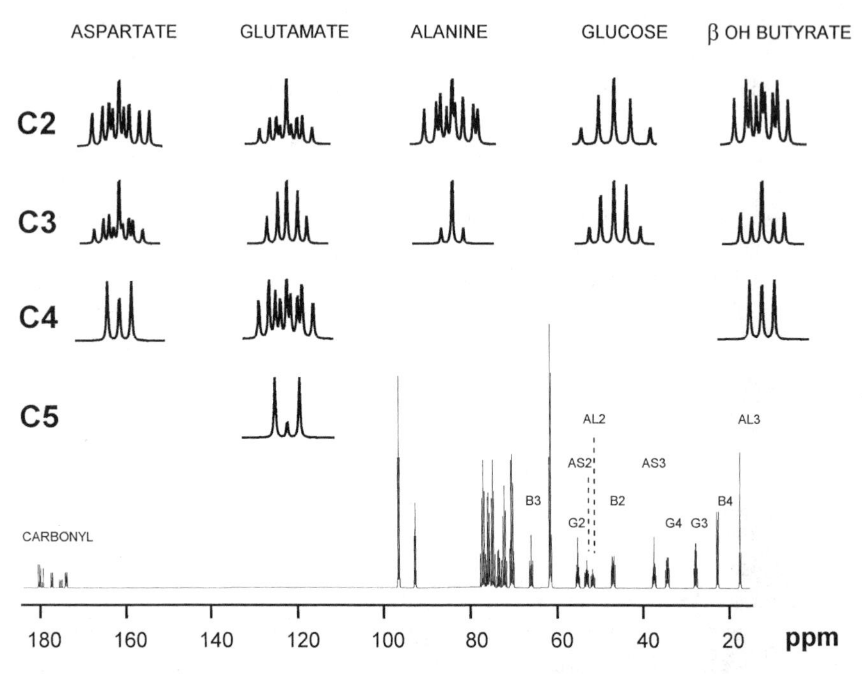

Figure 3. Simulated spectra of aspartate, glutamate, alanine, β-hydroxybutyrate, and glucose anticipated for a liver perfused with [3-^{13}C]lactate (F_{C2} = 0.5), [U-^{13}C]fatty acids (F_{C3} = 0.4), and ^{13}C-enriched bicarbonate using *tcaSIM*. The activities of pyruvate dehydrogenase, pyruvate carboxylase, and pyruvate kinase were set to 0.5, 1, and 1, respectively, relative to a critic acid cycle flux of 1.

tumble, thus producing equal enrichment in C1 and C4 (and C2 and C3) of all malate derived from succinyl-CoA. The system was assumed to be at metabolic and isotopic steady state and the relative concentrations of all metabolites were roughly equal to those typically found in extracts of rat livers. The complete simulated ^{13}C NMR spectrum is shown as well as expanded views of most aspartate, glutamate, alanine, glucose, and β-hydroxybutyrate multiplets.

Although a complete isotopomer analysis of this spectrum does return the metabolic indices used as input variables in *tcaSIM*, this spectrum can be interpreted to some extent by simple inspection. For example, the difference in total areas of aspartate C2 and C3 indicates that these carbons were not symmetrically enriched in the oxaloacetate pool, reflecting incomplete "backward scrambling" in fumarate/succinate. Under steady-state conditions, the multiplet pattern in aspartate C3 must be equivalent to the pattern in glutamate C2, as it is in this simulation. Since the only available exogenous substrates were [3-^{13}C]lactate (which yields [2-^{13}C]acetyl-CoA) and [U-^{13}C]fatty acids (which yields [1,2-^{13}C]acetyl-CoA], the small singlet in glutamate C5, indicating that [1-^{13}C]acetyl-CoA was generated metabolically, documents flux from labeled OAA → pyruvate → acetyl-CoA. The complex alanine spectrum confirms that some of the pyruvate pool exchanging with alanine must have been derived from oxaloacetate (and not simply pyruvate derived from [3-^{13}C]lactate), while the glucose multiplets (calculated assuming that J_{12}, J_{23}, and J_{34} were equivalent and first-order conditions applied) provide yet another window on oxaloacetate enrichment.

None of these methods formally includes the time dependence of ^{13}C distribution through each pathway. An elegant model, presented in 1983 by Edwin Chance and colleagues (Chance *et al.*, 1983), nicely illustrates the fourth type of model applied for analysis of carbon tracers: a kinetic analysis of the cycle. In this method, fluxes through each cycle step and through all reactions in exchange with intermediate pools were considered using conventional chemical kinetics. Each enzyme was treated as a "black box" and no assumptions were made about enzyme mechanisms. The system was assumed to be at metabolic steady state, yielding a network of reactions that could be described by a series of differential equations. The solution of this system coupled with fractional enrichment measurements over time using numerical methods allowed calculation of absolute flux through several pathways associated with the citric acid cycle. The advantage of this type of analysis is that it yields an estimate of absolute flux; the disadvantage is the requirement of repeated ^{13}C NMR measurements. A thorough discussion of kinetic models is given elsewhere in this book.

4. THE EVOLUTION OF ^{13}C ISOTOPOMERS IN THE CITRIC ACID CYCLE

As shown above, entry of a ^{13}C-enriched metabolite into the citric acid cycle of an actively respiring tissue can have multiple influences on the appearance of the ^{13}C NMR spectrum of that tissue. In an energy-demanding tissue like the heart, this entry most often occurs as a two-carbon condensation of acetyl-CoA with oxaloacetate to form citrate in the reaction catalyzed by citrate synthase. Because citrate is prochiral, the labeled two-carbon acetyl unit appears only in carbons 4 and 5 of aconitate, isocitrate, and α-ketoglutarate during the first span of the cycle. As these intermediates are typically present at concentrations well below that detectable by ^{13}C NMR, this information is normally recorded in the five-carbon amino acid, glutamate, formed by transamination of α-ketoglutarate. Thus, the appearance of the resonance multiplets in glutamate C4 and C5 should mirror the labeling pattern in acetyl-CoA entering the citric acid cycle at citrate synthase. This is the fundamental principle behind the "direct analysis" method described below for monitoring substrate selection in intact tissue. In those instances where ^{13}C spectra are collected as a function of time after introducing a labeled compound to a tissue, the amount of time necessary to encode the labeling pattern of the acetyl group of acetyl-CoA into glutamate C4 and C5 has never been fully tested. The common assumption that glutamate C4 and C5 accurately mirrors acetyl-CoA at all times requires that flux from acetyl-CoA to glutamate is limited only by citric acid cycle flux and that the pool of six-carbon intermediates (citrate, aconitate, and isocitrate) and α-ketoglutarate is small compared to the glutamate pool. The pool size requirement is easily met in most tissues, while the flux from acetyl-CoA to glutamate may in some cases be limited by exchange between α-ketoglutarate and glutamate (Robitaille *et al.*, 1993; Weiss *et al.*, 1995; Yu *et al.*, 1995; Zhao *et al.*, 1995). We found equivalent ^{13}C isotopomer populations in succinate and glutamate in ischemic heart tissue (Jones *et al.*, 1993) and concluded that both metabolites equilibrated with the same pool of α-ketoglutarate. Later, Lewandowski *et al.* (1996), using similar techniques, found that the ^{13}C isotopomers in succinate, α-ketoglutarate, and glutamate were equivalent in hearts perfused to steady state with [2-^{13}C]acetate, while succinate differed from α-ketoglutarate and glutamate in hearts perfused with [3-^{13}C]pyruvate.

An example of temporal evolution of multiplet structure in the ^{13}C NMR spectrum of glutamate is shown in Fig. 4. These spectra were collected on an

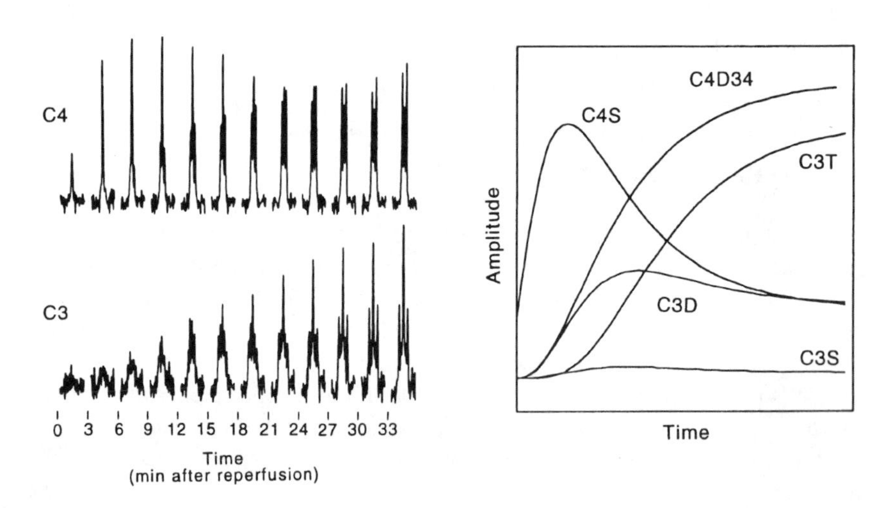

Figure 4. Time-dependent evolution of the ^{13}C NMR spectra of an intact rat heart perfused with [2-^{13}C]acetate (left panel). Spectra were acquired on a 11.75 Tesla magnet in 3 min blocks. Only glutamate C4 and C3 are shown (adapted from Malloy *et al.*, 1990b). Time-dependent evolution of glutamate C4 (C4S and C4D34) and C3 (C3S, C3D, and C3T) multiplets as calculated using *tcaSIM* (right panel). Estimates of citric acid cycle intermediate pool sizes similar to those expected for a rat heart perfused with [2-^{13}C]acetate were used in this simulation.

isolated rat heart perfused with 3 mM [2-^{13}C]acetate in a narrow bore 11.75 Tesla magnet as described in Malloy *et al.* (1990b) using broadband proton decoupling to remove all C–H spin–spin coupling. Thus, the multiplets observed in these spectra reflect the different isotopomers in glutamate C3 and C4 as the ^{13}C label originating in the methyl carbon of acetate entered those two positions of the glutamate molecule. Condensation of [2-^{13}C]acetyl-CoA with unenriched oxaloacetate yields [4-^{13}C]citrate, [4-^{13}C]isocitrate, [4-^{13}C]α-ketoglutarate, and hence [4-^{13}C]glutamate in the first span of the citric acid cycle. This is detected as a singlet in the glutamate C4 resonance in the first spectrum, signal averaged over 3 minutes after the addition of [2-^{13}C]acetate. The intensity of this singlet is substantially higher in the second and third spectra, reflecting production of even more [4-^{13}C]glutamate during these time periods. That portion of [4-^{13}C]α-ketoglutarate remaining in the cycle pools was converted to [3-^{13}C]succinyl-CoA and subsequently into equal amounts of [2-^{13}C]- and [3-^{13}C]succinate, fumarate, malate and oxaloacetate, assuming randomization of label between C2 and C3 in the symmetric intermediates, succinate and fumarate. Once the [2-^{13}C]oxaloacetate isotopomer is formed, this molecule can condense with another [2-^{13}C]acetyl-CoA to yield the first cycle intermediate with adjacent ^{13}C-enriched carbons, [3,4-^{13}C]citrate. This isotopomer is subsequently converted into [3,4-^{13}C]glutamate which, by the third acquisition, appears as a doublet (with a J_{CC} = 34 Hz) nearly centered around the glutamate C4 singlet (there is a small but measurable isotope chemical shift). This doublet gradually increased in intensity in subsequent spectra, reflecting distribution of ^{13}C into the C2 position of oxaloacetate and all other 4-carbon intermediates as the cycle turnover continued. Note that after about the fourth spectrum, the singlet in glutamate C4 also decreased in intensity with time because, as oxaloacetate C2 became more highly enriched, the probability of forming [4-^{13}C]citrate and [4-^{13}C]glutamate gradually decreased. Thus, any change in intensity of the C4 singlet and doublet is a complex function of metabolite pool sizes (including citric acid cycle intermediates and all metabolites in exchange with these intermediates), fraction of acetyl-CoA derived from [2-^{13}C]acetate (the F_{C2} parameter described below), citric acid cycle flux, and in certain cases other exchanges as well (Robitaille *et al.*, 1993; Weiss *et al.*, 1995; Yu *et al.*, 1995; Zhao *et al.*, 1995). Analysis of such time-dependent multiplet data provides one basis for quantitating citric acid cycle kinetics and related interchanges (Jeffrey, unpublished results).

Concomitant with these alterations in glutamate C4 multiplets, the C3 multiplets also evolve with time. The first multiplet to appear is a doublet reflecting the [3,4-^{13}C]glutamate isotopomer, described above. Next, a small singlet could arise from [3-^{13}C]glutamate, but the intensity of this is highly dependent on the contribution of [2-^{13}C]acetate to acetyl-CoA. For the spectrum shown in Fig. 4, this contribution is high so the probability of forming [3-^{13}C]glutamate is low and consequently the C3 singlet is low. As cycle turnover continues, isotopomers such

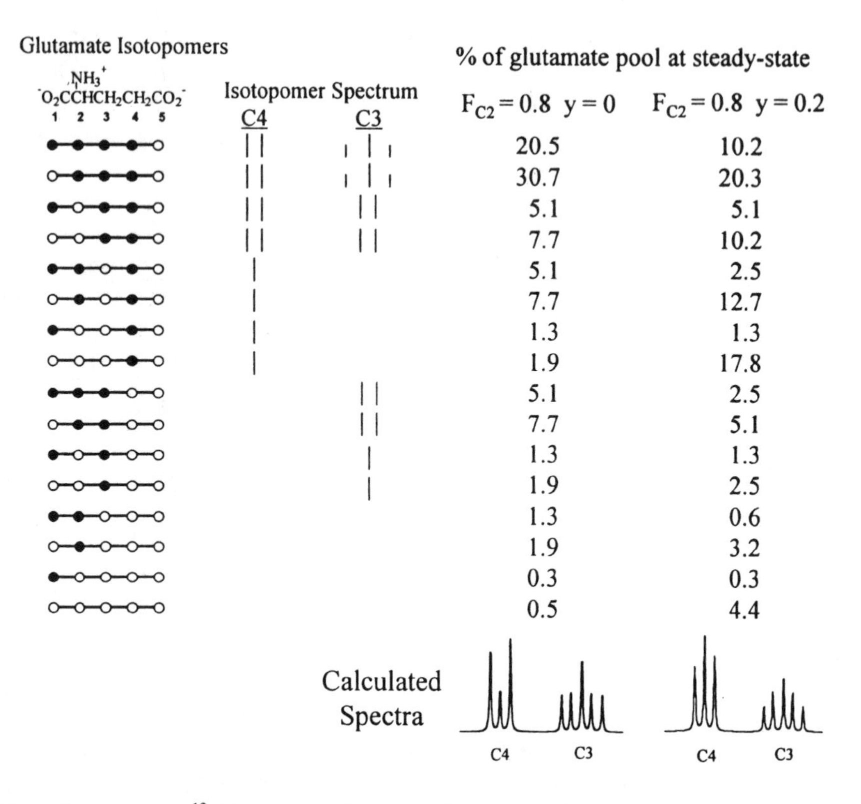

Figure 5. The sixteen ^{13}C isotopomers of glutamate shown in the left column will be generated under steady-state conditions by entry of [2-^{13}C]acetyl-CoA into the citric acid cycle. Each of those isotopomers contributes either a singlet or a doublet to the C4 resonance and either a singlet, doublet, or triplet to the C3 resonance. The two left columns indicate the relative concentrations of each isotopomer at steady state when the fraction of acetyl-CoA enriched in the methyl carbon (F_{C2}) is 80%, with or without activation of anaplerotic pathways ($y = 0$ or 20%). The spectra at the bottom demonstrate that glutamate C4 and C3 are both sensitive to anaplerosis.

as [2,3,4-^{13}C]glutamate containing three nearest-neighbor ^{13}C-enriched carbons are generated yielding a triplet in glutamate C3 (J_{CC} is approximately equal for C2–C3 and C3–C4 spin–spin coupling). With time, this multiplet becomes the dominant multiplet contributing to C3 in the experiment shown in Fig. 4. The time-dependent evolution of the C3 and C4 multiplets as predicted by *tcaSIM* are shown in the right panel of Fig. 4. As the rate of ^{13}C entry into the glutamate C3 (or C2) pool is also heavily dependent upon citric acid cycle flux (and the other parameters described above), the relative rates of enrichment of glutamate C4 vs. C3 have formed the basis of virtually all ^{13}C NMR citric acid cycle kinetic measurements to date, as described in Chapters 4–7 of this text.

After 30 minutes the relative intensities of these multiplets were no longer changing. At this point, the distribution of ^{13}C in citric acid cycle intermediates had reached steady-state and the relative concentrations of isotopomers no longer changed with time. The 16 glutamate isotopomers formed at steady state using [2-^{13}C]acetate as the sole enriched substrate are illustrated in Fig. 5 (half of the total 32 isotopomer of glutamate cannot occur because C5 was not enriched). The relative concentration of each isotopomer is easily calculated by *tcaSIM* or by following the fate of each labeled citrate formed by condensing oxaloacetate with acetyl-CoA as it equilibrates with the various cycle intermediate pools. Those concentrations depend upon (a) the ^{13}C fractional enrichment (equivalent to a ^{14}C specific activity) of the acetyl-CoA entering the cycle on each turn (the F_C variables, see below), and (b) flux of other molecules (either labeled or unlabeled) through the citric acid cycle via an anaplerotic reaction. As noted earlier, the ratio of anaplerotic flux to citric acid cycle flux has been given the label y. The figure illustrates that the population of glutamate isotopomers and the appearance of the glutamate C4 and C3 multiplets at steady state is influenced by both F_C and y.

5. THE STEADY-STATE ^{13}C ISOTOPOMER ANALYSIS

A ^{13}C NMR spectrum of an extract of an *in vivo* pig heart infused to steady state with a mixture of [1,2-^{13}C]acetate and [3-^{13}C]lactate is shown in Fig. 6. Note that there are more multiplets in the glutamate C4 resonance of this spectrum than in the spectrum shown in Fig. 4. The reason for this is quite simple: in this case, both acetate and lactate contributed to acetyl-CoA entering the citric acid cycle and the doubly labeled acetate forms additional ^{13}C isotopomers in the cycle that were not present in the previous example. For example, [1,2-^{13}C]acetyl-CoA formed from the acetate would condense in the cycle to yield [4,5-^{13}C]glutamate on the first turn and [3,4,5-^{13}C]glutamate on subsequent turns. These appear in the glutamate C4 resonance as a doublet (with J_{CC} = 52 Hz) and a doublet-of-doublets or quartet (resulting from spin–spin coupling between C4 and both C3 and C5), respectively. So, one can easily see by inspecting the glutamate C4 multiplet components that both acetate and lactate contributed to acetyl-CoA in this *in vivo* pig heart.

How does one obtain quantitative metabolic information from such a result? Before proceeding, we need to be able to describe enrichment patterns in the pathways feeding the citric acid cycle. For example, the two-carbon acetyl group might contain ^{13}C either singly labeled in the methyl carbon, singly labeled in the carbonyl carbon, labeled in both carbons, or completely unenriched. Thus, up to four different acetyl-CoA patterns (and therefore oxidation of four different labeled substrates) can be distinguished. The fraction each possible labeled oxidizable

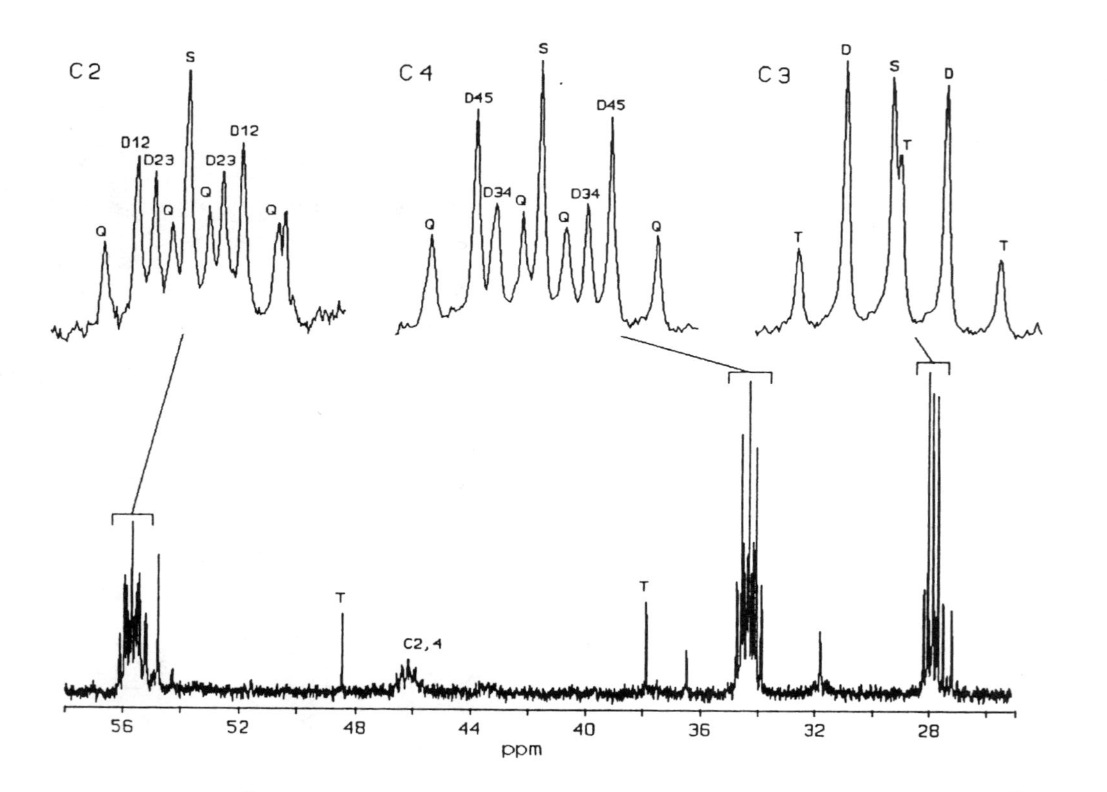

Figure 6. Proton decoupled ^{13}C NMR spectrum of an extract of an *in vivo* pig heart after steady-state oxidation of [1,2-^{13}C]acetate and [3-^{13}C]lactate. The C2, C3, and C4 glutamate resonances are expanded. Enriched citrate (C2,4) and natural abundance taurine (T) are also detected in the complete spectrum.

substrate contributes to acetyl-CoA entering the cycle at citrate synthase (the F_{Ci} parameters) is defined as follows:

F_{C0} = fraction of acetyl-CoA that is not enriched in ^{13}C above natural abundance levels.
F_{C1} = fraction of acetyl-CoA enriched in ^{13}C only in its carbonyl carbon.
F_{C2} = fraction of acetyl-CoA enriched only in its methyl carbon.
F_{C3} = fraction of acetyl-CoA enriched in both the carbonyl and the methyl carbons.

By definition, $F_{C0} + F_{C1} + F_{C2} + F_{C3}$ must always equal unity. Similarly, anaplerotic substrates might contain ^{13}C but here the number of possible isotope labeling patterns is greatly increased. Using pyruvate carboxylase as the prototype anaplerotic pathway, one could in principle have eight possible pyruvate isotopomers condensing with either $^{12}CO_2$ or $^{13}CO_2$ leading to 2^4 or 16 possible oxaloacetate isotopomers, requiring sixteen different F_A parameters for complete description. Rather than define all F_{Ai} fractions (where $i = 0\text{–}15$) here, we simply point out that once again ΣF_{Ai} must equal unity. In practice, the number of labeled substrates available for anaplerotic entry into the cycle is usually quite limited. For example, in a mammalian tissue exposed to a mixture of [1,2-^{13}C]acetate and [3-^{13}C]pyruvate, none of the ^{13}C originating in acetate can enter the cycle pools via anaplerosis while [3-^{13}C]pyruvate can only produce [3-^{13}C]- or [2-^{13}C]oxaloacetate (after equilibration through fumarase), assuming that the bicarbonate pool will not contain sufficient ^{13}C to affect the results unless $H^{13}CO_3^-$ is also provided.

We have previously reported equations (Malloy *et al.*, 1988, 1990a) which relate the metabolic indices of interest (the F_{Ci}, F_{Ai}, and y values) to the fractional multiplet areas of glutamate C2, C3, and C4 as measured in a single ^{13}C spectrum, derived using the input–output matrix method. For the example shown in Fig. 6, only F_{C0} (from endogenous substrates), F_{C2} (from [3-^{13}C]lactate), and F_{C3} (from [1,2-^{13}C]acetate) are possible, so $F_{C0} + F_{C2} + F_{C3} = 1$. Similarly, only [3-^{13}C]lactate could contribute as an anaplerotic substrate in this case so $F_{A0} + F_{A1} = 1$, where F_{A0} equals the fraction of anaplerotic substrate derived from unlabeled, endogenous sources and F_{A1} equals the fraction of derived [3-^{13}C]lactate. The equations appropriate to this metabolic circumstance are reproduced in Table 1 and the multiplet areas from the glutamate spectrum of Fig. 6 are summarized in the far right column. The solution of these equations for the metabolic indices of interest using a nonlinear least-squares algorithm gave the following results: $F_{C0} = 0.47$; $F_{C2} = 0.25$; $F_{C3} = 0.27$; $y = 0.13$; and $F_{A1} = 0$. This indicated that doubly labeled acetate and singly labeled lactate contributed 27 and 25%, respectively, to the acetyl-CoA oxidized by the Krebs citric acid cycle in this heart, total anaplerotic flux was 13% that of cycle flux, and none of the labeled lactate entered the cycle as an anaplerotic substrate. The identity of the unlabeled substrate sources could

Table 1
Relationships Between the Glutamate Fractional Multiplet Areas Measured from the ^{13}C NMR Spectrum of Fig. 2 and Metabolic Variables, F_{C0}, F_{C2}, F_{C3}, F_{A1}, and y[a]

Multiplet component	Metabolic variables	Multiplet area
C2S	$[(F_{C2}+yF_{A1})(2y+1)]/[F_{C2}+F_{C3}+yF_{A1})(2y+2)] + [F_{C0}(F_{C0}+y) - F_{C2}(y+1)]/[2(y+1)^2]$	0.26
C2D12	$[(F_{C3})(2y+1)]/[(F_{C2}+F_{C3}+yF_{A1})(2y+2)] + (F_{C0}F_{C2} + F_{C0}F_{C3})/[2(y+1)^2] - F_{C3}/(2+2y)$	0.26
C2D23	$[F_{C2}(y+1) + (F_{C2}+F_{C3})(F_{C0}+y)]/[2(y+1)^2]$	0.24
C2Q	$[F_{C3}(y+1) + (F_{C2}+F_{C3})^2]/[2(y+1)^2]$	0.24
C3S	$F_{C0}(F_{C0}+y)/(y+1)$	0.26
C3T	$(F_{C2}+F_{C3})^2/(y+1)$	0.27
C3D	$[(y+1) - F_{C0}(F_{C0}+y) - (F_{C2}+F_{C3})^2]/(y+1)$	0.47
C4S	$F_{C2}[2y+1-(F_{C2}+F_{C3}+yF_{A1})]/[(2y+1)(F_{C2}+F_{C3})]$	0.27
C4D34	$F_{C2}(F_{C2}+F_{C3}+yF_{A1})/[(2y+1)(F_{C2}+F_{C3})]$	0.22
C4D45	$F_{C3}[2y+1-(F_{C2}+F_{C3}+yF_{A1})]/[(2y+1)(F_{C2}+F_{C3})]$	0.26
C4Q	$F_{C3}(F_{C2}+F_{C3}+yF_{A1})/[(2y+1)(F_{C2}+F_{C3})]$	0.23

[a]F_{A1} is the fraction of any anaplerotic substrate that can yield ^{13}C enrichment in oxaloacetate C3 (or C2). Full scrambling in the succinate/fumarate pool was assumed. See text for an explanation of the multiplet areas in the right column. The fraction of glutamate enriched in C4 (C4F) is simply $(F_{C2} + F_{C3})$.

not be identified in this experiment, but the combination of 0.23 mM free fatty acids and 6–7 mM glucose found in the blood of these animals could certainly have accounted for the 47% of unlabeled acetyl-CoA seen entering the cycle pools. The source of the unlabeled anaplerotic substrate was not identified.

Jeffrey *et al.* (1995) have applied steady-state ^{13}C isotopomer principles to evaluate the effects of the antianginal drug, perhexiline, on substrate selection in a perfused working rat heart preparation. They exposed hearts to a physiological mixture of fatty acids, ketone bodies, glucose, and lactate and determined the percent contribution of each substrate to acetyl-CoA entering the citric acid cycle of control hearts and in hearts exposed to a clinical dose (2 μM) of perhexiline. They found that fatty acids, lactate, acetoacetate, and glucose contributed 54%, 11%, 23%, and 0%, respectively, to acetyl-CoA in control hearts, with the remaining 12% coming from endogenous substrates (likely triglycerides). This profile was altered in hearts exposed to perhexiline. In this case, the fatty acid contribution to acetyl-CoA decreased to 35%, the endogenous substrate contribution declined to zero (consistent with the endogenous source being triglycerides), while the contribution from lactate increased to 36%. The acetoacetate and glucose contributions remained unchanged in hearts exposed to the drug. This study provided the first quantitative evidence that perhexiline reduces fatty acid oxidation by the myocardium. The mechanism(s) of inhibition of fatty acid oxidation and stimulation of lactate oxidation by the drug have not been fully addressed, but information of this type raises some interesting metabolic questions. For example, how can oxidation of lactate be stimulated without a concomitant increase in glucose oxidation and, since one might expect NADH levels to rise upon stimulation of lactate oxidation, why did the drug not stimulate oxidation of acetoacetate as well?

An exciting application of the steady-state analysis was reported by Viallard and colleagues (Viallard *et al.*, 1993), who examined the effects of left ventricular hypertrophy on anaplerosis in the isolated rat heart. In control myocardium, anaplerosis was identical to earlier reports in hearts supplied with acetate, about 6% of citric acid cycle flux. An important observation was the significant increase in y to 16% in hearts from hypertensive animals. One might speculate that stimulation of anaplerosis is a marker for the biochemical state associated with hypertrophy, and may even precede myocardial hypertrophy. When ^{13}C NMR isotopomer analysis becomes practical in humans, it will be interesting to monitor anaplerosis during the evolution and regression of hypertrophy.

Steady-state isotopomer methods have been used by numerous other investigators to interrogate citric acid cycle metabolism in animal hearts under a variety of physiological conditions. In some reports, ^{13}C spectra of intact hearts have been collected during presteady-state labeling of glutamate by singly enriched substrates and the rates of ^{13}C entry into glutamate C4 vs. C2 (or C3) have been used as an index of citric acid cycle flux (Lewandowski, 1992; Weiss *et al.*, 1992, 1993; Chatham *et al.*, 1995; Yu *et al.*, 1995). These topics are discussed in detail elsewhere under dynamic methods, Chapters 4–7. A few interesting observations are in order, however. Lewandowski (1992) has shown in rabbit hearts perfused to steady state with [2-^{13}C]acetate that the fractional enrichment of acetyl-CoA derived from acetate (as reported by F_{C2}) did not differ between control hearts, hearts arrested with KCl (low workload), or hearts stimulated by isoproterenol (high workload). Anaplerosis, as reported by y from steady-state ^{13}C spectra, was low and no different in control and high workload hearts ($y = 0.09 \pm 0.02$) but significantly higher (0.32 ± 0.02) in hearts experiencing low workloads. Since y reports a ratio (anaplerotic flux/citrate synthase or TCA cycle flux), this indicates that absolute anaplerotic flux increased in parallel with increased O_2 demand between control and high workload hearts. However, O_2 consumption was 3.7-fold lower in KCl-arrested hearts than control hearts while y increased 3.6-fold. This interesting result indicates that absolute anaplerotic flux was similar in control and KCl-arrested groups and that the measured change in y actually reflects lower TCA cycle flux in this group (Lewandowski, 1992). This suggests that anaplerosis is always active in hearts and that there is basal flux through the citric acid cycle pools (likely multiple pathways) even when the myocyte membrane is depolarized and energy demand is minimal.

Damico *et al.* (1990) have used ^{13}C isotopomer analysis to examine age-related differences in substrate selection in newborn pig hearts. Comparing metabolism in hearts from piglets aged 5–7 days vs. 25–30 days, the ^{13}C spectra of heart extracts showed quite clearly that a dramatic switch in substrate selection away from pyruvate (a glucogenic substrate) toward acetate (as a fatty acid substitute) occurred during this 20–25 day maturation period. This shift in substrate oxidation away from glucogenic sources toward fatty acids may reflect age-related differences in control of mitochondrial synthesis of ATP.

Steady-state ^{13}C isotopomer methods have also been used to compare skeletal muscle metabolism in perfused hindlimbs from control vs. septic rats (Yang *et al.*, 1992). These authors reported that skeletal muscle in both control and septic animals derived about 80% of total acetyl-CoA from [2-^{13}C]acetate in the perfusate and that relative anaplerotic flux (y) was about 36% that of total citric acid cycle flux. More recently, our group (Szczepaniak *et al.*, 1996) monitored the time dependence of ^{13}C incorporation into skeletal muscle glutamate *in vivo* from [2-^{13}C]acetate infused into the blood of live rabbits. Once again, an isotopomer analysis of tissue extracts collected after the glutamate pool had reached steady state indicated that 87% (F_{C2}) of the skeletal muscle acetyl-CoA was derived from [2-^{13}C]acetate under these conditions (about the same as in hearts from the same animals) and that anaplerosis was indeed quite high ($y = 26\%$ compared to 3% in the *in vivo* heart). However, if one considers that O_2 consumption is about 20-fold lower in resting skeletal muscle compared to heart, then absolute anaplerotic flux in skeletal muscle (although high when compared to citric acid cycle flux) was about 4-fold lower in skeletal muscle than in heart.

A final example of the utility of a steady-state ^{13}C isotopomer analysis is given by the work of Jessen *et al.* (1993), who examined the effects of myocardial protection by aspartate and glutamate during reperfusion. Control hearts and hearts subjected to 25 min of global ischemia were perfused with a mixture of [1,2-^{13}C]acetate and [3-^{13}C]lactate ± 5 mM unlabeled aspartate and glutamate prior to freeze-clamping and ^{13}C NMR analyses. As found previously (Sherry *et al.*, 1992), hearts recovering from ischemia utilized more acetate and less lactate than controls and anaplerosis was significantly higher in the reperfused myocardium. As addition of aspartate and glutamate to cardioplegic solutions has been associated with superior recovery of function during reperfusion (Lazar *et al.*, 1980; Choong *et al.*, 1988), the finding that exogenous aspartate and glutamate had no effect on functional recovery in reperfused working hearts and did not alter the profile of substrate utilization was unanticipated. Further experiments with ^{13}C-enriched aspartate and glutamate showed that neither molecule appeared to enter the myocardium and neither contributed to the increased anaplerotic flux seen in the reperfused tissue. Studies such as this may prove useful in evaluating the effects of various oxidizable substrates on myocardial viability and in formulating strategies to improve cardiac performance after ischemia.

6. THE NONSTEADY-STATE ANALYSIS

There are many situations in active tissue where metabolic information such as that illustrated above would be quite valuable even though the system may not be at metabolic or isotopic steady state. Examples might be immediately after a change in heart workload, in regions of heart tissue where flow is restricted and O_2 consumption may be limiting, or *in vivo* where steady state simply cannot be confirmed under some conditions. In those circumstances, the equations presented above may not be valid and an alternative procedure is required. We have demonstrated elsewhere that the F_{Ci} variables can be obtained with confidence from the glutamate C4 and C5 multiplet areas and the total C4/C3 ratio at any time during isotopic presteady state *and* during changes in intermediate pool sizes (Malloy *et al.*, 1990b). The equations describing these relationships include:

$$\mathrm{C4Q} * (\mathrm{C4/C3}) = F_{\mathrm{C3}}$$
$$\mathrm{C4D34} * (\mathrm{C4/C3}) = F_{\mathrm{C2}}$$
$$(\mathrm{C5S/C5D}) * F_{\mathrm{C3}} = F_{\mathrm{C1}}$$
$$1 - F_{\mathrm{C1}} - F_{\mathrm{C2}} - F_{\mathrm{C3}} = F_{\mathrm{C0}}$$

Thus, if one made available to tissue a mixture of substrates that could produce [1-^{13}C]acetyl-CoA, [2-^{13}C]acetyl-CoA, [1,2-^{13}C]acetyl-CoA, and unenriched acetyl-CoA, all four F_{C} parameters could be derived from the glutamate C4 and C5 multiplet areas and the C4/C3 ratio (note that the resonance ratios may require corrections for NOE and T_1 differences, depending upon how the experimental data are collected; also, the C4/C2 ratio may be substituted for C4/C3 should C3 be obscured by other metabolite resonances). These relationships are independent of changing pool sizes, anaplerosis, and alterations in the isotopomeric composition of citric acid cycle pools because of the direct correspondence between ^{13}C enrichment in glutamate C4 and C5 and the C2 and C1 of acetyl-CoA. These two glutamate resonances sample the isotopomer population of the acetyl carbons in acetyl-CoA on each pass through citrate synthase, so the analysis is independent of cycle turnover or flux of other molecules through one or more of the intermediate pools. The only requirement is that sufficient ^{13}C must have entered the glutamate pool to allow detection by NMR. Note, however, the nonsteady-state analysis does not allow a determination of anaplerosis, y.

Like the other tools for analysis of the ^{13}C spectrum, the nonsteady-state analysis assumes that the distribution of isotope in any molecule reflects the distribution of isotope in its precursor. In this case, we assume that the enrichments in carbons 4 and 5 of glutamate are identical to enrichments in carbons 4 and 5 of α-ketoglutarate, which in turn are equivalent to carbons 2 and 1 of acetyl-CoA, respectively. Although this assumption is appropriate under steady-state conditions, it is conceivable that rapid changes in acetyl-CoA enrichment may not be instantaneously transmitted to glutamate C4 and C5. If flux through citrate synthase is slowed, for example, then a rapid change in substrate selection (reflected by changes in the labeling pattern of acetyl-CoA) may be detected in the ^{13}C spectrum of glutamate only after a short delay.

The utility of the nonsteady-state analysis was recently demonstrated in a study of perfused rat hearts exposed to either 0.225 mM H_2O_2 or 0.4 mM *tert*-butylhydroperoxide (Jones *et al.*, 1996) in the presence of 10 mM unenriched glucose, 1 mM [3-^{13}C]lactate, and 0.25 mM [1,2-^{13}C]acetate. As these concentrations of oxidants caused substantial decreases in contractility, an assumption of isotopic steady state could not be assured even though the hearts were presented with ^{13}C substrates 45 min prior to freeze-clamping (more than sufficient time to assure isotopic steady state during normoxia). Three glutamate C4 resonances are shown in Fig. 7: one from the ^{13}C spectrum collected on an extract of a control heart and two others from hearts perfused with either H_2O_2 or *tert*-butylhydroperoxide. Direct inspection of those resonances shows that acetate utilization was lower in both hearts exposed to the oxidants than in the control heart. When the C4D34 and C4Q multiplet areas were measured and the nonsteady-state analysis applied (multiplying by the total C4/C3 ratio), the lactate, acetate, glucose, and endogenous substrate contributions to acetyl-CoA in the control hearts were 38, 22, 5, and 35%, respectively (the glucose vs. endogenous substrate contributions were separated by performing separate experiments with [UL-^{13}C]glucose and unlabeled lactate and acetate). The substrate contributions to acetyl-CoA were substantially different in hearts exposed to oxidants. H_2O_2 nearly doubled the contribution from lactate (72%), slightly decreased the acetate contribution (18%), increased the glucose contribution (9%), and essentially inhibited oxidation of all endogenous substrates. *tert*-Butylhydroperoxide had less of an effect compared to controls on lactate utilization (41%) and acetate utilization (14%), but dramatically increased glucose utilization (44%). Once again, the contribution from endogenous substrates to acetyl-CoA was negligible in hearts exposed to *tert*-butylhydroperoxide. These data show that cardiac metabolism is altered in substantially different ways by these two oxidants and serve to illustrate that significant metabolic information may be obtained about various pathways feeding the acetyl-CoA pool even during metabolic conditions that cannot insure that the tissue is at ^{13}C isotopic steady state.

This same nonsteady-state isotopomer analysis has been used to evaluate the fractional contribution of [3-^{13}C]lactate or [3-^{13}C]pyruvate to acetyl-CoA (F_{C2}) from ^{13}C spectra of tissue extracts sampled from *in vivo* dog hearts after infusion

→

Figure 7. ^{13}C NMR spectra (C4 glutamate region only) of extracts of hearts perfused with 10 mM unenriched glucose, 1 mM [3-^{13}C]lactate, and 0.25 mM [1,2-^{13}C]acetate. Spectra are from control hearts (A), hearts perfused with 0.225 mM H_2O_2 (B), and hearts perfused with 0.4 mM *t*-butylperoxide. The peak labels refer to multiplets arising from entry of lactate (L) or acetate (A) into the citric acid cycle. The resonance designated S2/S2′ is from succinate.

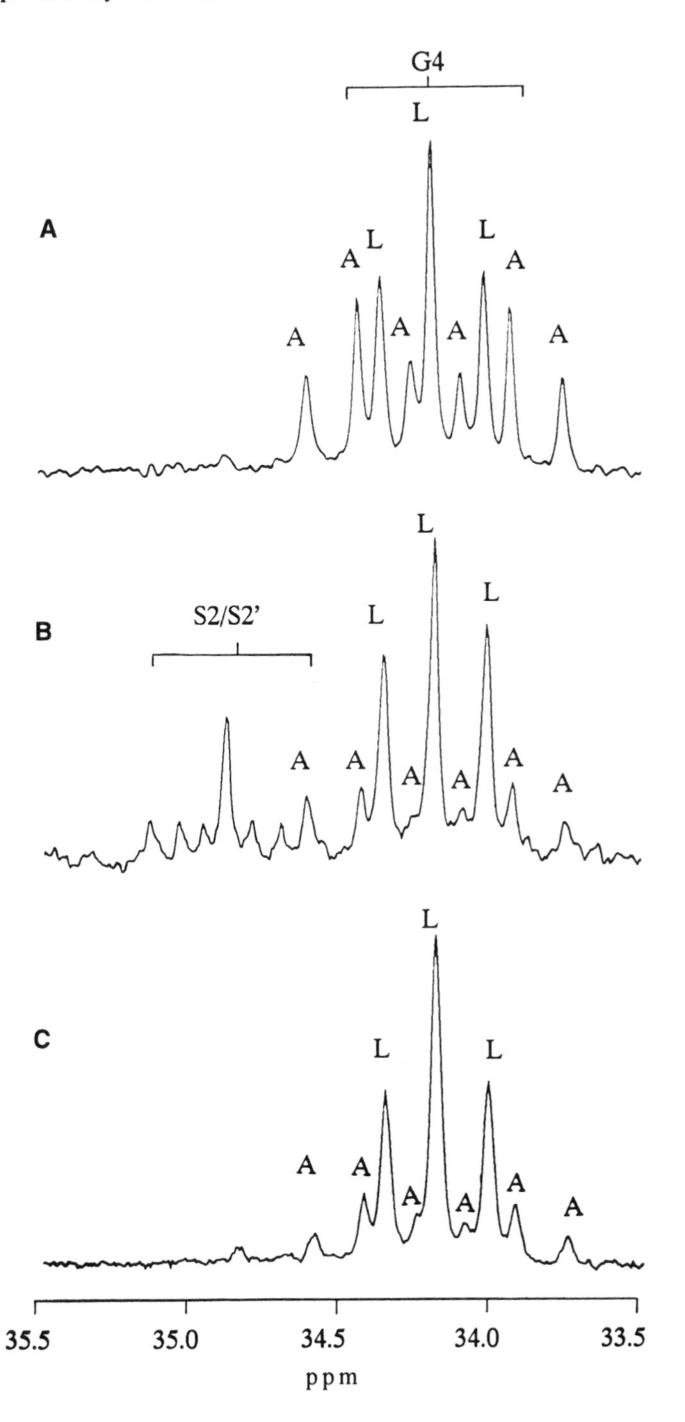
G4
L
A
A
L
L
A
A
A
A
A
L
B
S2/S2'
L
L
A
A
A
A
A
A
L
C
L
L
A
A
A
A
A
A
35.5
35.0
34.5
34.0
33.5
ppm

of either substrate into the blood of live animals (Laughlin *et al.*, 1993). The fractional contribution of labeled substrate (57% for [3-^{13}C]lactate and 38% for [3-^{13}C]pyruvate) to acetyl-CoA was equal to the ^{13}C fractional enrichment of the methyl carbon of alanine (as determined by ^{1}H NMR), indicating that alanine mirrored the pool of pyruvate entering the citric acid cycle in the *in vivo* myocardium and that high concentrations of exogenous lactate or pyruvate suppress oxidation of other endogenous substrates. Although these authors did not attempt to identify the source of the unlabeled substrate contributing to acetyl-CoA (F_{C0}), the observation that the ^{13}C fractional enrichment of alanine and acetyl-CoA were equal suggests that the source of the unlabeled substrate must have been glycolytic (if one can safely assume that alanine reflects cytosolic pyruvate). Interestingly, the ^{13}C fractional enrichment of the methyl carbon of lactate (again determined by ^{1}H NMR) was equal to that of alanine in dogs infused with [3-^{13}C]lactate but significantly less than alanine in dogs infused with [3-^{13}C]pyruvate. These observations suggested that myocardial lactate dehydrogenase had reduced activity in the presence of elevated levels of pyruvate.

7. THE DIRECT C4 ANALYSIS: A READOUT OF RELATIVE SUBSTRATE UTILIZATION

The nonsteady-state analysis described above requires that the citric acid cycle intermediate pools have "turned over" enough times after exposure to a ^{13}C-enriched substrate to enrich glutamate C3 sufficiently so that its resonance integral can be measured accurately and the multiplets, C4D34 and C4Q, are visible in the spectrum. This metabolic requirement cannot always be met, especially in circumstances where citric acid cycle activity might be compromised. In those instances, one may have no choice but to derive metabolic information from the glutamate C4 multiplets only. By inspecting the equations in Table 1, it is easy to see that there are some simple relationships between the C4 multiplets that can be useful. For example, both C4D34/C4Q and C4S/C4D45 directly report the ratio of F_{C2}/F_{C3}. This means that in metabolic situations where the glutamate C3 pool is not sufficiently enriched and C4D34 and C4Q may be small or not visible at all, the ratio F_{C2}/F_{C3} may still be obtained from the C4S/C4D45 ratio. Let us consider again a situation where three ^{13}C-enriched substrates and one unenriched substrates are presented to a tissue (Fig. 8). In this case, [1,3-^{13}C]acetoacetate would generate [1-^{13}C]acetyl-CoA, [3-^{13}C]lactate would generate [2-^{13}C]acetyl-CoA, uniformly enriched [UL-^{13}C]fatty acids would generate [1,2-^{13}C]acetyl-CoA, and natural abundance glucose plus any endogenous substrates would contribute to unenriched acetyl-CoA. Direct inspection of Fig. 8 shows that the C5S/C5D ratio would directly report the ratio of acetoacetate/fatty acids used by the tissue, while any of the ratios C4S/C4D45, C4D34/C4D45, or (C4S + C4D34)/(C4D45 + C4Q) would

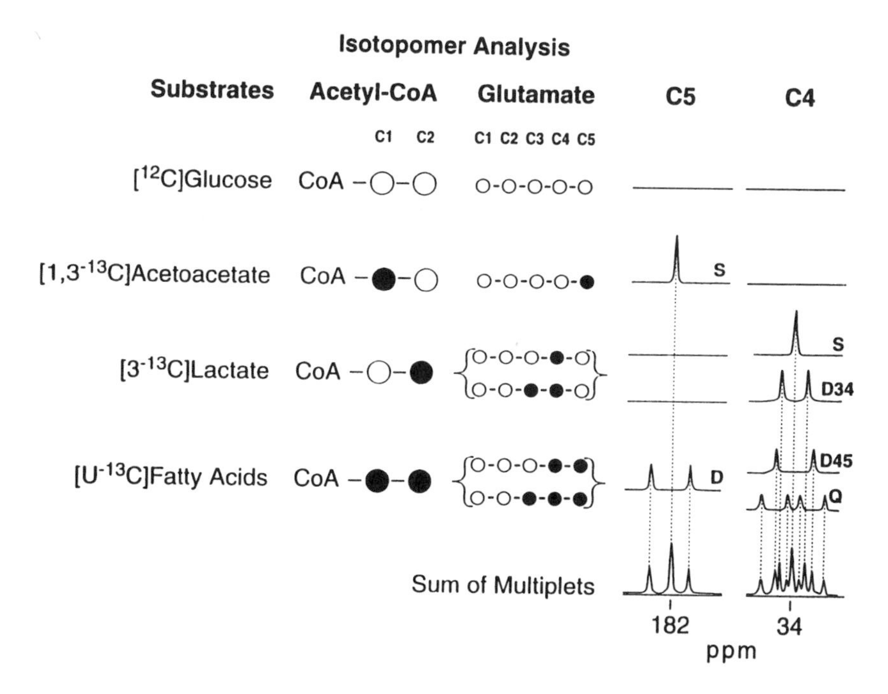

Figure 8. The direct analysis method for determining substrate competition. Substrates labeled as shown to the left can generate the four possible acetyl-CoA isotopomers. After entry of these four isotopomers into the citric acid cycle pools, this information is encoded directly in the glutamate C4 and C5 multiplet patterns. In this example, C5S can only be derived from [1,3-^{13}C]acetoacetate while C5D can only arise from [UL-^{13}C]fatty acids. Thus, the C5S/C5D ratio reports relative utilization of acetoacetate and fatty acids in the citric acid cycle. Similarly, C4S and C4D34 can arise only from entry of [3-^{13}C]lactate while C4D45 and C4Q arise only from [UL-^{13}C]fatty acids. The ratios C4S/C4D45, C4D34/C4Q, or (C4S + C4D34)/(C4D45 + C4Q) report relative utilization of lactate and fatty acids in the cycle. The contribution of unenriched glucose cannot be determined by direct inspection. Reproduced from Solomon *et al.* (1996).

report the lactate/fatty acid ratio. This method has been used recently to monitor substrate utilization in reperfused vs. control myocardium exposed to a physiological mixture of glucose (5.5 mM), lactate (1.2 mM), free fatty acids (0.35 mM), and acetoacetate (0.17 mM) (Solomon *et al.*, 1996). Rabbit hearts were first perfused to metabolic steady state with this same substrate mixture containing natural abundance levels of ^{13}C (1.1%). The left anterior descending coronary artery was then occluded for 15 min. After removing the occluder, the hearts were reperfused for 5 min with a substrate mixture containing the same amounts of enriched substrates, arrested with cold KCl, and dissected to separate ischemic from nonischemic tissue. Each portion was then freeze-clamped extracted and prepared for

^{13}C NMR. Figure 9 illustrates the glutamate C4 and C5 resonances from one such heart. Qualitatively, one can see by inspection that the relative acetoacetate/fatty acid utilization ratio (C5S/C5D) increased somewhat while the lactate/fatty acid utilization ratio (C4S/C4D45) decreased in the ischemic-reperfused region compared to the nonischemic region. Quantitatively, the acetoacetate/fatty acid utilization ratio increased by 55% while the lactate/fatty acid utilization ratio decreased by 56%. By including estimates of the total C3 resonance area (even though this signal intensity was low), a nonsteady-state analysis showed that indeed absolute fatty acid utilization did not differ in these two regions, but acetoacetate utilization was higher while lactate utilization was lower in the ischemic-reperfused tissue. This example illustrates that *relative* substrate utilization may be monitored by direct analysis of the glutamate C4 multiplets, even in circumstances where glutamate C3 may not be sufficiently enriched to measure its resonance area.

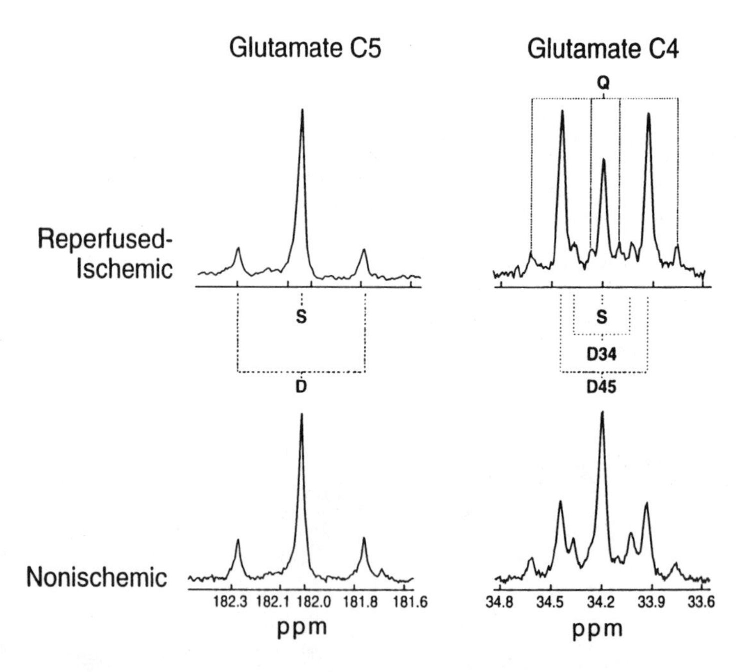

Figure 9. Glutamate C5 and C4 resonances from reperfused-ischemic (top) and nonischemic (bottom) regions of a rabbit myocardium provided with the substrates shown in Fig. 8. The C5 resonance shows an increase in C5S/C5D while the C4 resonance reports a decrease in C4S/C4D45 in the reperfused-ischemic compared to the nonischemic tissue. Reproduced from Solomon *et al.* (1996).

8. STEADY-STATE ANALYSIS UNDER NONSTEADY-STATE CONDITIONS

Two basic approaches to the design and analysis of ^{13}C NMR spectra are outlined in this book. Virtually all kinetic studies designed to measure flux in the citric acid cycle rely upon measurement of *multiple* spectra. Repeated measurement of ^{13}C enrichment in glutamate is most typical. The ^{13}C isotopomer approach emphasized in this chapter requires only a *single* spectrum to calculate the relative activities of metabolic pathways. Even if conditions are not at steady state, a detailed picture of the sources of acetyl-CoA can be obtained, repeatedly if necessary, as shown in Fig. 8. If steady state applies, then relative flux and labeling patterns in the anaplerotic pathways can also be determined.

It is also possible to obtain both kinetic data and measurements of anaplerosis if conditions are selected carefully. One feature of carbon flow in the citric acid cycle is that the methyl carbon of acetyl-CoA can ultimately enrich any of the carbons in glutamate except carbon 5, while the carbonyl of acetyl-CoA can only yield enrichment in carbons 1 or 5, but never carbons 2, 3, or 4. Therefore, ^{13}C enrichment in carbons 2, 3, and 4 is dependent on the fraction of acetyl-CoA that is *either* [1,2-^{13}C] *or* [2-^{13}C]. Furthermore, the ^{13}C NMR spectrum of glutamate C3 is not sensitive to enrichment in the C1 of acetyl-CoA. This feature of isotopomer analysis may prove quite useful for selected studies. For example, if a tissue is perfused to steady state with a substrate yielding [1,2-^{13}C]acetyl-CoA, the equations which describe the C3 multiplets in Table 1 plus the ratio C3/C4 can be used to calculate y. The labeling pattern in the available substrate could then be switched to one yielding [2-^{13}C]acetyl-CoA. Even before the distribution of ^{12}C achieves steady state in carbons 1 and 5 of glutamate, the steady-state equations can be used for analysis of the C3/C4 and the multiplets in C3.

These principles are illustrated in Fig. 10, which shows the calculated kinetics of ^{13}C enrichment in C4 and C3 (top panel) in a heart exposed to [1,2-^{13}C]acetate. These curves are similar to those published nearly 15 years ago by Chance's group. The bottom panel shows the evolution of multiplets in the C3 resonances which reach steady state. The C4 multiplets (middle panel) show a strikingly different pattern after the switch to [2-^{13}C]acetate: the C4Q and D45 disappear rapidly, as expected, and the C4S and C4D34 grow progressively. However, throughout this period the equations describing the C3 multiplets do not change: $C3S = F_{C0}(F_{C0} + y)/(y + 1)$, and $C3T = (1 - F_{C0})^2/(y + 1)$. Therefore, both y and F_{C0} (and therefore $F_{C2} + F_{C3}$) may be calculated throughout this period. With repeated sampling of spectra, citric acid cycle kinetics could be calculated simultaneously. This method of switching substrates has the advantage of a constant level of enrichment in carbons 2, 3, and 4. This means that signal-to-noise would not change during collection of kinetic data, in contrast to experiments when substrate enrichment is switched from ^{12}C to ^{13}C, or vice versa.

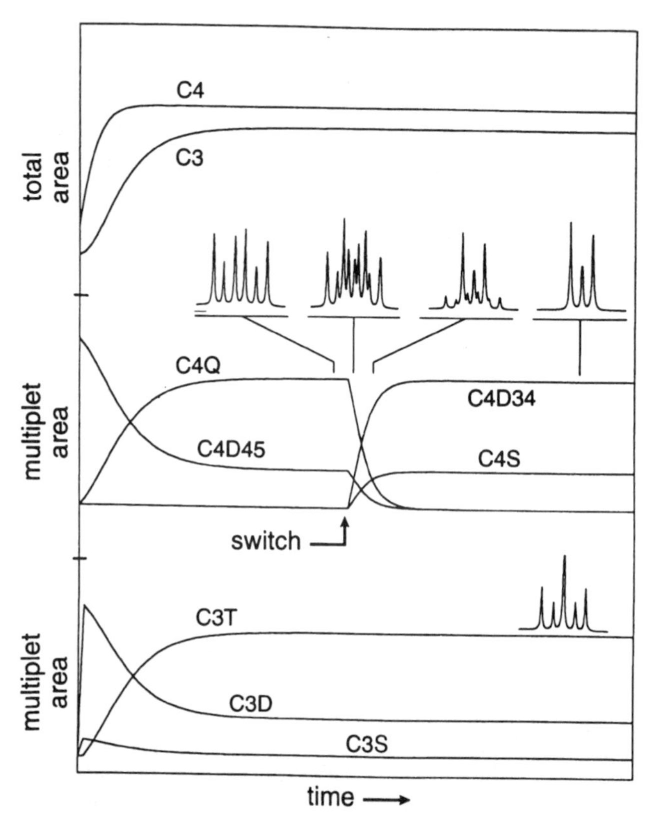

Figure 10. ^{13}C NMR spectra and time course of ^{13}C enrichment in glutamate. In this simulation, the acetyl-CoA was initially 90% [1,2-^{13}]acetyl-CoA and the remainder was not enriched. Anaplerosis was set to 8%. As the labeled acetyl-CoA was metabolized, ^{13}C enrichment increased in carbon 4 (C4) and, at a slower rate, carbon 3 (C3), as shown in the upper panel. The evolution of the multiplets in carbon 4 (C4Q and C4D45) and carbon 3 (C3T, C3D, and C3S) are shown in the lower panel. After steady-state conditions were achieved, the enrichment in acetyl-CoA was changed to 90% [2-^{13}C]acetyl-CoA. Note that the total enrichment in C3 and C4 did not change, and that the spectrum of C3 was not altered. However, the C4 resonance evolved dramatically. This simulation illustrates that certain resonances or features of the glutamate spectrum can be considered at steady state even when isotopic steady state does not apply. The simulation was created with *tcaSIM*.

9. ABSOLUTE METABOLIC FLUXES FROM ^{13}C ISOTOPOMER DATA

We have recently shown that the metabolically related F_{Ci} and F_{Ai} indices derived by ^{13}C isotopomer analysis can be combined with a single O_2 consumption measurement to determine the percent contribution of each individual substrate to total citric acid cycle flux (Malloy *et al.*, 1996). When the myocardium is oxidizing a single substrate, the relationship between citric acid cycle flux and O_2 consumption is substrate-dependent and readily derived. However, when a heart is exposed to a physiological mixture of substrates, the amount of each substrate oxidized by the myocardium will depend to some extent on neurohumoral conditions, relative concentrations of substrates, and the current metabolic state of the tissue (normoxia vs. ischemia, for example). Consequently, the precise relationship between O_2 consumption and citric acid cycle flux is sensitive to physiological factors which control the profile of substrates that are actually oxidized.

Given that complete oxidation of one mole of acetyl-CoA consumes 2 moles of molecular oxygen (one cycle turn nets 4 reducing equivalents or 8 electrons), one can easily derive the proportionality factor ($R_i = Q_i/C_i$) relating O_2 consumption (Q_i) to citric acid cycle flux (C_i) for any given substrate. For example, complete oxidation of one mole of acetate produces 4 reducing equivalents and consumes 2 moles of O_2; hence, $R_i = 2$ for acetate. Complete oxidation of glucose produces 2 additional reducing equivalents, one in glycolysis and another at the level of pyruvate dehydrogenase, for a total of 6; hence, $R_i = 3$ for glucose. The R_i values for other common substrates are summarized in Table 2. Other definitions include the following: Total O_2 consumption by the heart was defined as Q_t, while Q_0, Q_1, Q_2, and Q_3 refer to O_2 consumption resulting from oxidation of substrates 0, 1, 2, and 3, respectively. Thus, $Q_t = Q_0 + Q_1 + Q_2 + Q_3$. Similarly, citric acid cycle flux was defined as C_t, and C_0, C_1, C_2, and C_3 refer to cycle flux due to oxidation of substrates 0, 1, 2, and 3. Consequently, $C_t = C_0 + C_1 + C_2 + C_3$. Since the R_i factor for each substrate (Table 2) relates O_2 consumption to citric acid cycle flux, $Q_t = C_0R_0 + C_1R_1 + C_2R_2 + C_3R_3$. Given that the F_C variables are defined by the fraction any given substrate makes to total acetyl-CoA entering the citric acid cycle, $F_{C0} =$

Table 2
Calculated R_i Factors for Various Oxidizable Substrates

Substrate	R_i
Glucose	3.00
Lactate	3.00
Free fatty acids (of average chain length)	2.80
Pyruvate	2.50
Butyrate	2.50
β-Hydroxybutyrate	2.25
Acetoacetate	2.00
Acetate	2.00

C_0/C_t, $F_{C1} = C_1/C_t$, $F_{C2} = C_2/C_t$, and $F_{C3} = C_3/C_t$. These relationships can be combined to give

$$Q_t/C_t = F_{C0}R_0 + F_{C1}R_1 + F_{C2}R_2 + F_{C3}R_3 \quad (1)$$

Hence, citric acid cycle flux (C_t) can be determined from the F_{Ci} values measured by ^{13}C NMR, the R_i factor for each substrate, and a single measure of total O_2 consumption (Q_t). This equation applies only when anaplerosis can be ignored. Although flux through anaplerotic reactions is generally low in the perfused heart supplied by a single substrate (about 5% of citrate synthase flux), this activity can be stimulated significantly in some circumstances.

How can anaplerosis modify the relationship between O_2 consumption and citric acid cycle flux? Although carboxylation of pyruvate to form oxaloacetate does not produce reducing equivalents, a parallel disposal reaction must occur for entry of each pyruvate into the cycle pools at steady state. The most likely disposal reaction in heart tissue involves removal of 3 carbon units as pyruvate via the NADP-linked malic enzyme (Nuutinen *et al.*, 1981). During normoxia, oxidative flux far exceeds demands from biosynthetic pathways requiring NADPH, so one can assume that NADPH produced by the malic enzyme is converted to NADH by the nicotinamide nucleotide transhydrogenase system.

An inspection of Fig. 1 will show that metabolism of 1 mole of propionate through the citric acid cycle to pyruvate produces 2 moles of reducing equivalents (one mole at succinate dehydrogenase and one mole at the malic enzyme). Metabolism of pyruvate through pyruvate carboxylase is more complex. Carboxylation of exogenous pyruvate from the perfusate and subsequent decarboxylation via the NADP-linked malic enzyme produces no net reducing equivalents. Pyruvate derived from exogenous lactate or glucose, or endogenous glycogen, however, would produce one mole of NADH. Thus, activation of an anaplerotic reaction involving either propionate or pyruvate could increase O_2 consumption without increasing flux through citrate synthase. This contribution is given by yC_tR_A, where R_A is a substrate-dependent factor which relates the number of reducing equivalents produced by each anaplerotic pathway (R_A is 0 for carboxylation of exogenous pyruvate, 0.5 for carboxylation of pyruvate derived from glucose, lactate, or glycogen, and 1 for carboxylation of propionate). Thus,

$$Q_t = C_0R_0 + C_1R_1 + C_2R_2 + C_3R_3 + yC_tR_A \quad (2)$$

and

$$Q_t/C_t = F_{C0}R_0 + F_{C1}R_1 + F_{C2}R_2 + F_{C3}R_3 + yR_A \quad (3)$$

This analysis is most valuable for hearts exposed to a mixture of substrates, particularly when anaplerosis is active. Hearts perfused in the Langendorff mode were exposed to [3-^{13}C]L-lactate, [3-^{13}C]pyruvate, [U-^{13}C]fatty acids, [1,3-

^{13}C]acetoacetate, and unlabeled glucose, each at a concentration normally present in the plasma of a fed, rested rat (Remesy and Demigne, 1983). Typical ^{13}C NMR spectra of hearts utilizing this mixture of substrates with and without 5 mM unenriched propionate are shown in Fig. 11. Simple inspection of the C4 and C5 resonances demonstrates that lactate and pyruvate (labeled "L" in the C4 resonance), fatty acids (labeled "F" in the C4 and C5 resonances), and acetoacetate (labeled "A" in the C5 resonance) each contributed to acetyl-CoA. Although the glutamate C5 resonance in the upper spectrum is not shown because it overlapped with other carbonyl carbons, the multiplets of this resonance could be deconvoluted into a singlet and a doublet. A complete steady-state analysis, using methods outlined above, provided the relative contribution of each substrate to acetyl-CoA plus the activity and labeling pattern of anaplerotic substrates.

The sources of substrate contributing to unenriched acetyl-CoA could possibly include glucose in the perfusate, or, as noted above, glycogen or triglycerides. The rate of glycogenolysis in the presence of fatty acids is relatively low, about 0.5 mmole/gdw/min in working hearts and much less than this in Langendorff-perfused hearts (Crass *et al.*, 1969). Separate experiments (data not shown) using [U-^{13}C]glucose and the same mixture of substrates without ^{13}C enrichment showed no measurable glucose oxidation. Therefore, R_0 was assumed to be 2.8, the value representing a mixture of triglycerides with different chain lengths. Substrates contributing to F_{C1}, the fraction of acetyl-CoA labeled in the carbonyl carbon, could only be derived from acetoacetate in the perfusate; thus, R_1 was 2.00. Substrates contributing to F_{C2}, the fraction of acetyl-CoA labeled in the methyl carbon, could only have been derived from pyruvate and lactate in the perfusate. Assuming that pyruvate and lactate were oxidized in proportion to their relative concentrations, R_2 was calculated as $3(10/11) + 2.5(1/11) = 2.95$. Substrates contributing to F_{C3}, representing doubly labeled acetyl-CoA, could only have been derived from exogenous fatty acids. The calculated R_3 value for the particular mixture of exogenous fatty acids presented to this heart was 2.79.

In the absence of propionate, the sources of acetyl-CoA as derived from the ^{13}C spectrum were: endogenous lipids ($F_{C0} = 0.10 \pm 0.04$); acetoacetate ($F_{C1} = 0.23 \pm 0.03$); lactate plus pyruvate ($F_{C2} = 0.03 \pm 0.01$); and fatty acids ($F_{C3} = 0.65 \pm 0.02$). Oxygen consumption was 53 ± 13 μmole/min/gdw. The ^{13}C isotopomer analysis also indicated that anaplerosis was low ($y = 0.07 \pm 0.04$) and that none of the substrate entering the anaplerotic reactions was derived from ^{13}C-enriched sources. Substitution of these metabolic indices into Eq. (3) gave the contribution of each substrate to citric acid cycle flux. For this group of hearts, exogenous fatty acids, acetoacetate, and lactate/pyruvate contributed 11.4, 3.4, and 1.3 μmole/min/gdw, respectively, to a total citric acid cycle flux of 20.0 ± 4.9 μmole/min/gdw. The remaining citric acid cycle flux (3.9 μmole/min/gdw) could be attributed to endogenous unlabeled substrates.

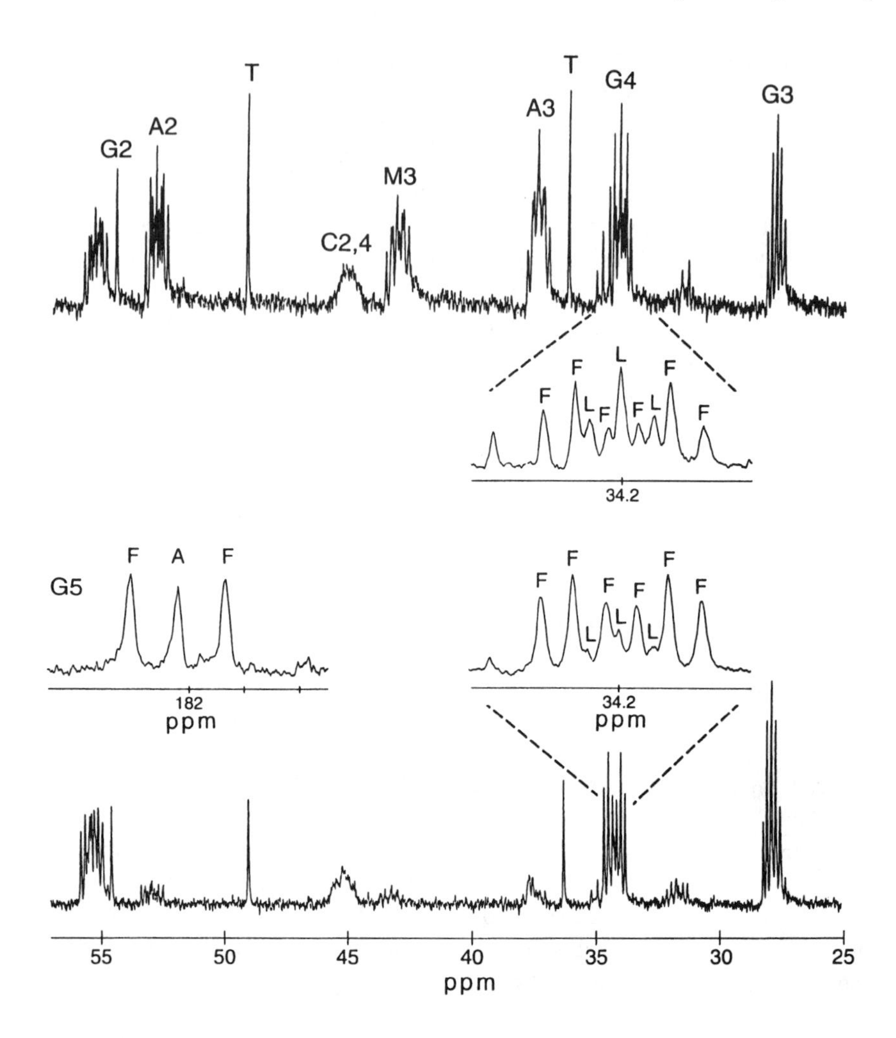

Figure 11. Proton decoupled ^{13}C NMR spectra of extracts from hearts supplied with [U-^{13}C] long-chain fatty acids, [3-^{13}C]L-lactate, [3-^{13}C]pyruvate, [1,3-^{13}C]acetoacetate, and glucose. Resonances from taurine (T) and glutamate carbons 2–5 (G2, G3, G4, G5) are shown (G5 is the insert) in 4B. The addition of 5 mM propionate increased the concentration of aspartate (A2 and A3), malate (M3), and citrate (C2, 4); these assignments are shown in 4A. Multiplets which are a consequence of entry of acetoacetate (A), lactate (L), and fatty acids (F) into the acetyl-CoA pool are identified in the expanded resonances. Note that in the presence of propionate, the contribution of lactate (L) increased relative to fatty acids (F), as indicated by a direct readout of the glutamate C4 resonance. The resonance from glutamate C5 in the upper spectrum is not shown because it overlapped with other carbonyl carbons. Reproduced from Malloy *et al.* (1996).

The addition of 5 mM propionate to the perfusate substantially altered the pattern of substrate flow into the citric acid cycle, as evidenced by the dramatic changes seen in the ^{13}C NMR spectrum (Fig. 11). A complete isotopomer analysis indicated that the sources of acetyl-CoA in this case included endogenous lipids ($F_{C0} = 0.08 \pm 0.04$), acetoacetate ($F_{C1} = 0.20 \pm 0.04$), lactate plus pyruvate ($F_{C2} = 0.30 \pm 0.01$), and fatty acids ($F_{C3} = 0.43 \pm 0.03$), while relative anaplerotic flux had now increased to 0.29 ± 0.05. Once again, F_{A1} was zero, so all of the anaplerotic substrate was assumed to have been derived from propionate. O_2 consumption in these hearts was identical to that of the control group, 55 ± 5 mmol/min/gdw. Interestingly, addition of propionate stimulated lactate/pyruvate oxidation by the citric acid cycle, but reduced the contributions of exogenous fatty acids and acetoacetate to citric acid cycle flux. Substituting the F_{Ci} parameters for this group of hearts into Eq. (3) showed that exogenous fatty acids, acetoacetate, and lactate/pyruvate contributed 8.5, 3.1, and 4.7 μmole/min/gdw, respectively, to a total citric acid cycle flux of 18.54 ± 1.46 μmol/min/gdw. Again, the remaining portion of citric acid cycle flux (2.24 μmole/min/gdw) resulted from endogenous substrates.

This study illustrated that it is possible to obtain a complete profile of flux through the citric acid cycle using a combination of ^{13}C NMR and O_2 consumption data. This includes individual flux contributions from multiple substrates through the oxidative portion of the citric acid cycle as well as flux through any anaplerotic pathways. O_2 consumption was identical in these two groups of hearts, so one might at first be tempted to conclude that citric acid flux was also identical. However, the data show that citric acid cycle flux tended to be lower in the propionate group (20.0 vs. 18.5 μmole/min/gdw) due to the additional reducing equivalents produced via entry and disposal of propionate into the cycle pools (see Fig. 1). Propionate also increased relative flux of the glycogenic substrates, lactate and pyruvate, and decreased relative flux of fatty acids by nearly an equivalent amount. This likely reflects direct stimulation of pyruvate dehydrogenase by propionate (Latipaa *et al.*, 1985).

A number of reports have appeared recently in which relationships between the rate of appearance of ^{13}C in glutamate and O_2 consumption in either isolated organs (Chance *et al.*, 1983; Weiss *et al.*, 1992) or *in vivo* (Fitzpatrick *et al.*, 1990) have been described. It is worth pointing out that any such relationship will be substrate-dependent and that changes in an NMR measurable index (such as half-time for enrichment of the C4 of glutamate, for example) does not necessarily require a change in O_2 consumption. This may be an important consideration *in vivo* whenever more than one ^{13}C-enriched substrate with differing R_i values are available for oxidation.

10. OTHER CONSIDERATIONS

Beyond the information provided by the multiplets in a ^{13}C NMR spectrum, there are also practical advantages of using ^{13}C as a metabolic tracer. First, all precautions related to radiation protection are eliminated. This allows experiments in working environments where radiotracer studies are simply unacceptable or at least cumbersome due to requirements for radiation containment. Perhaps less obvious, the use of ^{13}C-enriched precursors is helpful in that the experimentalist can easily determine if a compound with a complex labeling pattern is enriched in the correct positions, a task quite difficult for ^{11}C- or ^{14}C-enriched compounds. Furthermore, a variety of ^{13}C-enriched compounds are now commercially available, including compounds with complex labeling patterns such as alternately labeled fatty acids, that are normally difficult to obtain if the isotope is ^{11}C or ^{14}C. There are also advantages that result from the NMR method itself. For example, purification, chemical degradation, or derivitization of metabolic products is not required. Most importantly, ^{13}C NMR and the methods described in this chapter are directly applicable in humans (Gruetter *et al.*, 1994).

For those investigators with access to suitable equipment, the most exciting application of ^{13}C NMR is direct observation of intact tissues. However, a valuable strategy is to begin by working with perchloric acid extracts of almost any tissue (bacteria, yeast, isolated cells, perfused organs, etc.). The experiment should be designed in such a way that at least 25% or more of the acetyl-CoA pool will be labeled in the C2 position ([2-^{13}C]acetyl-CoA, [1,2-^{13}C]acetyl-CoA, or both). This requires that at least one significant source of acetyl-CoA is highly enriched. The experiment should continue to near-steady state which typically requires 30 minutes in Langendorff rat hearts, and somewhat longer in tissues with lower oxygen consumption. Perchloric acid extracts of freeze-clamped tissue are prepared in the usual fashion and neutralized with KOH, not K_2CO_3 (Sherry *et al.*, 1990). The freeze-dried extracts are reconstituted in deuterated water in a 5 or 10 mm NMR tube. A drop of unenriched dioxane provides sufficient ^{13}C signal (natural abundance) to provide an internal chemical shift standard at 67.4 ppm. The nonsteady-state or steady-state equations (where applicable) are used to determine the F_{Ci} parameters. If one is uncertain whether the steady-state assumptions apply, an agreement between the two evaluations suggests that the system is at steady state, and that anaplerosis can also be determined.

Finally, one should be aware that experimental conditions may differ for a metabolism study involving ^{13}C vs. ^{11}C or ^{14}C. As the sensitivity of both NMR and MS is low compared to measurement of ^{14}C radioactivity, it has been suggested that the biological relevance may be compromised because the mass of the added compounds is often not negligible. This is certainly a concern, since substrate loads could definitely influence metabolic profiles. We have found, however, that administration of [3-^{13}C]alanine or [3-^{13}C]lactate intravenously at physiological blood levels yields spectra suitable for isotopomer analysis in extracts of hepatic tissue. Similarly, mixtures of ^{13}C-enriched fatty acids, lactate, pyruvate, ketones, and glucose in physiological concentrations yield beautiful ^{13}C spectra from isolated rat hearts. This indicates that if the experiment is designed carefully, physiologically relevant metabolic data may be obtained with confidence using the techniques outlined in this chapter.

ACKNOWLEDGMENTS. The authors thank Drs. Piyu Zhao, John Jones, and Mike Soloman for providing some of the figures and Dr. Mark Jeffrey for providing *tcaSIM*. We also acknowledge grant support from the National Institutes of Health (HL-34557 and RR-02584) and the Department of Veterans Affairs.

REFERENCES

Chance, E. M., Seeholzer, H., Kobayashi, K., and Williamson, J. R., 1983, *J. Biol. Chem.* **258**:13785.

Chatham, J. C., Forder, J. R., Glickson, J. D., and Chance, E. M., 1995, *J. Biol. Chem.* **270**:7999.

Choong, Y. S., Gavin, J. B., and Armiger, L. C., 1988, *J. Mol. Cell. Cardiol.* **20**:1043.

Cohen, D. M., and Bergman, R. N., 1995, *Am. J. Physiol.* **268**:E397.

Cohen, S. M., 1983, *J. Biol. Chem.* **258**:14294.

Cohen, S.M., Rognstad, R., Shulman, R. G., and Katz, J., 1981, *J. Biol. Chem.* **256**:3428.

Crass, M. F., McCaskill, E. S., and Shipp, J. C., 1969, *Am. J. Physiol.* **216**:1569.

Damico, L. A., Closter, J., Jan, L., and Clark, B. J., 1990, The ontogeny of substrate choice in the neonatal piglet heart, *Ninth Annual Meeting of the Society of Magnetic Resonance in Medicine*, Book of Abstracts, p. 179.

Davis, E. J., Spydevold, O., and Bremer, J., 1980, *Eur. J. Biochem.* **110**:255.

Eakin, R. T., Morgan, L. O., Gregg, C. T., and Matwiyoff, N. A., 1972, *FEBS Lett.* **28**:259.

Exton, J. H., and Park, C. R., 1967, *J. Biol. Chem.* **242**:2622.

Fitzpatrick, S. M., Hetherington, H. P., Behar, K. L., and Shulman, R. G., 1990, *J. Cereb. Blood Flow Metab.* **10**:170.

Goebel, R., Berman, M., and Foster, D., 1982, *Fed. Proc.* **41**:96.

Gruetter, R., Novotny, E. J., Boulware, S. D., Mason, G. F., Rothman, D. L., Shulman, G. I., Prichard, J. W., and Shulman, R. G., 1994, *J. Neurochem.* **63**:1377.

Haut, M., and Basickes, S., 1967, Comput. Biomed. Res. **1**:139.

Jeffrey, F. M. H., Rajagopal, A., Malloy, C. R., and Sherry, A. D., 1991, *Trends Biochem. Sci.* **16**:5.

Jeffrey, F. M. H., Alvarez, L., Diczku, V., Sherry, A. D., and Malloy, C. R., 1995, *J. Cardiovasc. Pharmacol.* **25**:469.

Jeffrey, F. M. H., Storey, C. J., Sherry, A. D., and Malloy, C. R., 1996, *Am. J. Physiol.* **271**:E788.

Jessen, M. E., Kovarik, T. E., Jeffrey, F. M., Sherry, A. D., Storey, C. J., Chao, R. Y., Ring, W. S., and Malloy, C. R., 1993, *J. Clin. Invest.* **92**:831.

Jones, J. G., Sherry, A. D., Jeffrey, F. M. H., Storey, C. J., and Malloy, C. R., 1993, *Biochemistry* **32**:12240.

Jones, J. G., Le, T. H., Storey, C. J., Sherry, A. D., Malloy, C. R., and Burton, K. P., 1996, *Free Rad. Biol. Med.* **20**:515.

Katz, J., 1985, *Am. J. Physiol.* **248**:R391.

Katz, J., Lee, W.-N. P., Wals, P. A., and Bergner, E. A., 1989, *J. Biol. Chem.* **264**:12994.

Kelleher, J. K., 1986, *Am. J. Physiol.* **250**:E296.

Kornberg, H. L., 1966, *Essays Biochem.* **2**:1.

Krebs, H., 1970, *Perspect. Biol. Med.* **14**:154.

Latipaa, P. M., Peuhkurinen, K. J., Hiltunen, J. K., and Hassinen, I. E., 1985, *J. Mol. Cell. Cardiol.* **17**:1161.

Laughlin, M. R., Taylor, J., Chesnick, A. S., DeGroot, M., and Balaban, R. S., 1993, *Am. J. Physiol.* **264**:H2068.

Lazar, H. L., Buckberg, G. D., Manganaro, A. J., Becker, H., and Maloney, J. V., Jr., 1980, *Surgery* **88**:702.

Lewandowski, E. D., 1992, *Circ. Res.* **70**:576.

Lewandowski, E. D., and Hulbert, C., 1991, *Magn. Reson. Med.* **19**:186.

Lewandowski, E. D., Doumen, C., White, L. T., LaNoue, K. F., Damico, L. A., and Yu, X., 1996, *Magn. Reson. Med.* **35**:149.

London, R. E., 1988, *Prog. NMR Spectrosc.* **20**:337.

Magnusson, I., Chandramouli, V., Schumann, W. C., Kumaran, K., Wahren, J., and Landau, B. R., 1987, *J. Clin. Invest.* **80**:1748.

Malloy, C. R., Sherry, A. D., and Jeffrey, F. M. H., 1988, *J. Biol. Chem.* **263**:6964.

Malloy, C. R., Sherry, A. D., and Jeffrey, F. M. H., 1990a, *Am. J. Physiol.* **259**:H987.

Malloy, C. R., Thompson, J. R., Jeffrey, F. M. H., and Sherry, A. D., 1990b, *Biochemistry* **29**:6756.

Malloy, C. R., Jones, J. G., Jeffrey, F. M. H., Jessen, M. E., and Sherry, A. D., 1996, *MAGMA* **4**:35.

Moreland, C. G., and Carroll, F. I., 1974, *J. Magn. Reson.* **15**:596.

Nuutinen, E. M., Peuhkurinen, K. J., Pietilainen, E. P., Hiltunen, J. K., and Hassinen, I. E., 1981, *Biochem. J.* **194**:867.

Remesy, C., and Demigne, C., 1983, *Ann. Nutr. Metab.* **27**:57.

Robitaille, P.- M. L., Rath, D. P., Skinner, T. E., Abduljalil, A. M., and Hamlin, R. L., 1993, *Magn. Reson. Med.* **30**:262.

Sherry, A. D., Malloy, C. R., Roby, R. E., Rajagopal, A., and Jeffrey, F. M., 1988, *Biochem. J.* **254**:593.

Sherry, A. D., Malloy, C. R., Jeffrey, F. M. H., Chavez, F., and Srere, H. K., 1990, *J. Magn. Reson.* **89**:391.

Sherry, A. D., Malloy, C. R., Zhao, P., and Thompson, J. R., 1992, *Biochemistry* **31**:4833.

Solomon, M. A., Jeffrey, F. M. H., Storey, C. J., Sherry, A. D., and Malloy, C. R., 1996, *Magn. Reson. Med.* **35**:820.

Strisower, E. H., Kohler, G. D., and Chaikoff, I. L., 1952, *J. Biol. Chem.* **198**:115.

Sundqvist, K. E., Hiltunen, J. K., and Hassinen, I. E., 1989, *Biochem. J.* **257**:913.

Szczepaniak, L., Babcock, E. E., Malloy, C. R., and Sherry, A. D., 1996, *Magn. Reson. Med.* **36:**451.

Tornheim, K., and Lowenstein, J. M., 1972, *J. Biol. Chem.* **247**:162.

Viallard, J. F., Dos Santos, P., Raffard, G., Tariosse, L., Gouverneur, G., Besse, P., Canioni, P., and Bonoron-Adele, S., 1993, *Arch. Mal. Coeur Vaiss.* **86**:1123.

Walker, T. E., Han, C. H., Kollman, V. H., London, R. E., and Matwiyoff, N. A., 1982, *J. Biol. Chem.* **257**:1189.

Walsh, K., and Koshland, D. E., 1984, *J. Biol. Chem.* **259**:9646.

Weinman, E. O., Strisower, E. H., and Chaikoff, I. L., 1957, *Physiol. Rev.* **37**:257.

Weiss, R. G., Gloth, S. T., Kalil-Filho, R., Chacko, V. P., Stern, M. D., and Gerstenblith, G., 1992, *Circ. Res.* **70**:392.

Weiss, R. G., Kalil-Filho, R., Herskowitz, A., Chacko, V. P., Litt, M., Stern, M. D., and Gerstenblith, G., 1993, *Circulation* **87**:270.

Weiss, R. G., Stern, M. D., de Albuquerque, C. P., Vandegaer, K., Chacko, V. P., and Gerstenblith, G., 1995, *Biochim. Biophys. Acta* **1243**:543.

Williamson, D. H., Lund, P., and Krebs, H. A., 1967, *Biochem. J.* **103**:514.

Yang, X., Song, S., Ackerman, J. J. H., and Hotchkiss, R. S., 1992, Evaluation of the effect of sepsis on TCA cycle flux and glucose metabolism in skeletal muscle using ^{13}C NMR spectroscopy, *Eleventh Annual Meeting of the Society of Magnetic Resonance in Medicine*, Book of Abstracts, p. 2742.

Yu, X., White, L. T., Doumen, C., Damico, L. A., LaNoue, K. F., Alpert, N. M., and Lewandowski, E. D., 1995, *Biophys. J.* **69**:2090.

Zhao, P., Wiethoff, A. J., Sherry, A. D., and Malloy, C. R., 1995, ^{13}C NMR spectra of perfused rat hearts in the presence of the transaminase inhibitor, aminooxyacetate, *Abstracts of the Proceedings of the Society of Magnetic Resonance and the European Society for Magnetic Resonance in Medicine and Biology*, Nice, France, August, p. 301.

3

Determination of Metabolic Fluxes by Mathematical Analysis of ^{13}C-Labeling Kinetics

John C. Chatham and Edwin M. Chance

1. INTRODUCTION

A long-term goal in the study of biochemistry and metabolism has been the measurement of reaction rates and fluxes in intact biological systems. From the 1940s onward the development of techniques for measuring $^{14}CO_2$ production from ^{14}C-labeled substrates and techniques for separation of ^{14}C-labeled intermediates lead to tremendous progress in this direction. Recently, the application of NMR spectroscopy to biological systems, in particular ^{31}P and ^{13}C NMR spectroscopy, has furthered our understanding of how different metabolic processes are regulated *in vivo*. Despite these advances, the goal of obtaining quantitative metabolic fluxes

---

John C. Chatham • Department of Radiology and Radiological Sciences, Division of NMR Research, The Johns Hopkins University School of Medicine, Baltimore, Maryland 21205-2195. **Edwin M. Chance** • Department of Radiology and Radiological Sciences, Division of NMR Research, The Johns Hopkins University School of Medicine, Baltimore, Maryland 21205-2195 and Department of Biochemistry and Molecular Biology, University College London, Gower Street, London WC1E 6BT, England.

Biological Magnetic Resonance, Volume 15: In Vivo Carbon-13 NMR, edited by L. J. Berliner and P.-M. L. Robitaille. Kluwer Academic / Plenum Publishers, New York, 1998.

has remained elusive and much of our knowledge of metabolism and metabolic regulation is still based on studies of tissue extracts and isolated enzymes.

In principal, ^{13}C NMR spectroscopy should provide significant insight into *in vivo* metabolism; however, a major limitation in its widespread use has been the lack of readily usable metabolic flux models for the analysis of labeling kinetics. In 1983 Chance and co-workers published a mathematical model to analyze ^{13}C NMR kinetic data from rat hearts perfused with [2-^{13}C]acetate and [3-^{13}C]pyruvate (Chance *et al.*, 1983). We have recently extended the original model to include glycolysis, the TCA cycle (Chatham *et al.*, 1995), and mitochondrial transport processes. We present here a more detailed description of the methods used in the analysis of glutamate labeling kinetics.

The foundation of our work rests on the early studies of Krebs and co-workers who described the individual reactions comprising the TCA cycle (Krebs, 1940; Krebs and Eggleston, 1940; Krebs and Johnson, 1937). Subsequent isotope studies described specific mechanisms of each step; of particular importance were the studies which described the stereospecific formation of the symmetric molecule citrate and its asymmetric conversion to α-ketoglutarate (Ogston, 1951; Martius and Schorre, 1950; Potter and Heidelberger, 1949; Ogston, 1948; Wood *et al.*, 1941) as well as the asymmetric behavior of fumarase (Farrar *et al.*, 1957; Englard and Colowick, 1956; Fisher *et al.*, 1955). An excellent description of the work leading to our understanding of the reactions in the TCA cycle can be found in the series *Metabolic Pathways* edited by David M. Greenberg (Lowenstein, 1967; Krebs and Lowenstein, 1960).

In the 1950s and 1960s mathematical descriptions of intermediary metabolism were developed in order to analyze data arising from the metabolism of ^{14}C-labeled substrates (Wood *et al.*, 1963; Weinman *et al.*, 1957; Strisower *et al.*, 1952). Indeed the work by Strisower and colleagues (Weinman *et al.*, 1957; Strisower *et al.*, 1952) probably represents the first attempt at using a model of the TCA cycle for the analysis of isotope-incorporation data. The measurement of $^{14}CO_2$ production may provide a direct indication of the rate of substrate oxidation in an intact biological system; however, analysis of the intermediate steps is limited since it represents the end product of a large number of reactions about which there is no direct information. Furthermore, there are a variety of sources of CO_2 which may not directly reflect oxidative metabolism of the substrate of interest, or there may be several different oxidative pathways that yield $^{14}CO_2$. For example, in the case of [U-^{14}C]glucose, in addition to $^{14}CO_2$ formation from the TCA cycle, the pentose phosphate pathway is another source of $^{14}CO_2$. It is also not possible to distinguish between oxidation via PDH or pyruvate carboxylase. There are also data to suggest that $^{14}CO_2$ production may not accurately reflect fatty acid oxidation, with $^{14}CO_2$ contributing 20–70% of total fatty acid oxidation (Veerkamp *et al.*, 1986). One solution to this is to follow the ^{14}C isotope through the various intermediates by separating the different intermediates. However, using this approach one obtains specific activities rather than actual mass of the isotope incorporation at a single time point. Consequently, such data cannot be used to determine reaction rates or fluxes unless multiple experiments are carried out for varying times. Such a process, although possible, is clearly laborious and fraught with errors. In addition, it is also problematic to determine the specific carbon atoms that are labeled, thus it is difficult to ascertain the precise pathways involved in the labeling reactions.

One of the advantages of ^{13}C NMR spectroscopy is that using ^{13}C-labeled substrates one can measure continuously and nondestructively the actual mass of ^{13}C-label incorporation into specific carbon atoms of different metabolic intermediates. Thus, in principle, it is possible to obtain a comprehensive profile of the fate of the ^{13}C tracer over time. Unfortunately, due to the low sensitivity of NMR spectroscopy, the intermediates of glycolysis and the TCA cycle are below the limit of detection in many biological systems. However, glutamate is usually present in sufficient concentration for detection by NMR, and is in exchange with the TCA cycle intermediate α-ketoglutarate, via two transaminase reactions. There have been many studies that have analyzed the steady-state ^{13}C-labeling of glutamate to determine the relative contribution of various substrates to the overall TCA cycle flux (Lewandowski, 1992; Malloy *et al.*, 1990a; Malloy *et al.*, 1990b; Malloy *et al.*, 1988; Malloy *et al.*, 1987). These approaches, however, are limited in that they do not provide estimates of absolute metabolic fluxes, which can be obtained by analysis of ^{13}C-labeling kinetics (Chatham *et al.*, 1995; Chance *et al.*, 1983).

If glutamate is in rapid exchange with the TCA cycle, measurement of the time course of ^{13}C-label incorporation into glutamate by ^{13}C NMR spectroscopy may provide a means for noninvasively determining the oxygen consumption of an organ *in vivo*. Consequently, there has been much interest in analyzing the kinetics of isotopic incorporation into glutamate in order to estimate TCA cycle fluxes *in vivo*. Such calculations of absolute fluxes cannot be made from the measurement of steady-state enrichment data and external fluxes alone, due to the complexity of the reaction networks involved. It was only in the late 1970s that the computational techniques required to analyze large reaction networks became available (Chance *et al.*, 1977; Curtis, 1976; Curtis and Chance, 1972; Chance and Curtis, 1970) and in the 1980s that development of NMR hardware and methodology enabled ^{13}C NMR studies of intact biological systems to be possible (Dickinson *et al.*, 1983; Bailey *et al.*, 1981; den Hollander *et al.*, 1981).

2. APPROACH TO ANALYZING LABELING KINETICS

When analyzing kinetic data there are fundamentally two different approaches: (1) use the simplest possible mathematical function to adequately describe the kinetics or (2) use the knowledge we have regarding the chemistry of the system, describe this mathematically in as much detail as possible, and fit this to the experimental data. Although the first option is clearly the simplest from a computational point of view, it is difficult to interpret the data in terms of the biochemistry involved. For example, the kinetics of labeling of C_4-glutamate with ^{13}C can be relatively well described mathematically by a single exponential function; however, this process, at it simplest, is a result of the rate of TCA cycle flux, transaminase reactions, and mitochondrial transport systems. Thus a rate constant obtained from a simple exponential fit is comprised of an unknown combination of these different processes.

The alternative approach, of using as much information about the system that is available to construct a mathematical description of the metabolic pathways involved in the labeling process, is clearly complex. Until recently this was indeed a limitation as the methods required the use of a main frame, large-scale, sequential processor. However, improvements in both computer software and hardware over the past 15 years have enabled the solution of large networks to be carried out on a 32-bit personal computer in only a few minutes. When constructing a network to describe the labeling of glutamate, one has the option of considering either a subset of exchange reactions only involving intermediates labeled in specific positions arising from specific ^{13}C-labeled substrates or describing every possible carbon–carbon exchange reaction that may take place. While the latter option is more computationally intense, the resulting network is flexible and can be easily adapted to any number of possible substrates with variable labeling configurations.

3. FORMULATION OF MODEL

In Fig. 1 a schematic of the reaction network used to analyze glutamate labeling kinetics is presented. The network includes the basic reactions of the TCA cycle as well as the malic enzyme, cytoplasmic transaminase and malate-aspartate shuttle reactions. Oxygen consumption (MVO_2) was calculated from the algebraic sum of the fluxes of oxidizing and reducing equivalents (i.e. the sum of the fluxes through the TCA cycle, malate-aspartate shuttle, glycerol phosphate shuttle, glycolysis, β-oxidation, and proteolysis) divided by two.

The model requires that intermediate pool sizes be defined in order for fluxes to be calculated. While it was evident that the glutamate pool size would effect the flux calculations, it was not clear what effect differences in the pool sizes of TCA cycle intermediates would have on these calculations. An extensive sensitivity analysis showed that, provided the intermediate pool sizes were small relative to glutamate, changes in pool size had no effect on the calculated fluxes. Pool sizes for most of the TCA cycle intermediates were taken from the literature (Chance *et al.*, 1983) and, for succinate, fumarate, and oxaloacetate, an arbitrary value of 0.1 μmole/g wet weight was used.

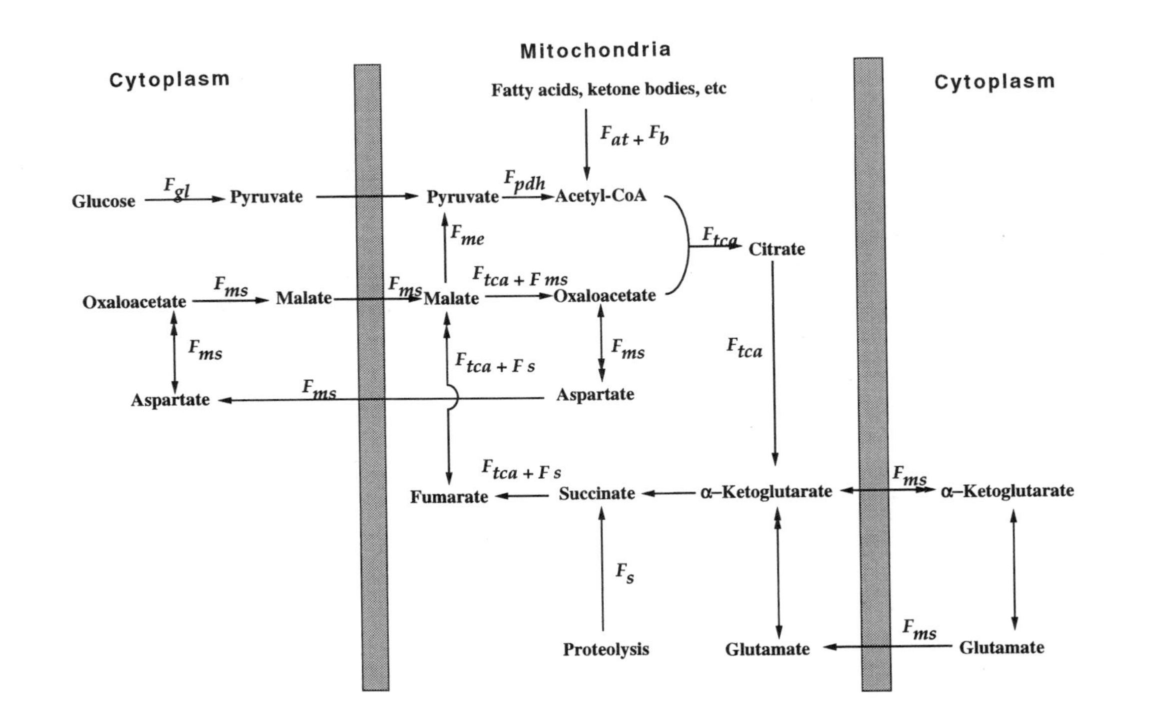

Figure 1. Reaction network used to analyze the kinetics of glutamate labeling. The abbreviations of steady-state fluxes are as follows: Fat = oxidation, of exogenous mitochondrial substrates (i.e., labeled, fatty acids, ketone bodies, acetate); Fb = β-oxidation—from endogenous sources (i.e., unlabeled, from triglyceride breakdown); Fgl = glycolysis—includes flux from both exogenous (i.e., labeled) and endogenous (i.e., unlabeled, from glycogen) sources; Fme = malic enzyme; Fms = malate–aspartate shuttle; Fpdh = pyruvate dehydrogenase; Fs = proteolysis; Ftca = TCA cycle.

Another unknown is the distribution of metabolites between the mitochondrial and cytosolic compartments. To determine whether this distribution was important in calculating the fluxes, the ratio of these compartments was varied between 5 and 40%. This covers the range of values for the fraction of the total cell volume occupied by the mitochondria in skeletal and cardiac muscle (Bers, 1991). Analysis indicated that the relative sizes of these two compartments over this range did not have a significant effect on the calculated fluxes.

The ^{13}C-labeled substrates enter the network as influxes and are assumed to be the principal sources of energy and of ^{13}C-label in the system. In order to account for our observation that endpoint enrichments of glutamate were significantly less than the maximum possible enrichment, additional unlabeled endogenous influxes had to be included in the network. Isotopic dilution at C_4-glutamate originates from influxes into the acetyl-CoA pool. Two principal pathways that could be responsible for isotopic dilution of acetyl-CoA are glycolysis (Fgl), and β-oxidation of endogenous fatty acids (Fb). Depending on the perfusion conditions, Fgl could arise from unlabeled glucose or from glycogen breakdown; it is not possible to distinguish between these pathways in our network.

Entry of unlabeled substrates into the TCA cycle, other than via acetyl-CoA, such as metabolism of amino acids, was represented by a single unlabeled influx (Fs) at succinyl-CoA. It should be noted that since succinyl thiokinase is freely reversible, both succinate and succinyl-CoA were treated as a single pool in our model. Although Fs was restricted to influx at succinyl-CoA, the present network cannot discriminate between unlabeled substrate influx here or via other anaplerotic pathways; the net result is the same, i.e., dilution of C_3-glutamate enrichment relative to C_4.

In order to maintain constant pool sizes of the TCA cycle intermediates and to allow the possibility of achieving a steady state, a branch point in the network at malate via malic enzyme was included. This opens an alternative pathway from malate to citrate with different fates for the individual carbon atoms. In the absence of malic enzyme, additional flux from proteolysis would lead to the flux into acetyl-CoA being less than cycle flux and the flux into oxaloacetate being greater than TCA cycle flux. The result of this would be a decrease in acetyl-CoA and malate pools and an increase in oxaloacetate and citrate; thus the system would not be at steady state. Malic enzyme has been shown to be active in rat heart mitochondria by a number of workers (Nagel *et al.*, 1980; Newsholme and Williams, 1978; Nolte *et al.*, 1972; Brdiczka and Pette, 1971). We have also recently obtained independent evidence of flux through this enzyme in the intact heart (Chatham and Forder, in preparation).

All the reactions in the network are considered to be irreversible with the exception of the transaminases, and fumarase. The reverse flux of the cytosolic transaminase was varied as an unknown parameter. The best agreement between the experimental and calculated results was obtained when the reverse reaction was slow compared to the forward reaction, indicating that the cytosolic transaminase is out of equilibrium. High forward and backward rates for fumarase relative to the TCA cycle were required to maintain a positive velocity and to achieve and maintain the fumarate and malate pool sizes at their equilibrium values.

The labeling reactions and enzymes used in the model are listed in Table 1. The network is constructed from differential equations describing each reactant in the network; one differential equation is used for every possible ^{13}C-labeled reactant. A list of the reactants and the number of differential equations required is given in Table 2. The reason for this large number of equations is that the number of possible labeled species for any reactant is 2^n, where n is the number of carbon atoms. For example, a total of 32 differential equations are required to describe the interconversion between α-ketoglutarate and glutamate.

Table 1
Summary of Metabolite Pools and Labeling Reactions Used in Model[a]

Labeling reactions[b]	Enzyme
Mitochondrial	
[Act] → [Cit]	Citrate synthase
[Cit] → [Ket]	Isocitrate dehydrogenase
[Ket] → [Suc]	α-ketoglutarate dehydrogenase
[Suc] → [Fum]	Succinate dehydrogenase
[Fum] ↔ [Mal]	Fumarase
[Mal] → [Oxa]	Malate dehydrogenase
[Oxa] → [Asp]	Transaminase
[Glu] ↔ [Ket]	Transaminase
[Pyr] → [Act]	Pyruvate dehydrogenase
$[Asp]_m \leftrightarrow [Asp]_c$	Mitochondrial transport
Cytoplasm	
[Asp] ↔ [Oxa]	Transaminase
[Oxa] → [Mal]	Malate dehydrogenase
[Ket] ↔ [Glu]	Transaminase
$[Pyr]_c \leftrightarrow [Pyr]_m$	Mitochondrial transport
$[Mal]_c \leftrightarrow [Mal]_m$	Mitochondrial transport
$[Glu]_c \leftrightarrow [Glu]_m$	Mitochondrial transport

[a]Double-headed arrows are those in which both forward and reverse fluxes are included in the network; all other reactions are irreversible.

[b]Oxa = oxaloacetate; Act = acetyl-CoA; Cit = citrate; Ket = α-ketoglutarate; Suc = succinate; Fum = fumarate; Mal = malate; Glu = glutamate; Pyr = pyruvate. Subscripts "c" and m refer to cytoplasm and mitochondria, respectively.

Table 2
List of reactants, their number of carbon atoms (n), and the number of differential equations used in network[a]

Reactant	n	Number of differential equations
Mitochondria		
Act	2	4
Cit	6	64
Ket	5	32
Suc	4	16
Fum	4	16
Mal	4	16
Oxa	4	16
Glu	5	32
Pyr	3	8
Cytoplasm		
Asp	4	16
Oxa	4	16
Ket	5	32
Mal	4	16
Glu	5	32

[a]See Table 1 for abbreviations.

The reason for the apparent complexity and size of the network is that no assumptions regarding the chemistry of the system are needed in order to simplify the mathematics. However, there is a great deal of symmetry in the system which simplifies the mathematics; for example, although a total of 64 reactions are required to describe all the possible carbon–carbon exchanges between citrate and α-ketoglutarate, these are clearly not 64 independent reactions. Furthermore, since we assume that the system is at metabolic steady state, then the flux into any one node of the network must equal the flux out of that node. As a result, the reactions describing the flux from oxaloacetate to citrate are dependent on the reactions describing the flux from citrate to α-ketoglutarate as well as the flux from malate to oxaloacetate. This interdependence occurs throughout the network and is a direct result of the steady-state constraint. Consequently, the internal fluxes (i.e., enzyme activities) depend on the external influxes and the only unknowns are the exogenous and endogenous influxes and the reverse flux of the transaminases. Thus, despite the complexity of the network, the number of unknown parameters is relatively small.

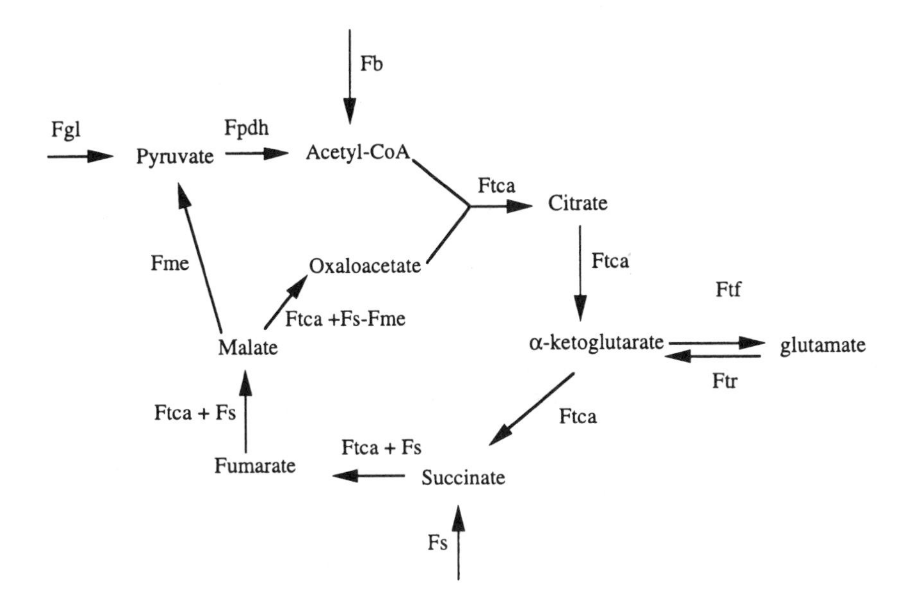

Figure 2. Simplified reaction network to demonstrate the simplification resulting from the assumption of metabolic steady state. The abbreviations of steady-state fluxes are the same as those in Fig. 1, except for: Ftf = forward transaminase; Ftr = reverse transaminase. Under conditions of metabolic steady state: Ftf = Ftr = unknown; Fs = Fme = unknown; Fb = unknown; Fgl = unknown; Ftca = Fpdh + Fb; Fpdh = Fgl + Fme; MVO2 = 1/2(Ftca + Fgl + Fb + Fs).

An example of how this complex network reduces to such a relatively small number of unknowns is demonstrated in Fig. 2, where a simplified network is shown; this network represents the situation that would exist in the absence of any mitochondria transport processes. Despite the fact that the total number of differential equations used to describe such a network is 204, there are only a total of 8 separate fluxes. Furthermore, because of the steady-state constraint (i.e., intermediate pool sizes remain constant for the duration of the experiment), the flux into any one intermediate must equal the flux out; consequently, these 8 fluxes are not independent of each other. For example, if we consider the situation at oxaloacetate where the flux entering oxaloacetate is (Ftca + Fs – Fme) and the flux leaving oxaloacetate is Ftca, then if the concentration of oxaloacetate is to remain constant Ftca must equal (Ftca + Fs – Fme), which means that Fms = Fs. Using this approach the total of 8 fluxes reduces to only 4 unknown parameters, namely, the substrate influxes, which in this network are glycolysis (Fgl), β-oxidation (Fb), the rate of α-ketoglutarate-glutamate exchange (i.e., transaminase flux, where Ftf = Ftr), and proteolysis (Fs = Fme). So in this simple case there are only 4 unknowns parameters and a minimum of 2 known parameters, i.e., the kinetics of labeling of C_4- and C_2-

(or C_3-) glutamate. It should be noted that in the absence of any evidence to the contrary the labeling of C_2- and C_3-glutamate is considered to be indistinguishable. If oxygen consumption is also measured, then a third constraint can be imposed on the network, since $MVO_2 = 1/2(Ftca + Fgl + Fb + Fs)$ would have to be satisfied.

The complete network that we have used (Fig. 1), is clearly more complex than that described in Fig. 2. The addition of mitochondrial transport processes results in the addition of at least four more unknowns, namely, the malate shuttle flux (Fms), the cytoplasmic transaminase flux, the α-ketoglutarate translocase flux, and the glycerol shuttle flux. While such a model by its very nature will never be overdetermined, additional parameters, other than glutamate labeling kinetics, can be used to provide additional constraints on the system. The measurement of MVO_2 we believe is an essential parameter, since this flux is the principal determinant of the TCA cycle flux. Furthermore, measurements of ^{13}C spectra from tissue extracts can be used to provide estimates of Fs, as well as the relative contributions of Fb and Fpdh to acetyl-CoA formation (Malloy *et al.*, 1990a; and Chapter 3). In addition, measurement of rates of substrate consumption, metabolite efflux (i.e., lactate efflux) will also provide additional estimates of Fgl and Fb, etc.

4. NUMERICAL METHODS

The majority of the metabolic network described here was constructed using FACSIMILE 3.0 or below (Chance *et al.*, 1977; Curtis, 1976). This program was originally developed by ARC Scientific (257 Woodstock Road, Oxford, England OX2 7AE) for AEA Technology (Harwell, Didcot, England OX11 0RA), for the general solution of large networks of simultaneous, ordinary differential equations of which this study is one application. However, current versions of this code do not run on higher releases of FACSIMILE; consequently, software specifically designed for this purpose is being developed and will be available in the near future from Differential Kinetics Inc. (1052 Corning Hill Parkway, Annapolis MD 21401) in conjunction with ARC Scientific.

The full dependence matrix required for the solution of a network of differential equations consists of n rows and n columns, where n is the total number of reactants in the system. The elements of the dependence matrix consist of 1s where there is dependence between reactants and 0s elsewhere. As a result many of the elements are zero, and most of the nonzero elements are along the diagonal. With a large number of reactants, as is the case here, the matrix becomes unwieldy and solving these systems with traditional techniques is time-consuming, since the amount of computation required to calculate the number of Newton iterations increases by about n^3. This problem is overcome by the use of sparse matrix techniques, which substantially reduce the number of nonzero elements that have to be considered. This approach reduces the amount of computation required to calculate the Newton iterations to about $3n$ (Chance and Curtis, 1970). In other words, for an n of 300 (approximately the number of elements in the network used here), using traditional techniques, the number of Newton iterations is about 2.7×10^7; however, the use of sparse matrices decreases this to approximately 900. In other words, the amount of computation is reduced by at least four orders of magnitude. The basic numerical features have already been described in detail (Curtis, 1976; Curtis and Chance, 1972).

Another essential facility is the ability to handle vectors of variables, enabling operations of entire arrays to be performed as a single instruction. This is especially useful in coding the TCA cycle reactions where there is a great deal of symmetry in the network (excluding the randomization at fumerase). Consequently, relatively large sections of the code can be written with relatively few array instructions. For example, there are 32 possible ^{13}C-labeled species of α-ketoglutarate, thus there are 32 differential equations required to describe the exchange between α-ketoglutarate and glutamate, as shown in Table 3; however, to encode all these equations involves only the following set of instructions:

```
ARRAY<32> WS;
%KFn%KBn: KG = GL;
ARRAY;
```

where ARRAY<32> operates on vectors of declared length 32; K_{Fn} defines the forward, and K_{Bn} the reverse, scalar rate constants for the exchange (i.e., the transaminase reaction) between all the ^{13}C-labeled species (n) of α-ketoglutarate (KG) and glutamate (GL). From this simple instruction FACSIMILE constructs all 32 differential equations describing all possible exchanges between α-ketoglutarate

Table 3
All Possible Exchange Reactions Between α-Ketoglutarate (KG) and Glutamate (GL)[a]

$KG_0 \rightarrow GL_0$	$KG_{12} \rightarrow GL_{12}$	$KG_{123} \rightarrow GL_{123}$	$KG_{1234} \rightarrow GL_{1234}$
$KG_1 \rightarrow GL_1$	$KG_{13} \rightarrow GL_{13}$	$KG_{124} \rightarrow GL_{124}$	$KG_{1235} \rightarrow GL_{1235}$
$KG_2 \rightarrow GL_2$	$KG_{14} \rightarrow GL_{14}$	$KG_{125} \rightarrow GL_{125}$	$KG_{1245} \rightarrow GL_{1245}$
$KG_3 \rightarrow GL_3$	$KG_{15} \rightarrow GL_{15}$	$KG_{134} \rightarrow GL_{134}$	$KG_{1345} \rightarrow GL_{1345}$
$KG_4 \rightarrow GL_4$	$KG_{23} \rightarrow GL_{23}$	$KG_{135} \rightarrow GL_{135}$	$KG_{2345} \rightarrow GL_{2345}$
$KG_5 \rightarrow GL_5$	$KG_{24} \rightarrow GL_{24}$	$KG_{145} \rightarrow GL_{145}$	$KG_{12345} \rightarrow GL_{12345}$
	$KG_{25} \rightarrow GL_{25}$	$KG_{234} \rightarrow GL_{234}$	
	$KG_{34} \rightarrow GL_{34}$	$KG_{235} \rightarrow GL_{235}$	
	$KG_{35} \rightarrow GL_{35}$	$KG_{245} \rightarrow GL_{245}$	
	$KG_{45} \rightarrow GL_{45}$	$KG_{345} \rightarrow GL_{345}$	

[a]Subscripts indicate position of ^{13}C-label. These 32 exchange reactions can be encoded by only 3 lines of instructions, as described in the text.

and glutamate (Table 3). The rate equations for the forward and reverse reactions are:

$$K_{Fn} = \text{FORWARD TRANSAMINASE FLUX}/M_{KG}$$

$$K_{Bn} = \text{REVERSE TRANSAMINASE FLUX}/M_{GL}$$

respectively, where M_{KG} and M_{GL} are the masses of KG and GL, respectively. Under these conditions the net flux through the transaminase reaction is $K_{Fn}M_{KG} - K_{Bn}M_{GL}$. It is apparent from this that under conditions of metabolic steady state (i.e., M_{KG} and M_{GL} are constant) K_{Fn} and K_{Bn} are independent of the labeling pattern and the net forward flux is the same for all 32 differential equations.

A major advantage of taking this approach is that we are able to follow the fate of any single carbon atom in the network. This flexibility allows the use of any combination of labeled substrates and enables the addition of new pathways with minimal changes in the code. As one equation is required for each reactant, additions to the network will only increase the number of equations if new reactants are added. Consequently, the addition of a new pathway with the current set of reactants will only increase the number of terms in the relevant equations, not the total number of equations. For example, the addition of pyruvate carboxylase to the network will not increase the number of equations, since the reactants—pyruvate and oxaloacetate—are already included in the network; however, as the reaction is unidirectional it will add one extra term in the equation describing the flux from pyruvate to oxaloacetate (if the reaction was reversible it would add two extra terms to the equation). On the other hand, if we include lactate dehydrogenase to the network this will add an additional 8 differential equations, since we are adding an extra 3-carbon reactant.

Stringent criteria are applied to the fitting of the kinetic data to determine the adequacy of the network we have established. These criteria are: (1) the standard deviation for the minimum least-squares fitting is within the error of the primary data, (2) the distribution of residuals is random, and (3) the calculated flux parameters are statistically well determined (Clore and Chance, 1978).

5. RESULTS AND DISCUSSION

The reaction network shown in Fig. 1 was used in the analysis of glutamate labeling kinetics from hearts perfused with [1-^{13}C]glucose alone, with [4-^{13}C]β-hydroxybutyrate plus glucose and [2-^{13}C]acetate alone, and with glucose as substrates. In all cases, using the C_2- and C_4-glutamate labeling data and the experimentally determined MVO_2 as input parameters for the model, there was a

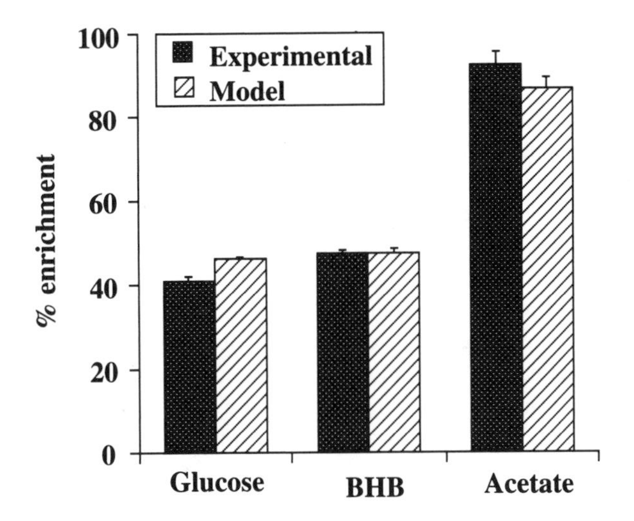

Figure 3. Comparison of fractional enrichment of glutamate determined experimentally and α-ketoglutarate calculated from the model. The enrichment of glutamate and α-ketoglutarate should be the same at steady state. Error bars represent the range of two experiments.

good fit between the calculated and experimentally determined glutamate labeling kinetics and oxygen consumption rates (Chatham *et al.*, 1995). We also compared the calculated enrichment at C_4-α-ketoglutarate with the measured enrichment at C_4-glutamate (Fig. 3); at steady state these should be equal. In all the experiments the measured values were within the 90% confidence limits of the calculated values. For example, in the β-hydroxybutyrate and glucose experiments, where the maximum fractional enrichment is 50%, the calculated enrichment was 47 ± 1% compared with 48 ± 2% determined from the heart extracts.

In order for the network to adequately describe the experimental results, both the calculated kinetics of glutamate labeling and the calculated MVO_2 had to agree with the experimentally measured values. Initially, the calculated MVO_2 rates were systematically lower than the measured values for this parameter, with the discrepancy being greatest with glucose as sole substrate and least for acetate. This indicated that the glutamate labeling rates were too slow in order to account for the measured oxygen consumption. As the current model utilizes the malate–aspartate shuttle as the sole mechanism of labeling of the cytosolic glutamate pool, this suggested that another mechanism of transporting reducing equivalents into the mitochondria was required. In the absence of any other evidence, we have assumed that this additional redox shuttle is the glycerol phosphate shuttle and the inclusion of the glycerol phosphate shuttle results in excellent agreement between the calculated and measured MVO_2 for all the experiments.

In many of the experiments, there was a small but significant dilution at C_4-glutamate, which was accounted for primarily by unlabeled influx from glycolysis. The contribution of proteolysis to the overall TCA cycle was not statistically well determined by the network, although it was significantly greater than zero and ranged between 0.02–0.12 μmole/min/g wet wt. This is in good agreement with other studies of proteolysis in the isolated perfused rat heart which reported rates in the range of 0.01–0.05 μmole/min/g wet wt (Pisarenko *et al.*, 1986; Peuhkurinen *et al.*, 1983; Takala *et al.*, 1980).

Despite the excellent agreement between the calculated and experimentally determined parameters, the results of this study raise several important issues that need to be addressed. For example, we have assumed that the malate–aspartate shuttle is the principal mechanism of labeling cytosolic glutamate; thus, using mitochondrial substrates, such as acetate or ketone bodies, it was necessary to reverse the direction of the malate–aspartate shuttle. On thermodynamic grounds there is no reason why the individual reactions of the malate–aspartate shuttle could not be reversible, and early reports indicated that efflux of glutamate from the mitochondria did take place (Azzi *et al.*, 1967). Subsequently, it was found that this happened only under de-energized conditions, when the mitochondrial membrane is collapsed (Williamson *et al.*, 1980). Experimentally, it has been shown that in isolated mitochondria the malate–aspartate shuttle is unidirectional with glutamate influx and aspartate efflux (LaNoue *et al.*, 1974). In other words, it functions only in the direction of transporting NADH from the cytosol to the mitochondria. If this is the case, then using the current model, there would be no labeling of the glutamate pool with acetate or β-hydroxybutyrate as substrates because this requires reversal of the shuttle. An alternative solution would be if the mitochondrial α-ketoglutarate translocase were freely reversible and the principal glutamate labeling reaction was the cytoplasmic rather than the mitochondrial transaminase.

We have also found it necessary to include a second mechanism for transporting NADH from the cytosol to the mitochondria in order to obtain good agreements between both MVO_2 and glutamate labeling kinetics. We have proposed that this is the glycerol phosphate shuttle; however, the precise nature of the shuttle is unimportant as long as the net effect is to transfer reducing equivalents from the cytosol to the mitochondria. Isaacs *et al.* (1969) provided evidence that the glycerol phosphate shuttle was operational in cardiac tissue; in contrast, Safer et al. (1971) suggested that hydrogen flux through this shuttle was limited under normal conditions by low concentration of glycerol-3-phosphate. However, low tissue concentration of glycerol-3-phosphate need not in itself preclude shuttle activity. It should be noted that the incorporation of ^{13}C-label from [1-^{13}C]glucose into glycerol-3-phosphate has been observed under similar conditions (Chacko and Weiss, 1993), indicating that at least the cytoplasmic portion of this shuttle is operative. It is clear that further studies are necessary to determine the importance of the glycerol phosphate shuttle; however, one of the difficulties in addressing this issue is that this shuttle is not directly involved in any of the labeling reactions. It is possible that there may be an alternative mechanism of glutamate labeling that could result in good agreement between the calculated and experimental MVO_2 without the need for an additional redox shuttle.

6. CONCLUSIONS

Using the network described here, we have excellent agreement between the calculated and experimentally determined oxygen consumption rates, the time courses of enrichment of C_4- and C_3-glutamate, and the fractional enrichment of the glutamate pools for three different substrates—glucose, acetate, and β-hydroxybutyrate (Chatham *et al.*, 1995). We believe that this was the first metabolic model for the analysis of glutamate labeling kinetics to include glycolysis, the TCA cycle, as well as mitochondrial and cytosolic compartmentation, and to confirm the results from the model with experimental data. More recently, Yu and colleagues (Yu *et al.*, 1995) have also started to address the issue of mitochondrial transport processes in the modeling of ^{13}C-kinetic data. It is important to note that in the case of large network models when a good fit to the data is obtained, it is only possible to conclude that the model is a good candidate for describing the data; it is not necessarily a unique solution. Thus, the model described here is only one possible description of reality and, as discussed above, changes may be necessary in order for the model to be more consistent with the known biochemistry of the system.

In order to obtain greater confidence in any one model, improved experimental data and the addition of other experimental constraints will be valuable. Probably the most significant limitation in accurately fitting the glutamate labeling curves is the quality of the NMR data. The early time points are especially important for defining the shape of the labeling curve, and yet these data points inevitably have the lowest signal-to-noise ratio. The advent of increased field strengths and the use of appropriately labeled substrates designed to maximize the enrichment of the glutamate pool should significantly improve the quality of the experimental data.

The work described here used the labeling kinetics of C_2- and C_4-glutamate and oxygen consumption as the only input parameters for the model; however, there is no reason why endpoint measurements of enrichment from high-resolution ^{13}C NMR spectra of tissue extracts cannot be used to provide additional constraints on the system. For example, although proteolysis was required by the model to fit the data, the value of this flux was poorly determined as the difference in steady-state enrichment of C_2- and C_4-glutamate in the intact heart was small. However, an estimate of this flux relative to total TCA cycle flux can be obtained using the methods of Malloy and colleagues (Malloy *et al.*, 1987) from spectra of tissue extracts where acquisition times are not limited by the viability of the preparation. This estimate could then be used to provide a constraint on the flux determined by the model. The model can also predict the enrichment of TCA cycle intermediates that are too low to be detected by NMR spectroscopy. It is possible with GC-MS to obtain enrichment data for many of these low concentration intermediates (Comte *et al.*, 1997; Des Rosiers *et al.*, 1995); thus GC-MS could provide a wealth of additional experimental data to support the model.

It is important to recognize that the apparently simple process of transferring ^{13}C-label from α-ketoglutarate to glutamate is still poorly understood. The majority of glutamate in the heart is cytosolic and the ^{13}C-label must be transferred from the mitochondrial α-ketoglutarate pool to the cytosolic glutamate pool. Currently, we do not know whether the primary labeling reaction is via the mitochondrial transaminase with subsequent transport of glutamate out of the mitochondria, or whether α-ketoglutarate is transported to the cytosol and labeling takes place via the cytoplasmic transaminase. If glutamate is first labeled and is then transported out of the mitochondrial compartment, we need to know how this is achieved. In either case it is possible that transport of label out of the mitochondria may be the rate-determining process in labeling the cytoplasmic glutamate pool (Yu *et al.*, 1995). Consequently, alternative mechanisms for mitochondrial transport of glutamate, such as glutamate–hydroxyl exchange or glutamate–glutamine exchange, need to be considered. Clearly, the precise mechanism for labeling the glutamate pool may be tissue-specific, and thus a model that accurately describes the labeling kinetics of one organ system may not be directly applicable to another.

Clearly, there are many possible approaches to the problem of modeling intermediary metabolism and the analysis of ^{13}C-labeling kinetics. The approach outlined here, we believe, is one of the most comprehensive and is founded upon experimental information. The combination of experimental and theoretical studies is essential for the development of a robust and comprehensive network. Furthermore, the issue of intracellular compartmentation, particularly metabolite exchange between the mitochondria and cytosol, is an integral part of our model and, until recently (Yu *et al.*, 1995), this had been ignored by most investigators. It is important to note that we perceive the development of this network not simply as a tool for the analysis of ^{13}C NMR spectra, but rather as a general approach to the investigation of metabolic processes in intact biological systems.

ACKNOWLEDGMENTS. Dr. John R. Forder of The Johns Hopkins University School of Medicine played a vital role in the development of the model described here. We also wish to acknowledge the contribution of University College London, England to these studies. This work was supported in part by grant R29 HL48789 from the National Institutes of Health (JCC) and a NATO Collaborative Research Grant No. 930206 (EMC).

REFERENCES

Azzi, A., Chappell, J. B., and Robinson, B. H., 1967, *Biochem. Biophys. Res. Commun.* **29**:148.
Bailey, I. A., Gadian, D. G., Matthews, P. M., Radda, G. K., and Seeley, P. J., 1981, *FEBS Lett.* **123**:315.
Bers, D. M., 1991, *Excitation–Contraction Coupling and Cardiac Contractile Force*, Kluwer Academic Publishers, Norwell, MA.
Brdiczka, D., and Pette, D., 1971, *Eur. J. Biochem.* **19**:546.
Chacko, V. P., and Weiss, R. G., 1993, *Am. J. Physiol.* **264**:C755.
Chance, E. M., and Curtis, A. R., 1970, *FEBS Lett.* **7**:47.
Chance, E. M., Curtis, A. R., Jones, I. P., and Kirby, C. R., 1977, *FACSIMILE: A Computer Program for Flow and Chemistry Simulation, and General Initial Value Problems,* H.M. Stationary Office, London.
Chance, E. M., Seeholzer, S. H., Kobayashi, K., and Williamson, J. R., 1983, *J. Biol. Chem.* **258**:13785.
Chatham, J. C., Forder, J. R., Glickson, J. D., and Chance, E. M., 1995, *J. Biol. Chem.* **270**:7999.
Clore, G. M., and Chance, E. M., 1978, *Biochem. J.* **173**:799.
Comte, B., Vincent, G., Bouchard, B., and Des Rosiers, C., 1997, *J. Biol. Chem.* **272**:26117.
Curtis, A. R., 1976, *Biochem. Soc. Trans.* **4**:364.
Curtis, A. R., and Chance, E. M., 1972, *FEBS Symp.* **25**:39.
den Hollander, J. A., Behar, K. L., and Shulman, R. G., 1981, *Proc. Natl. Acad. Sci. U.S.A.* **76**:2693.
Des Rosiers, C., Di Donato, L., Comte, B., Laplante, A., Marcoux, C., David, F., Fernandez, C., and Brunengraber, H., 1995, *J. Biol. Chem.* **270**:10027.
Dickinson, J. R., Dawes, I. W., Boyd, A. S. F., and Baxter, R. L., 1983, *Proc. Natl. Acad. Sci. U.S.A.* **80**:5847.
Englard, S., and Colowick, S. P., 1956, *J. Biol. Chem.* **221**:1019.
Farrar, T. C., Gutowsky, H. S., Alberty, R. A., and Miller, W. G., 1957, *J. Am. Chem. Soc.* **79**:3978.
Fisher, H. F., Frieden, C., McKee, J. S. M., and Alberty, R. A., 1955, *J. Am. Chem. Soc.* **77**:4436.
Isaacs, G. H., Sacktor, B., and Murphy, T. A., 1969, *Biochim. Biophys. Acta* **177**:196.
Krebs, H. A., 1940, *Biochem. J.* **34**:775.
Krebs, H. A., and Eggleston, L. V., 1940, *Biochem. J.* **34**:442.
Krebs, H. A., and Johnson, W. A., 1937, *Enzymologia* **4**:148.
Krebs, H. A., and Lowenstein, J. M., 1960, The tricarboxylic acid cycle, in *Metabolic Pathways* (D. M. Greenberg, ed.), Academic Press, New York, pp. 129–203.
LaNoue, K. F., Bryla, J., and Bassett, D. J. P., 1974, *J. Biol. Chem.* **249**:7514.
Lewandowski, E. D., 1992, *Biochemistry* **31**:8916.
Lowenstein, J. M., 1967, The tricarboxylic acid cycle, in *Metabolic Pathways* (D. M. Greenberg, ed.), Academic Press, New York, pp. 146–270.
Malloy, C. R., Sherry, A. D., and Jeffery, F. M. H., 1987, *FEBS Lett.* **212**:58.
Malloy, C. R., Sherry, A. D., and Jeffrey, F. M. H., 1988, *J. Biol. Chem.* **263**:6964.
Malloy, C. R., Sherry, A. D., and Jeffrey, F. M. H., 1990a, *Am. J. Physiol.* **259**:H987.
Malloy, C. R., Thompson, J. R., Jeffrey, F. M. H., and Sherry, A. D., 1990b, *Biochemistry* **29**:6756.
Martius, C., and Schorre, G. Z., 1950, *Z. Naturforsch.* **5b**:170.
Nagel, W. O., Dauchy, R. T., and Sauer, L. A., 1980, *J. Biol. Chem.* **255**:3849.
Newsholme, E. A., and Williams, T., 1978, *Biochem. J.* **176**:623.
Nolte, J., Brdiczka, D., and Pette, D., 1972, *Biochim. Biophys. Acta* **284**:497.
Ogston, A. G., 1948, *Nature* **162**:963.
Ogston, A. G., 1951, *Nature* **167**:693.
Peuhkurinen, K. J., Takala, T. E. S., Nuutinen, E. M., and Hassinen, I. E., 1983, *Am. J. Physiol.* **244**:H281.
Pisarenko, O. I., Solomatina, S., and Studneva, M., 1986, *Biochim. Biophys. Acta* **885**:154.

Potter, V. R., and Heidelberger, C., 1949, *Nature* **164**:180.
Safer, B., Smith, C. M., and Williamson, J. R., 1971, *J. Mol. Cell. Cardiol.* **2**:111.
Strisower, E. H., Kohler, G. D., and Chaikoff, I. L., 1952, *J. Biol. Chem.* **198**:115.
Takala, T., Hiltunen, J. K., and Hassinen, I. E., 1980, *Biochem. J.* **192**:285.
Veerkamp, J. H., van Moerkerk, H. T. B., Glatz, J. F. C., Zuurveld, J. G. E. M., Jacobs, A. E. M., and Wagenmakers, A. J. M., 1986, *Biochem. Med. Metabol. Biol.* **35**:248.
Weinman, E. O., Strisower, E. H., and Chaikoff, I. L., 1957, *Physiol. Rev.* **37**:252.
Williamson, J. R., Hoek, J. B., Murphy, E., and Coll, K. E., 1980, *Ann. N. Y. Acad. Sci.* **341**:593.
Wood, H. G., Katz, J., and Landau, B. R., 1963, *Biochem. Zeit.* **338**:809.
Wood, H. G., Werkman, C. H., Hemingway, A., and Nier, A. O., 1941, *J. Biol. Chem.* **139**:483.
Yu, X., White, L. T., Doumen, C., Damico, L. A., LaNoue, K. F., Alpert, N. A., and Lewandowski, E. D., 1995, *Biophys. J.* **69**:2090.

4

Metabolic Flux and Subcellular Transport of Metabolites

E. Douglas Lewandowski

1. INTRODUCTION

^{13}C NMR spectroscopy is becoming increasingly useful for monitoring metabolic turnover within intact tissues and thereby enables studies of the metabolic support of physiological function. ^{13}C nuclei are only 1.1% naturally abundant and are 1.6% as sensitive as protons for NMR detection. However, the low natural abundance enables the use of exogenous compounds that are enriched at specific sites with ^{13}C to be used for monitoring intermediary metabolism. With selective enrichment of carbon sites within metabolites of interest, the otherwise inherent insensitivity of ^{13}C NMR detection is actually well-suited to precise targeting of labeling metabolites in intact tissues. Such isotopic enrichment combined with proton decoupling and nuclear Overhauser enhancement (NOE) can be used to counter the low sensitivity of ^{13}C NMR observations to provide a powerful tool for examining cell metabolism.

A distinct advantage of ^{13}C detection is apparent for NMR studies on mixed samples of ^{13}C-enriched compounds and metabolites, as found in biological sys-

---

E. Douglas Lewandowski • NMR Center, Massachusetts General Hospital, Harvard Medical School, Bldg. 149, 13th Street, Charlestown, Massachusetts 02129.

Biological Magnetic Resonance, Volume 15: In Vivo Carbon-13 NMR, edited by L. J. Berliner and P.-M. L. Robitaille. Kluwer Academic / Plenum Publishers, New York, 1998.

tems, due to the relatively large chemical shift range of the ^{13}C nucleus as compared to that of ^{1}H (~220 ppm vs. ~15 ppm). The more favorable spectral resolution results from the larger number of electrons surrounding the carbon nucleus. This permits ^{13}C resonances to be identified which would otherwise be difficult to resolve in the proton spectrum, even with current applications of heteronuclear cross-polarization. The selective enrichment of metabolites and the ability to identify resonances signals over a wide chemical shift range are attractive features of ^{13}C NMR in addressing the metabolic ambiguities of intact tissues under normal and disease states.

The ability to obtain NMR data from intact tissues in a nondestructive manner also enables serial observations of the progressive incorporation of ^{13}C into metabolic intermediates. In this manner, the dynamics of tissue metabolism can be followed in response to physiological function and pathophysiological condition for regional or whole organ studies. As will be discussed below, most recent findings indicate that the separate compartmentation of metabolites in the cell provides the means to monitor subcellular metabolite transport across the mitochondrial membrane in coordination with flux through oxidative metabolism. Therefore, ^{13}C NMR not only provides unique and detailed information about the metabolic fate of a ^{13}C-enriched precursor, but also can be used to observe dynamic processes that relate to the physiochemical regulation of the cell such as metabolic flux, enzyme activity, and the exchange of metabolic intermediates between intracellular compartments.

However, for such measurements to be accurate, careful consideration of experimental factors must be considered. These considerations are: (1) fractional enrichment and metabolite concentration effects on signal intensity, (2) compartmentation of ^{13}C-enriched metabolites, (3) the distinctions between static and dynamic measures of metabolism, and (4) the correct match of kinetic analysis to the desired observations of metabolic processes. This chapter addresses each of these practical components of the carbon isotope kinetics in introducing the utility and application of dynamic-mode ^{13}C NMR spectroscopy.

2. THE GENERAL UTILITY OF DYNAMIC-MODE ^{13}C NMR: LESSONS FROM THE HEART

An example of the type of experimental data addressed in this chapter is shown in Fig. 1, which displays a series of sequential, proton decoupled ^{13}C spectra with digital subtraction of the natural abundance signals. Each spectrum was acquired in one minute, from an isolated rabbit heart oxidizing [2-^{13}C]acetate. The identifiable resonances demonstrate the progressive incorporation of the ^{13}C label from the metabolic substrate, or fuel, into the carbon chain of the glutamate molecule. This labeling of glutamate is fundamental to many ^{13}C NMR studies of metabolic flux

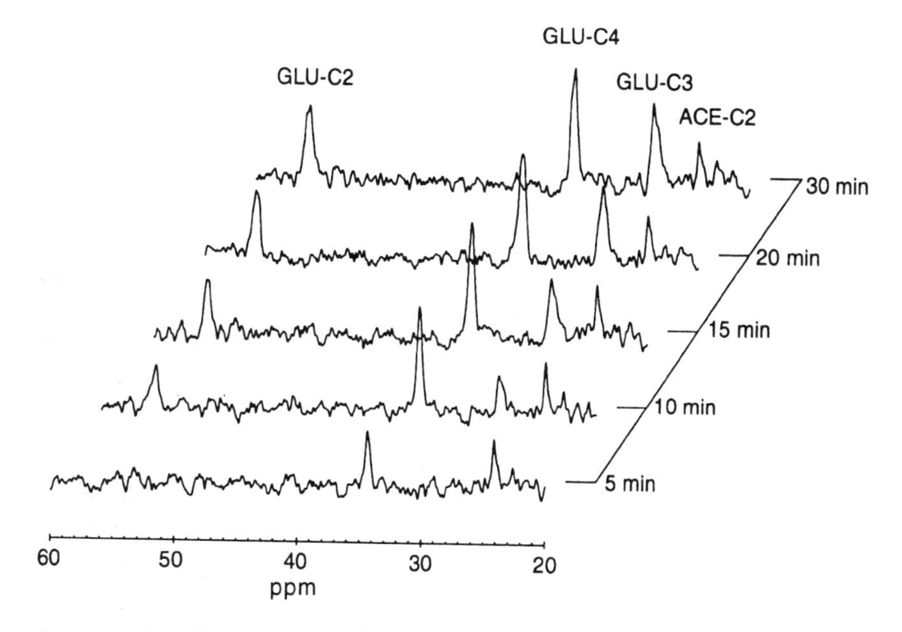

Figure 1. Sequential ^{13}C NMR spectra obtained from an isolated rabbit heart oxidizing [2-^{13}C] acetate. The progressive enrichment of the 2-, 3-, and 4-carbon positions of glutamate can be used to assess flux through oxidative metabolism. Identified resonance peaks are the initial site of labeling at the 4-carbon (GLU C4) and the secondary labeling sites, the 2-carbon (GLU C2) and the 3-carbon (GLU C3).

due to the information provided by the relative rates of ^{13}C enrichment at each of the identified carbon sites within the glutamate molecule. From NMR detection of the rates of carbon isotope turnover within the observed glutamate pool, dynamic metabolic processes within the cells of intact tissues can be determined. Therefore, the term "dynamic carbon-13 NMR spectroscopy" refers to the NMR detection of dynamic processes occurring within the cell. This experimental approach can provide a valuable index of metabolic flux and metabolite exchange between subcellular compartments in response to the physiological state of a tissue or whole organ.

Metabolic activity has long been known to be a correlate to physiological function. Studies to examine this link have often relied on biochemical and physiological evaluation of experimental heart preparations. Fundamental mechanisms can then be studied in isolated perfused hearts, without the confounding variables of organs contributing to whole body metabolism along with neural and hormonal effects on metabolism, albeit under much less physiological conditions. Isolated, perfused hearts have also been used to establish the biological basis for conditions of ^{13}C NMR detection in the intact organ (Damico *et al.*, 1996; Lewandowski *et al.*, 1996; Yu *et al.*, 1996; Chatham *et al.*, 1995; Lewandowski *et al.*, 1995a,b; Yu *et al.*, 1995; Lewandowski, 1992a,b; Weiss *et al.*, 1992; Lewandowski *et al.*, 1991b; Lewandowski and Hulbert, 1991; Lewandowski and Johnston, 1990; Weiss *et al.*, 1989; Malloy *et al.*, 1988; Sherry *et al.*, 1988; Sherry *et al.*, 1985; Chance *et al.*, 1983; Bailey *et al.*, 1981). From the understanding of basic mechanisms gained by isolated heart studies, *in vivo* evaluations with ^{13}C NMR can then be better understood under more complex physiological conditions.

The heart has traditionally been very useful for studies of metabolic regulation and flux by offering a fairly homogeneous cell population. In the heart, the vast majority of cell types are myocytes, which weight the NMR-observed metabolic activity. The heart also displays a relatively simple oxidative intermediary metabolism in comparison to other organ systems. One easily obtained index of organ function, the mechanical work performed by the heart, provides the correlate for understanding metabolic demand. The mechanical work output of the heart is known to be linearly related to the rate of oxidative metabolism which supports this function, as evidenced by concurrent measurements of the contractile function of the heart and oxygen consumption rates (Zimmer *et al.*, 1990). Given this direct relationship between metabolic flux and function in the heart, the heart serves as a very handy, easily manipulated engine for studying work output vs. energy production/utilization. Under such conditions, the heart can be used as a "flux phantom" for characterizing the links between metabolic rate and physiological function.

Therefore, advantages for establishing ^{13}C NMR protocols using the heart as a basic model are: (1) NMR signals from intact cardiac tissue are predominated by the large percentage of myocytes, or cardiac muscle cells, (2) contractile activity can be easily manipulated for altering metabolic flux rates, and (3) the metabolism is relatively simple to facilitate studies on metabolic regulation. It is easy to see then that many fundamental issues in the biochemistry of the cell can be learned from the simple lessons provided by NMR studies of cardiac metabolism.

Comprehensive metabolic analyses of whole organs are unfortunately not well suited to the concurrent measurement of physiological function due to the requirements of tissue sampling procedures. The destructive nature of such metabolic evaluations have to date been limiting factors in applicability to the basic physiology of intact myocardium and the clinical evaluation of diseased myocardium. Early examinations of the role of metabolic processes in supporting cardiac contractile activity were directed toward the measurement of static concentrations of key metabolites in experimental heart preparations (Peuhkurinen *et al.*, 1983; Taegtmeyer *et al.*, 1980; Neely *et al.*, 1973). Although enlightening, such snapshots of tissue content of extractable metabolites were limited in the perspective of metabolic turnover and concurrent physiological function. While absolute concentration measurements may provide an index of the metabolic or energetic state of the myocardium (Kupriyanov *et al.*, 1988; Taegtmeyer *et al.*, 1985; Reimer *et al.*, 1983; Neely *et al.*, 1973), such classical biochemistry has not demonstrated a strict correlation to cardiac performance. These more traditional methods of metabolite assays are further limited because sampling disrupts the intact system, halting the dynamic processes of metabolism and function.

Alternative methods of isotope uptake and release kinetics, such as radiolabeled carbon dioxide production from ^{14}C-labeled metabolic precursors, have served to demonstrate that metabolic flux through specific intermediary pathways is related to both contractile performance and the pathophysiological state of myocardium (Lopaschuk *et al.*, 1990; Renstrom *et al.*, 1989). However, such radiolabeling methods are restricted to measurements of metabolite efflux and washout from the tissues and vasculature. Unless biochemical assays of tissue samples are ultimately performed, the nondestructive analysis by radioisotopes is not specific for intracellular events alone. Therefore, cardiovascular studies are greatly enhanced by the nondestructive method of ^{13}C NMR spectroscopy that allows for repeated or continual metabolic evaluation of the same section of myocardium. The ability to obtain concurrent functional measurements and "on-line" monitoring of metabolic activity in the intact cell is the most effective means for relating tissue biochemistry to the "bottom line" of overall physiological function.

To best meet these requirements for observing metabolic processes within functioning organs, two methodologies fulfill these criteria to provide nondestructive, kinetic information: positron emission tomography (PET) and nuclear magnetic resonance (NMR) spectroscopy. While PET provides information on the uptake and spatial distribution of a positron emitting nucleus of an exogenous metabolic agent which can then be related to pathophysiological state (Schneider and Taegtmeyer, 1991; Buxton *et al.*, 1988), chemically discreet analysis of the metabolic fate of the label or carrier is unavailable. For example, the positron emitting isotope of carbon, ^{11}C, has been used to monitor rates of substrate oxidation based on isotope retention in the active tissue (Buxton *et al.*, 1988), but the method does not provide the chemical specificity that is afforded by the chemical shift information of NMR spectroscopy. Furthermore, the radiolabeled agent is not always a direct probe of the metabolic process of interest, such as the case of ^{18}F fluorodeoxyglucose for studying glucose uptake and glycolysis. In contrast, NMR spectroscopy is uniquely available to provide specific chemical information on the metabolism of exogenous ^{13}C-labeled substrates, although at the expense of spatial information. However, as discussed below, the inherent insensitivity of the NMR experiment, in comparison to radioisotope methods, is that signal detection from ^{13}C enrichment requires concentrations of labeled agent in large excess of the tracer levels used for PET. Thus, the potential exists for altering the metabolism of interest by supplying high concentrations of a specific ^{13}C-enriched metabolic precursor.

By virtue of this chemical specificity, in combination with the inherent insensitivity of the NMR technique, such NMR data require careful interpretation. The potential for ^{13}C NMR methods to provide kinetic information of tissue biochemistry in intact myocardium can be applied to specific targeting of metabolic events associated with pathophysiological changes (Lewandowski and White, 1995; Weiss *et al.*, 1993; Johnston and Lewandowski, 1991; Lewandowski *et al.*, 1991a,b; Lewandowski and Johnston, 1990). However, systematic analyses of the intracellular events influencing dynamic changes in ^{13}C NMR spectra from hearts are only now being performed. The most recent analytical studies suggest that dynamic ^{13}C NMR methods may be more powerful than were even originally anticipated, indicating that the isotope kinetics are sensitive to not only metabolic flux but metabolite transport as well. For example, monitoring metabolite exchange between the mitochondria and cytosol of intact hearts has now been demonstrated with ^{13}C NMR spectroscopy. This enables studying the regulation of oxidative metabolism, which was previously restricted to the much more artificial model of isolated mitochondria.

First applied to cell suspensions to follow the metabolism of exogenous ^{13}C-enriched substrates in 1972 (Eakin *et al.*, 1972), ^{13}C NMR spectroscopy has been used to study a wide range of biological systems including isolated cells (Cohen *et al.*, 1979), perfused liver (Cohen, 1983) and heart (Damico *et al.*, 1996; Lewandowski *et al.*, 1996; Yu *et al.*, 1996; Chatham *et al.*, 1995; Lewandowski *et al.*, 1995a,b; Yu *et al.*, 1995; Lewandowski, 1992a,b; Weiss *et al.*, 1992; Lewandowski *et al.*, 1991b; Lewandowski and Hulbert, 1991; Lewandowski and Johnston, 1990; Malloy *et al.*, 1988; Sherry *et al.*, 1988; Sherry *et al.*, 1985; Chance *et al.*, 1983; Bailey *et al.*, 1981), and *in vivo* organs (Mason *et al.*, 1995; Laughlin *et al.*, 1993; Robitaille *et al.*, 1993a,b; Fitzpatrick *et al.*, 1990; Weiss *et al.*, 1989). Ever since Bailey *et al.* (1981) and others (Sherry *et al.*, 1985) demonstrated that NMR spectra of isotopically enriched metabolites could be obtained from isolated, perfused hearts that were supplied with ^{13}C-enriched fuels, the utility of these stable isotope methods for cardiac studies has progressively increased. A formative study by Chance and co-workers (Chance *et al.*, 1983) using time-resolved changes in high-resolution ^{13}C spectra from hearts frozen at varied intervals during perfusion with ^{13}C-enriched materials alerted researchers to the notion that NMR spectroscopy could provide relative or absolute measures of flux through intermediary metabolism.

The study of cardiac metabolism has been bolstered by a series of analyses by Malloy and co-workers (Malloy *et al.*, 1988; Malloy *et al.*, 1990) and Sherry and co-workers (Sherry *et al.*, 1988) that take advantage of $J_{^{13}C-^{13}C}$-coupling resolved ^{13}C NMR spectra of glutamate to allow calculations of the fractional contributions of competing fuels and entry points for carbon influx into oxidative intermediary metabolism. This high-resolution analysis of complete isotope isomer formation is discussed in detail in another chapter. However, the generally low-resolution conditions of *in vivo* NMR experiments often precludes the timely detection of such well-resolved multiplet structures. Thus, direct on-line observation of ^{13}C entry and turnover within metabolic intermediates of the intact, functioning heart (Fig. 1), as shown by Lewandowski and co-workers (Lewandowski and Johnston, 1990; Lewandowski, 1992a; Yu *et al.*, 1995; Yu *et al.*, 1996) and Weiss and co-workers (Weiss *et al.*, 1989; Weiss *et al.*, 1992), provided the basis for examining metabolic flux under lower spectral resolution conditions with rapid temporal resolution from intact tissues. Subsequent work by Lewandowski and Hulbert (1991) and co-workers (Yu *et al.*, 1995) on the application of dynamic data from the low resolution conditions of ^{13}C NMR of intact hearts demonstrated the independence of these methods from inconsistencies in isotope enrichment levels. The initial observation enabled further analysis by Robitaille and co-workers (Robitaille *et al.*, 1993a; Robitaille *et al.*, 1993b) of dynamic ^{13}C NMR spectra under the more physiological conditions of the *in vivo* canine heart.

However, the true potential of dynamic ^{13}C NMR spectroscopy of the heart has yet to be fully realized. Recent work indicates that ^{13}C NMR spectroscopy is sensitive not only to the chemical exchange of the isotope, but also to the metabolic communication between subcellular organelles, as the metabolic demands of the myocyte are translated to the mitochondria via the two individual transporters of the malate–aspartate shuttle (Yu *et al.*, 1996). This is a particularly exciting new area for investigation in intact, functioning organs. Additionally, targeting of specific enzyme complexes with selected ^{13}C-enriched substrates indicates that ^{13}C NMR spectroscopy is also sensitive to altered enzyme activity in the stunned myocardium (Lewandowski and Johnston, 1990; Johnston and Lewandowski, 1991; Lewandowski and White, 1995; O'Donnell *et al.*, 1996). Thus, as the numerous physiological functions of intact organs are dynamic processes, static measures of metabolites may aid description of pathophysiological state, but do not characterize the levels of metabolic control and regulation that support the physiology of the cell. Rather, dynamic measures must be performed to truly characterize the intracellular processes that support the metabolic demands of physiological function.

3. ^{13}C NMR AND METABOLIC ACTIVITY

Dynamic ^{13}C NMR spectroscopy allows absolute metabolic flux along a particular metabolic pathway to be determined. As ^{13}C-enriched substrate is metabolized, ^{13}C label is transferred to various metabolic intermediates. The labeling pattern is dependent on the chemistry of the enzyme reactions, and the rate of labeling is determined by flux through the pathway and the pool sizes of the intermediates. From the ^{13}C NMR spectra, resonances attributable to specific carbon atoms of particular intermediates are identified by their chemical shifts. The area of the resonance peak, when corrected for NOE effects and partial saturation, is proportional to the ^{13}C content of the particular carbon site, and the fractional ^{13}C enrichment can be calculated if the intermediate pool size is known. The type of information obtained on the formation and distribution of the resulting isotope isomers is similar to that obtained by radiolabeling studies with ^{14}C (Kelleher, 1985; Weinman *et al.*, 1957; Strisower *et al.*, 1952) or mass spectrometry with ^{13}C (Des Rosiers *et al.*, 1994). However, the NMR observation holds an advantage in that the metabolites of interest need not be isolated, purified, and chemically degraded to determine ^{13}C enrichment at each position within the carbon chain of a metabolic intermediate. Limitations in NMR sensitivity can be overcome by *in vitro* NMR analysis of tissue extracts, in combination with chemical isolation of low level intermediates (Lewandowski *et al.*, 1996). More importantly, the ^{13}C NMR technique permits data to be obtained continuously and nondestructively while metabolism is in progress. Hence, integrated biochemical and physiological response of the tissue to external interventions may be explored.

An early study with ^{13}C NMR (Bailey *et al.*, 1981) showed that it is possible to detect enriched aspartate and glutamate signals and follow the time dependence of ^{13}C incorporation into these pools in hearts that are oxidizing the ^{13}C-enriched substrate. Chance and co-workers (1983) measured the time dependence of ^{13}C fractional enrichment at each carbon position in these same pools in freeze-clamped extracts of perfused hearts and fit the data to a mathematical model of a key oxidative metabolic pathway, the tricarboxylic acid cycle, to obtain carbon flux information. (Such mathematical treatments are discussed below.) Neurohr and co-workers (Neurohr *et al.*, 1983) demonstrated the feasibility of detecting ^{13}C-enriched glutamate signals from the heart of an open-chested guinea pig during intravenous infusion of [2-^{13}C]acetate.

An important application of ^{13}C NMR is the evaluation of oxidative metabolic activity within a tissue, as an index of the functional activity in tissue types which do not offer functional parameters that are as readily obtained as the mechanical function of the heart (i.e., brain). For the fundamental understanding of how ^{13}C studies can be applied to such systems to monitor metabolic flux, the data from studies in the heart have provided particular insight.

Specifically, the progressive oxidative of carbon-based fuels (i.e., carbohydrates, fats, and ketone bodies) within the mitochondria occurs via the tricarboxylic acid (TCA) cycle. Administration of ^{13}C-enriched fuels for oxidative metabolism then enables the activity of the TCA cycle to be monitored in the intact heart (Figs. 1 and 2). The net result of this cycle, which conserves carbon mass, is the generation of electrons, as reducing equivalents, to fuel the oxidative production of energy (Lewandowski and Ingwall, 1994). With molecular oxygen as the ultimate electron acceptor, TCA cycle flux and oxygen consumption are both linked through the respiratory rate that supports the energy demands of metabolically active tissues.

Oxygen consumption measurements are available in invasive studies of intact organs, but are not directly measured within the intracellular environment. Rather, the oxygen consumption of a region or whole organ is frequently determined by the differences in oxygen content between the arterial and venous (A–V difference)

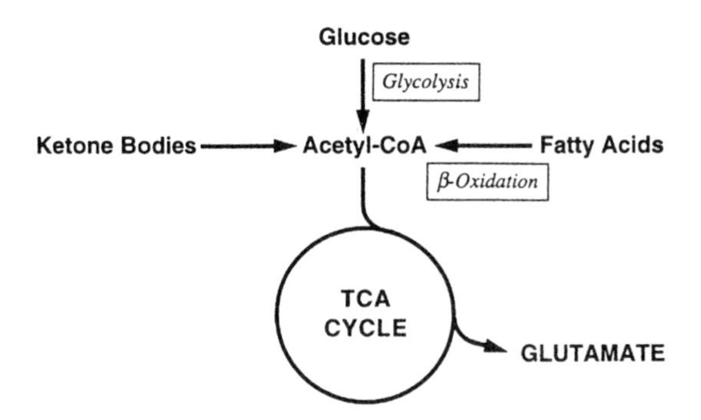

Figure 2. Diagram of carbon-based fuel entry into the tricarboxylic acid cycle and NMR observable amino acid pool of glutamate.

nutrient supplies (blood or perfusion medium). The measurement of oxygen use is dependent on the delivery of oxygen from the vasculature to the cell and then to the mitochondria where it is utilized. Thus, current invasive measurements of oxygen are actually quite remote from the intramitochondrial environment where oxygen serves as an electron acceptor.

Adding to this problem is the difficulty of measuring oxygen consumption noninvasively, where clinical or *in vivo* studies may dictate that the access to the organ for A–V difference measurements is extremely limited. In this instance, ^{13}C NMR observations of metabolic flux can provide a truly intracellular index of the oxidative rate within an intact, functioning tissue. With continued development of ^{13}C NMR detection schemes and improved understanding of factors that influence the observations of such isotopic enrichment rates, ^{13}C NMR will provide a powerful means for noninvasively assessing the metabolic rates of *in vivo* systems.

Of course, one commonly understood disadvantage of NMR detection is the inherently low sensitivity. Hence, resonance signals cannot be detected unless the concentration of ^{13}C is relatively high, in the millimolar range, and the concentrations of many key intermediates of oxidative metabolism, such as the TCA cycle, are well below the detectable range for NMR studies. Fortunately, the ability to detect the relatively large intracellular pool of ^{13}C-enriched glutamate makes ^{13}C NMR particularly well suited for nondestructive evaluations of oxidative metabolism. Although most ^{13}C-enriched TCA cycle intermediates cannot be observed directly by NMR, current ^{13}C NMR methods rely on the observation glutamate, which is in constant exchange with the TCA cycle intermediate, α-ketoglutarate. However, as glutamate itself is not a TCA cycle intermediate, glutamate labeling remains an indirect indicator of ^{13}C labeling of the TCA cycle intermediates. This distinction requires careful consideration for accurate representations of metabolic flux from observation of glutamate enrichment with ^{13}C NMR spectroscopy. These considerations are discussed in the following sections.

4. ^{13}C-ENRICHMENT PATTERNS AND OXIDATIVE METABOLISM

During the oxidation of a ^{13}C-enriched fuel or substrate, the incorporation of label into the glutamate pool is readily detected by ^{13}C NMR of the intact heart. The appearance of resonance peaks that correspond to ^{13}C enrichment at specific carbon positions within the glutamate pool have been described in detail in the literature (Cohen, 1987; Chance *et al.*, 1983). The general labeling pattern of the TCA cycle intermediates and amino acid pools is shown in Fig. 3.

Interpretation of ^{13}C NMR signal from glutamate in the heart is based on the relatively high concentration of intracellular glutamate that is in chemical exchange with the TCA cycle intermediate, α-ketoglutarate. This exchange occurs via the equilibrium reaction catalyzed by glutamate–oxaloacetate transaminase. From substrates which contribute to the formation of [2-^{13}C]acetyl-CoA, the initial incorporation of label into the TCA cycle intermediates occurs from the condensation of labeled oxaloacetate with [2-^{13}C]acetyl-CoA. This reaction catalyzed by citrate synthase introduces the ^{13}C label at the 4-carbon site of citrate which is eventually metabolized to [4-^{13}C]α-ketoglutarate. The chemical equilibrium between α-ketoglutarate and glutamate results in the appearance of label at the 4-carbon position of glutamate which can then be detected by NMR spectroscopy.

As the ^{13}C label is recycled within the TCA cycle, the symmetry of the four carbon succinate molecule results in incorporation of ^{13}C at the 2- and 3-carbon positions of succinate with equal probability. The ^{13}C label then appears at the 2- and 3-carbon positions of the subsequent metabolites of the TCA cycle, fumarate, malate and then oxaloacetate. Although some investigators have reported asymmetric labeling of malate in various mammalian tissues perfused with ^{13}C-labeled propionate (Sherry *et al.*, 1994), this phenomenon has not been observed in NMR experiments involving functioning tissues oxidizing substrates that enter the TCA cycle via acetyl-CoA (Yu *et al.*, 1995). The ^{13}C-enriched intermediates of the second span of the TCA cycle are then progressively metabolized, as the ^{13}C at 2- and 3-carbon sites of oxaloacetate is ultimately reintroduced into citrate at the 2- and 3-carbon positions. With a continued supply of ^{13}C-enriched substrate, newly condensed acetyl groups carry the ^{13}C label into the 4-carbon position of the TCA cycle intermediates and thereby glutamate, while recycled ^{13}C label appears at the 2- and 3-carbon positions of the glutamate molecule.

This progressive labeling of the NMR detectable glutamate pool can be monitored in intact tissues. Thus, the rate of citrate synthase activity and flux between the first and second spans of the TCA cycle are reflected by the appearance

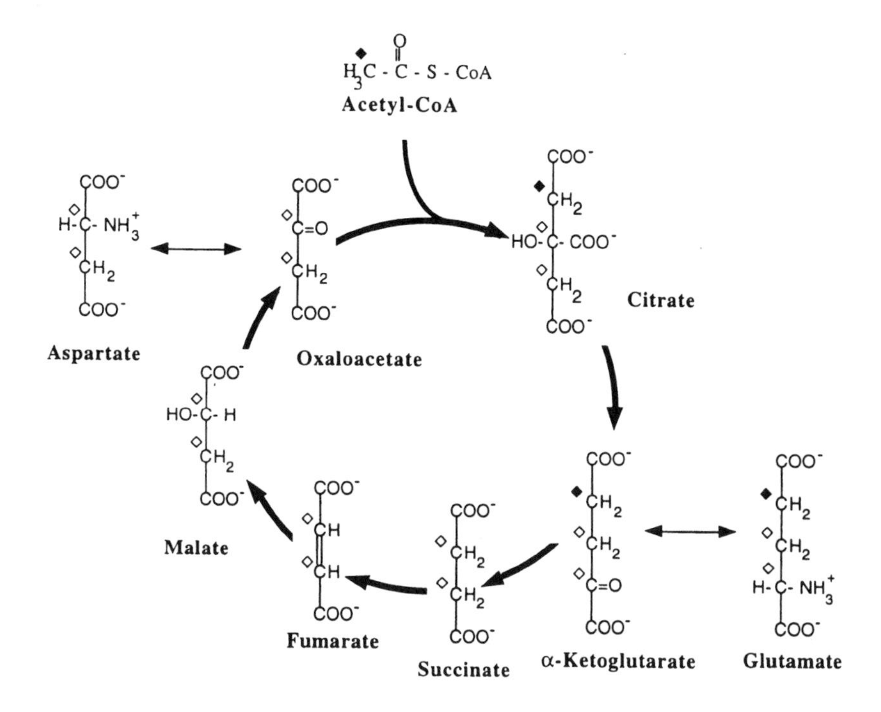

Figure 3. General labeling pattern of TCA cycle intermediates and amino acid pools from ^{13}C-enriched substrates that enter the cycle as [2-^{13}C]acetyl-CoA. Label first enters the first span of the TCA cycle, labeling the five carbon intermediates at the 4-carbon position. Solid diamond shapes represent the initial labeling at the 4-carbon position. Due the symmetry of the succinate molecule, the four carbon intermediate span of the TCA cycle is labeled with equal probability at the 2- and 3-carbon positions. Open diamond shapes represent the secondary labeling at the 2- and 3-carbon positions. With recycling of the label back to the first span of the cycle, at citrate synthase, the five carbon intermediates then are also labeled at the 2- and 3-carbon with equal probability. Chemical equilibrium between α-ketoglutarate and the large pool of glutamate allows the NMR-observed labeling pattern of glutamate to reflect the ^{13}C labeling of the TCA cycle intermediates.

of ^{13}C into glutamate. Eventually, the continued supply of ^{13}C-enriched fuels will result in a steady-state isotopic enrichment of the TCA cycle intermediates and glutamate. In the isolated perfused heart this steady state is reached within the first twenty to thirty minutes (Lewandowski, 1992a; Lewandowski and Johnston, 1990; Malloy *et al.*, 1988). The pre-steady-state incorporation of ^{13}C into glutamate can then be monitored to provide an index of the metabolic activity of the TCA cycle. The time to steady-state ^{13}C-enrichment has been directly related to the rate of oxidative metabolism of the heart, as shown in Fig. 4 and described in detail by Lewandowski (1992a) and Weiss and co-workers (Weiss *et al.*, 1992). The approach to steady-state enrichment is now known to be dependent on the relative rates of

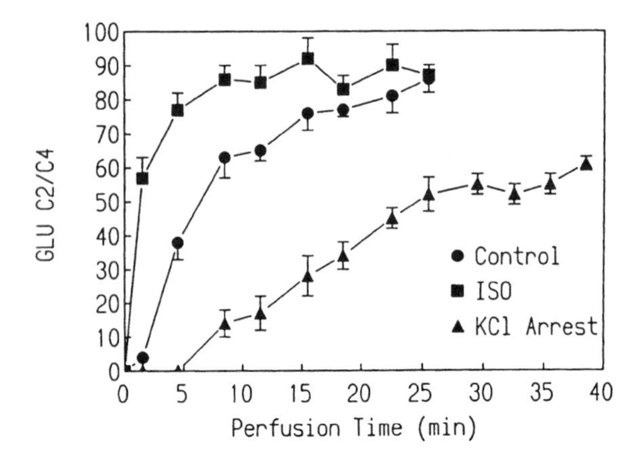

Figure 4. Relative rates of ^{13}C enrichment at the 2- and 4-carbon positions of glutamate in the isolated, beating rabbit heart oxidizing [2-^{13}C]acetate. Shown are the signal intensity ratios of the NMR signals from the 2-carbon site to the 4-carbon site. Note the increase in enrichment rate with increasing metabolic rates due to mechanical work in the heart. Shown are the ratios from control heart, hearts stimulated to high workloads with isoproterenol (ISO), and arrested, nonbeating hearts at basal metabolic rate (KCl Arrest). Reprinted with permission from E. D. Lewandowski, 1992, Nuclear magnetic resonance evaluation of metabolic and respiratory support of work load in the intact rabbit heart, *Circ. Res.* **70**:576. Copyright 1992 American Heart Association.

appearance of label between the 4-carbon and 2,3-carbon sites of glutamate, while being independent of the actual level of isotopic enrichment, or fractional enrichment, of the metabolites that are being labeled (Yu *et al.*, 1995).

Relative assessment of TCA cycle can be monitored from the relative rates of ^{13}C incorporation at the initial site of ^{13}C enrichment into glutamate, the 4-carbon, and at the secondary, 2- and 3-carbon, sites. Thus, a simplistic view allows us to consider the 4-carbon site in glutamate as a marker of the entry of labeled substrate into the first span of the TCA cycle, with ^{13}C enrichment at the 2- and 3-carbons as a "lap marker," indicating the entry of that label into the second span (the four carbon chains) and recycling into the first span of the cycle. To take advantage of the relative labeling rates as an index of TCA cycle flux, an early study examined the time-based evolution of the ratio of ^{13}C NMR signal intensity emanating from label at the 2-carbon of glutamate to that of the 4-carbon position. Although the endpoint ratio was independent of TCA cycle flux rates and was determined by the ratio of flux from unlabeled anaplerosis to that of citrate synthase, careful control for matching anaplerosis between experimental conditions allowed comparison of relative differences in TCA cycle flux. The differences in the relative rates of approach to a steady-state ratio of 2- to 4-carbon NMR signal ratios provides an easy relative marker of differences in oxidative metabolic activity (Lewandowski and Johnston, 1990; Lewandowski, 1992a). An example of the differences in the evolution of this ratio at three different rates of metabolic activity is shown in Fig. 4. However, quantitative methods can also be easily applied to the same experimental conditions (Yu *et al.*, 1995; Chatham *et al.*, 1995; Weiss *et al.*, 1992).

The data shown in Fig. 4 are from three different groups of hearts, each at a different rate of oxygen consumption. Note that direct comparison of the development of the signal intensity ratio over time is possible between the control hearts and the hearts operating at higher metabolic rates, due to the similar endpoint ratio of the 2- to 4-carbon signal intensities. This similarity in the steady-state ratio is indicative of similar rates of anaplerosis in comparison to TCA cycle activity. However, the group at very low metabolic rates showed a decreased steady-state ratio due to a different comparative rate of anaplerosis to citrate synthase. Thus, the ratio method does not allow for as direct a comparison for relative TCA cycle flux between the control and KCl arrested groups.

From such work, described above, a great deal of emphasis has been placed on detection of the glutamate signals with NMR spectroscopy. But as shown by these studies, the biological basis for detection must also be closely examined, since glutamate is an indirect indicator of the ^{13}C enrichment of the TCA cycle intermediates. In particular, the enzymes for the TCA cycle are inside the mitochondria, while at least 90% of the glutamate is located in the cytosol (Yu *et al.*, 1995; LaNoue *et al.*, 1974; LaNoue *et al.*, 1970). Therefore, the potential for differences in enrichment between the TCA cycle intermediates and the observed glutamate pool exists, depending on the metabolic conditions of a given experiment. For example, the fractional enrichment of glutamate with ^{13}C has been found to be lower than the enrichment of acetyl-CoA which is contributing to formation of the labeled TCA cycle intermediates, and the magnitude of the difference is substrate-dependent (Lewandowski, 1992a; Lewandowski, 1992b; Lewandowski and Hulbert, 1991).

Recent work has tested the assumption that the steady-state isotopomer distributions in the TCA cycle intermediates pools were accurately reflected by the high-resolution ^{13}C NMR signals from glutamate (Lewandowski *et al.*, 1996). Isolating both α-ketoglutarate and succinate on anion exchange columns from acid extracts of healthy myocardium documented for the first time the labeling patterns of these TCA cycle intermediates for comparison to glutamate. The resonance signals from the 4-carbon of α-ketoglutarate and the 2,3-carbons of succinate are shown in comparison to the glutamate 4-carbon resonance from the same hearts in Fig. 5. Under the simple experimental condition of the labeled substrate entering the TCA cycle exclusively via acetyl-CoA formation in healthy myocardium, the *J*-coupling multiplet patterns of glutamate were found to closely match that of α-ketoglutarate and the analysis of the multiplet structure of succinate at steady state. Fractional enrichment of the acetyl-CoA entering the TCA cycle from [2-^{13}C]acetate could be calculated from either the glutamate or succinate signals

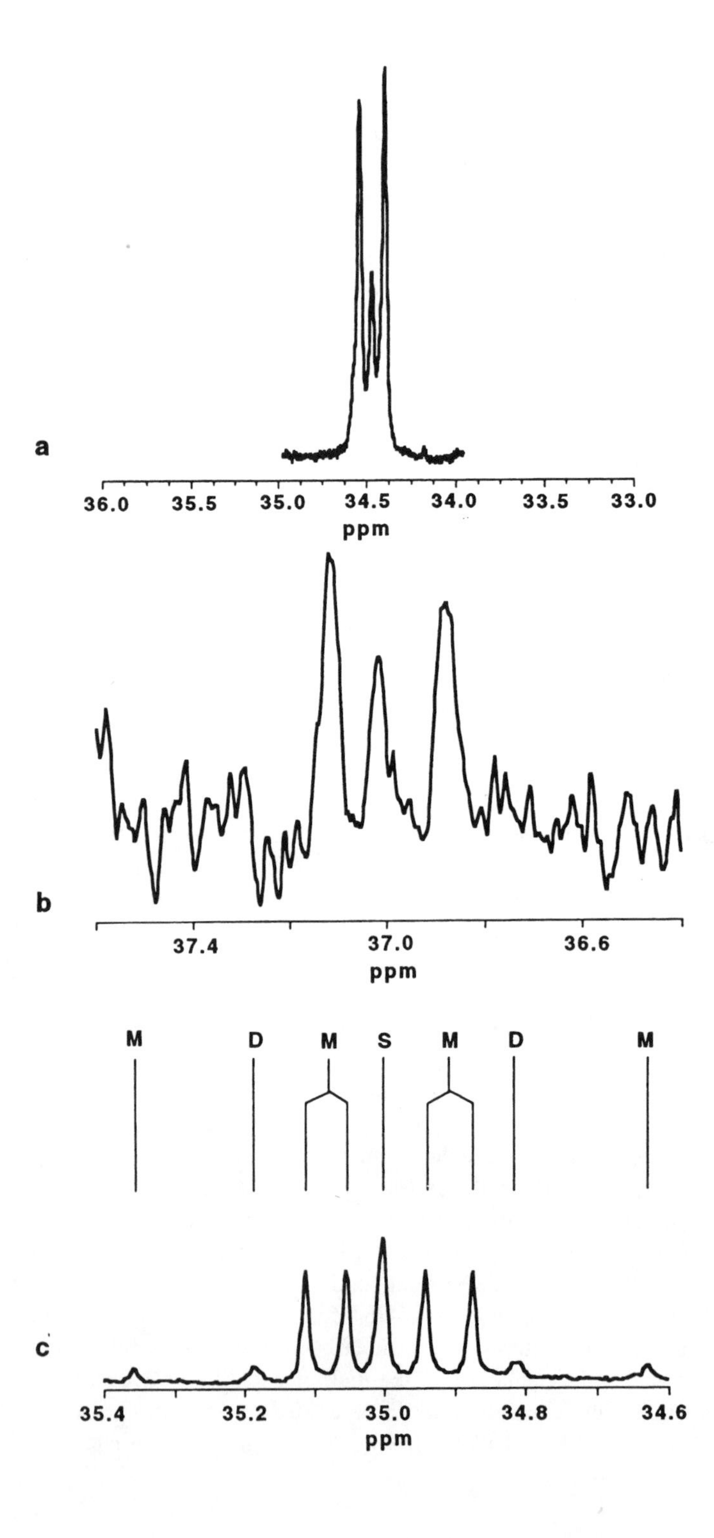
a
36.0
35.5
35.0
34.5
34.0
33.5
33.0
ppm
b
37.4
37.0
36.6
ppm
M
D
M
S
M
D
M
c
35.4
35.2
35.0
34.8
34.6
ppm

Figure 5. *In vitro* ^{13}C NMR resonance signals from isolated metabolites of hearts oxidizing [2-^{13}C]acetate. Multiplet structures arise from *J*-coupling between adjacent ^{13}C nuclei. Shown are the signals from (a) the 4-carbon of glutamate, (b) the 4-carbon of α-ketoglutarate, (c) the 2,3-carbons of succinate, where S denotes singlet, M multiplet, and D doublet. Reprinted with permission from E. D. Lewandowski *et al.*, 1996, Multiplet structure of ^{13}C NMR signals from glutamate and direct detection of tricarboxylic acid (TCA) cycle intermediates, *Magn. Reson. Med.* **35**:149. Copyright 1996 Williams and Wilkins.

(Lewandowski *et al.*, 1996; Jones *et al.*, 1993) to provide the same value of 0.9. However, under the same experimental conditions the fractional enrichment level of succinate, as determined from a proton spectrum, did not appear to match the fractional enrichment of glutamate, leading to additional questions regarding the compartmentation of these metabolites.

The importance of comparing the enrichment of succinate to that of glutamate is due to the metabolism of succinate vs. that of α-ketoglutarate when either metabolite traverses the mitochondrial membrane into the cytosol. While α-ketoglutarate may be the intermediate which is in direct chemical exchange with glutamate, this reversible transaminase reaction occurs both in the mitochondrial compartment, where a small fraction of the glutamate resides, and in the cytosolic compartment which is largely the glutamate detected by ^{13}C NMR. On the other hand, succinate produced in the mitochondria is not actively metabolized once in the cytosol. Therefore, the isotope exchange between α-ketoglutarate and glutamate can occur in isolation from the TCA cycle, making the labeling of α-ketoglutarate much less specific for intramitochondrial ^{13}C enrichment of the TCA cycle than that of succinate (Lewandowski *et al.*, 1996). Therefore, recent work indicates that under simple experimental conditions, where the entry of carbon isotope into the TCA cycle occurs overwhelmingly at the citrate synthase step, the steady state isotope isomer distribution within the glutamate pool closely matches that of the intramitochondrial TCA cycle pool. The basic principles and implications of this subcellular compartmentation for measurements of metabolic flux are discussed in the next section.

5. FRACTIONAL ENRICHMENT AND ^{13}C NMR OF METABOLIC FLUX

Metabolic flux is a measurement of a quantity of mass per unit time, generally weighted by the mass or volume of tissue within the region of detection. Such a measurement provides a truly quantitative correlate to the measured physiological function of an intact tissues or organ. While other nonquantitative measurements can be used to evaluate relative changes in metabolic rates, these are not to be confused with metabolic flux. Sometimes confused with metabolic flux rates are the relative measures of carbon entry points into the TCA cycle, as pioneered by Malloy and Sherry (Malloy *et al.*, 1988). These measures provide a handy means for using the isotopomer distributions in the glutamate pool as an index of acetyl-CoA enrichment and the relative proportions of differently labeled substrates that are contributing to flux through citrate synthase. However, if the TCA cycle rate slows, and citrate synthase flux slows to levels that reduce availability of carbon skeletons for the second span of the TCA cycle, anaplerosis will provide a proportionately higher percentage of carbon influx into the TCA cycle than will citrate synthase. This effect has been observed when comparing the contribution of anaplerosis in control hearts to nonbeating, KCl-arrested hearts. In the arrested hearts, the relative contribution from anaplerosis is higher than for control hearts, yet the absolute flux through both the TCA cycle and anaplerosis remains lower than under control conditions (Lewandowski, 1992a). The result of such a shift in carbon influx is that the measure of relative anaplerosis to citrate synthase will increase, even if anaplerotic flux is actually slower than under control conditions. In this case, we must be careful not to confuse a shift in the relative contributions of carbon influx into the TCA cycle with the actual flux through anaplerosis.

Fundamental to NMR studies that obtain actual, quantitative metabolic flux in one or more pathways (mass per unit time), as opposed to comparative measures of activity between multiple pathways (which is faster or slower), is the requirement of collecting two or more sequential spectra during the evolution period of the experiment. This is true whether the NMR experiment involves isotopic enrichment as with ^{13}C or a spin label, as in magnetization transfer studies. In the case of ^{13}C studies, the spectra must be collected during the pre-steady-state incorporation of the label into the NMR-observed metabolites. The progressive incorporation of the ^{13}C can then be quantified based on the observed increase in signal intensity over time, as the label becomes incorporated into a specific site on the targeted metabolite.

As in the case of all isotope studies, the fractional enrichment of the metabolite must also be considered. An important question for quantification is whether an increase in intensity from a particular ^{13}C resonance signal is due to an increase in the enrichment level of that carbon site within the metabolite pool or to an increase in the metabolite pool size at the same fractional enrichment. Both scenarios can occur due to subtle differences in substrate availability and enzyme activity. For example, when hearts oxidize [3-^{13}C]pyruvate during stimulation of the enzyme pyruvate dehydrogenase (PDH) the fractional enrichment of glutamate in the cell increases. This increase in the amount of label entering the glutamate pool is observed as an increase in glutamate signals in the ^{13}C spectrum. However, when hearts oxidize [3-^{13}C]lactate during identical conditions of PDH stimulation, the glutamate fractional enrichment does not increase while the actual amount of glutamate in the cell increases significantly. Again, an increase in ^{13}C NMR signals from glutamate occurs with PDH stimulation in the presence of [3-^{13}C]lactate, but for the opposite reason than the increase observed with PDH stimulation in the presence of [3-^{13}C]pyruvate.

Even with the administration of metabolic precursors that are nearly 100% enriched with ^{13}C at a specific site, some metabolic contributions from endogenous sources of ^{12}C will persist, unrecognized in the ^{13}C-observed spectrum (Lewandowski, 1992a; Lewandowski, 1992b). Although such contributions can be minimized, accuracy demands that the fractional enrichment of the metabolites must be considered for quantification of flux. In the *in vivo* setting, where whole body metabolism is able to contribute a vast source of endogenous metabolic precursors, high fractional enrichments are more difficult to obtain, and absolute flux requirements demand some consideration of the contribution to metabolic flux from the unlabeled metabolites.

Nevertheless, when the lower resolution ^{13}C spectra from intact tissues are used to define the time course of labeling at the 4- and 2,3-carbon sites of glutamate, without the full analysis of isotopomer fractions from multiplet structures, the kinetics of ^{13}C turnover within the glutamate pool are seen to occur independent of fractional enrichment (Yu *et al.*, 1997; Yu *et al.*, 1995; Lewandowski and Hulbert, 1991). Therefore, the fractional enrichment must be known for quantitation, but not for the time course of the enrichment rates. This independence from fractional enrichment of the substrate entering the TCA cycle at citrate synthase was first shown by Lewandowski and Hulbert (1991) using the time course of the simple ratio of 2-carbon glutamate enrichment to that of the 4-carbon of glutamate. In that study, the enrichment curves of glutamate were not affected whether hearts were oxidizing 99.9% [2-^{13}C]acetate or 50% [2-^{13}C]acetate.

A more recent study using ^{13}C-enriched butyrate, a four-carbon, short-chain fatty acid, with improved temporal resolution supports the original finding. In the more recent study, butyrate was supplied to hearts in either the single labeled, [2-^{13}C] form corresponding to nearly 50% enrichment of acetyl groups produced, or the double labeled, [2,4-^{13}C] form, accounting for over 90% of the acetyl groups produced for entry into the TCA cycle (Yu *et al.*, 1995). The results of this simple study indicated that the time constant for the resulting enrichment curves of both the 4-carbon and 2-carbon of glutamate were essentially the same for either level of enrichment. The fractional enrichment of the substrate entering the TCA cycle does not affect the observed enrichment rates, only the actual enrichment levels. Demonstrating this independence from fractional enrichment, Table 1 displays the time constants for both enrichment conditions.

Simply put, it takes the same amount of time for 50% of the glutamate pool to reach steady-state isotope enrichment from a 50% enriched substrate supply as it does for 100% of the glutamate pool to reach steady-state enrichment from a 100% enriched substrate supply. Thus, fractional enrichment is important for quantifying metabolic, but does not confound isotope enrichment rates for the ^{13}C NMR evaluation of metabolic flux. As stated in a previous section, these findings enable

Table 1
Time Constants for ^{13}C Enrichment of 4- and 2-Carbon Sites of Glutamate Molecule[a]

		Time constant (min)	
Substrate enrichment	F_{C2} (%)	Glu 4-carbon	Glu 2-carbon
[2-^{13}C]Butyrate	93	10	19
[2,4-^{13}C]Butyrate	48	11	20

[a]Values are shown for two different fractional enrichments of the oxidative substrate. F_{C2} denotes mean ^{13}C enrichment at 2-carbon of acetyl-CoA, entering the TCA cycle at citrate synthase. Note the lack of effect enrichment of entering substrate on the time to steady-state enrichment rates of glutamate.

the use of ^{13}C NMR spectroscopy to measure TCA cycle flux rates under *in vivo* conditions, where endogenous carbon sources dilute the ^{13}C enrichment of glutamate.

6. MODELS AND PARAMETERS OF GLUTAMATE ENRICHMENT

Several studies have employed modeling schemes to explore TCA cycle flux and its regulation from ^{13}C NMR spectra of intact organs (Yu *et al.*, 1996; Chatham *et al.*, 1995; Mason *et al.*, 1995; Yu *et al.*, 1995; Weiss *et al.*, 1992). Each method has been devised to address a specific set of experimental aims, and in sum demonstrate how such analysis can be stylized for a given application. For analysis of ^{13}C NMR data, an important early work, by Chance and co-workers (Chance *et al.*, 1983), applied a kinetic model to the analysis of ^{13}C labeling within glutamate as detected in *in vitro* spectra of tissue extracts at different time points. The approach consisted of the simultaneous solution of nearly 200 differential equations with a single compartment for each individual ^{13}C isotopomer. The model was recently expanded to 340 equations to provide a more comprehensive examination of related metabolic pathways (Chatham *et al.*, 1995). From a more recent analysis of ^{13}C NMR data from intact hearts, the expanded model suggests that the relative contributions of the glycerol phosphate shuttle and the malate–aspartate shuttles are different from previous experimental findings of Safer and co-workers (Safer, 1975; Safer and Williamson, 1973). In addition to this comprehensive method, several other simpler models have also been developed for specific applications.

For monitoring fast metabolic pathways it may be preferable to sacrifice spectral resolution for temporal resolution, to utilize data obtained under low resolution conditions, where the multiplet structure of the glutamate resonance signals is not available. Lewandowski and Johnston (1990) initially used an empirical time constant of glutamate 2-carbon/4-carbon (C2/C4) curve as an index of the TCA cycle flux. This approach demonstrated that, under widely varying workloads, the myocardium displayed consistent end point fractional enrichments and glutamate pool sizes, and yet very different time constants for C2/C4 curves (Lewandowski, 1992a). Although attractive in simplicity, as mentioned above, the approach provides only relative differences in metabolic rates and requires experimental conditions where anaplerotic contributions into the TCA cycle are matched. However, the ratio of signal intensities from the 2-carbon vs. 4-carbon positions in glutamate do appear to be less sensitive to moderate changes in the size of the glutamate pool than are recent quantitative measures of actual TCA cycle flux (Yu *et al.*, 1995).

Quantifying this time difference between the rates of 4- and 2-carbon enrichment in glutamate, Weiss and co-workers (Weiss *et al.*, 1992) used an empirical flux parameter K_T as an index of the TCA cycle flux (131), defined as

$$K_T = \frac{\sum {}^{13}\text{C NMR-detected TCA metabolites } (\mu\text{moles/g})}{F_C \times \text{glutamate}\Delta t_{50}\ (\text{min})}$$

where glutamate t_{50} is the time difference for the glutamate 4- and 2-carbons to reach half of their steady-state enrichment levels. ^{13}C-NMR-detected TCA metabolites include glutamate, aspartate, and citrate. Such an approach is based on the observations that the TCA cycle flux rate is inversely proportional to the time difference between ^{13}C appearance in the 4- and 2-carbon positions of glutamate (glutamate t_{50}). Although K_T has the unit of flux parameter, it may not always be equal to the TCA cycle flux if the time difference of 4- and 2-carbon labeling is influenced by factors other than net TCA cycle flux, such as changes in α-ketoglutarate dehydrogenase activity. Nevertheless, the approach demonstrates the close relation between metabolic activity and the turnover rates of isotope within the glutamate pool that has been discussed in the previous section.

Robitaille and co-workers have proposed post-steady-state analysis as a complementary method to the pre-steady-state method (Robitaille *et al.*, 1993a; Robitaille *et al.*, 1993b). The advantages of using post-steady-state analysis is that a closed-form solution of the rate of label washout from the 4-carbon of glutamate can be obtained and that the potential variability in the time curve of acetyl-CoA enrichment under *in vivo* conditions is avoided.

Mason and co-workers (Mason *et al.*, 1995; Mason *et al.*, 1992) have also developed a kinetic model for the analysis of glucose metabolism in cerebral energy production (81, 82). ^{1}H-observed/^{13}C-edited NMR spectroscopy technique was used in data acquisition to improve the sensitivity of NMR measurements in brain *in vivo* (102, 103). Their findings suggest that exchange between α-ketoglutarate and glutamate becomes fast in comparison to the TCA cycle. This may be the case for very glycolytic organs, such as the brain. Using this model to analyze ^{13}C enrichment of glutamate in the brain, the authors have proposed a yet to be discovered, third shuttle system for reducing equivalent transport into the mitochondria (Mason *et al.*, 1995).

Even more recently, a model has been developed by Yu and co-workers (Yu *et al.*, 1997; Yu *et al.*, 1995) to include the branch point between α-ketoglutarate and glutamate in utilizing a simple set of nine differential equations that describe isotope flux through the metabolic compartments of the TCA cycle, aspartate, and glutamate. This model was originally developed to explore the potential rate-limiting step of exchange between α-ketoglutarate, produced in the mitochondria, and the largely cytosolic glutamate pool. With nine linear differential equations in the model, a least-squares fitting of the model to the experimental NMR data of glutamate labeling then provides the two output variables of TCA cycle flux and the interconversion rates between α-ketoglutarate and glutamate. An optional feature of this model is the ability to include oxygen consumption measurements in least-squares data-fitting routines to constrain the variable search to TCA cycle flux rates within the range of experimentally observed oxygen consumption. The elements of this model are described in more detail in the next section.

An advantage of this approach that has been recently shown (Yu *et al.*, 1997) is that the least-squares fitting of the model to the data can be modified to enable each individual experiment to provide kinetic analysis rather than the grouped, mean values, by using the noise level in the NMR spectra in the cost function, rather than the standard error of the mean values for a group of hearts. Using the noise level of the spectra as the source of error in the fit, allows the model to be applied to each individual experiment, so that statistical mean and standard deviations can be determined for grouped data for statistical comparisons.

An additional consideration of the use of kinetic models for two-parameter fits to glutamate enrichment data is the covariance between the two fitted parameters, generally that for TCA cycle flux and the interconversion rate between α-ketoglutarate and glutamate. This issue has recently been dealt with in general terms by Yu *et al.* (1997), recognizing this intrinsic limitation in the currently applied models. Due to the fact that the central intermediate α-ketoglutarate is involved in both flux through the TCA cycle and the exchange of label with the NMR-observed glutamate pool, there exists a certain degree of uncertainty about the fitted parameters unless other experimental data can be used to constrain one of the parameters. In relatively simple experiments where substrate utilization is known, oxygen consumption can be used for this purpose. To increase the level of accuracy, the introduction of oxygen consumption, measured from the same experimental preparation, as a penalty function, can effectively reduce the covariance between the fitted values for TCA cycle flux and the interconversion between α-ketoglutarate and glutamate to as low as 60%.

7. DIRECT KINETIC ANALYSIS OF DYNAMIC ^{13}C NMR SPECTRA

^{13}C NMR spectra, whether obtained sequentially from the intact heart or from tissue extracts, contain various sorts of information that require mathematical treatment for a comprehensive evaluation. In the case of simply analyzing the ^{13}C-enrichment kinetics as an index of metabolic flux, all the mathematical model need do is serve as an analytical tool that is sensitive to rate-limiting or rate-determining steps in the flow of isotope between metabolite pools. For a mathematical model of isotope flux through established metabolic compartments or pools to serve this function, several basic principles must be considered. Using the analogy of dye dripping sequentially into a series of containers of water, these concepts can be visualized, as in Fig. 6. By this analogy, what we may wish to know is the concentration of the dye in any given container of water, or the concentration of isotope in any metabolite pool, at any particular point in time. To best characterize this system and obtain our answer, a set of linear differential equations can then be used to describe the concentration history of the dye in each compartment throughout the dynamic process that is being observed. The amount of water in each container and the flow rates will determine the concentrations over time.

Our ability to detect the rate-determining steps, as shown in Fig. 6, depends on which containers we are able to observe. A compartment with a large dilution factor, that is, the amount of water in it, will exert greater influence on the observed concentration over time than will a small container. This is analogous to a large metabolite concentration. A slow flow rate will prove limiting, analogous to a slowed biochemical reaction or one with significant reverse flux. A physical barrier, such as a mesh screen, will delay transfer of dye between containers as a rate-limiting step. This barrier is analogous to a membrane which restricts the exchange of isotope between metabolites in different sections of a cell. To recognize these rate-determining steps we must have some knowledge of the input rate and some information on the rate of change in a downstream compartment. We can then simplify our model by eliminating the elements are not rate-determining, such as a small container with a low dilution factor, a container downstream of a very rapid flow rate, or a container directly upstream from a very slow flow rate. If the flow rates are all the same, such as may be the case with a steady-state TCA cycle flux rate, then the concentration changes among the containers are influenced more by the size of the containers. Thus, we have established a simple compartment model which can be described by linear differential equations. It is always important to remember that there is no substitute for experimental data to provide constraints to an analytical model and that the model must be consistent with known experimental data. With that, we shall proceed to the kinetic analysis of dynamic ^{13}C NMR data using these principles.

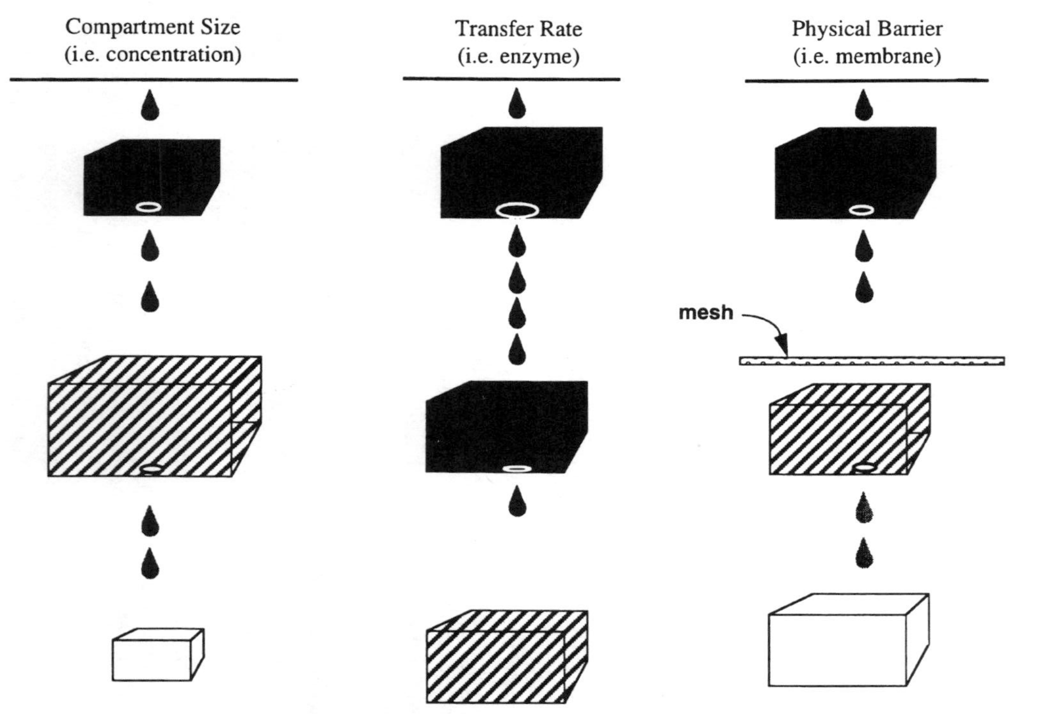

Figure 6. Illustration of dye entering and leaving a series of containers of water, demonstrating dilution factor and flow rate effects on the concentration of dye in each subsequent container. Containers are analogous to metabolite pools that are initially unlabeled with label entering from the previous compartment. Factors that effected the rate of change in the concentration of dye over time in each container are as shown: the size of the container, rate-limiting flow between two containers, and impeded flow between containers (as through a mesh screen).

A simple, direct kinetic model for the pre-steady-state isotope turnover within the key metabolite pools is derived from the simplified metabolic compartment model that includes the intermediates of the TCA cycle and the appropriate metabolite compartments that branch for interconversion with glutamate and aspartate. Analysis of ^{13}C enrichment in each metabolic compartment is carried out at both steady-state metabolic flux and constant intermediate pool (compartment) sizes. Because of the symmetry in 2- and 3-carbon labeling, the model only considers the labeling of 2-carbon while regarding the labeling of 3-carbon as the same as that of 2-carbon. The compartments of citrate, α-ketoglutarate, and glutamate are further divided into 2 subcompartments to represent ^{13}C labeling at the 2- and 4-carbon positions of each intermediate. The model below meets the purposes of analyzing experimental data, without extending beyond the capabilities of NMR detection of glutamate metabolism. This compartment model is represented in Fig. 7.

The simplification of eliminating the small succinate and fumarate pools and the other secondary labeling site (3-carbon of glutamate) was justified by comparing results from the reduced model and a more comprehensive model that included succinate and fumarate pools along with equations to describe both 2- and 3-carbon labeling (8 compartments, 19 differential equations) (Yu *et al.*, 1995). These two models showed no difference in characterizing ^{13}C-labeling kinetics of glutamate.

Effects of substrate utilization and anaplerosis are also accounted for by incorporating data measured from high resolution ^{13}C NMR as parameters in the model. Incorporation of unlabeled intermediate through anaplerosis was considered to enter the TCA cycle at malate while efflux of carbon mass (cataplerosis) occurred through malic enzyme (Russell and Taegtmeyer, 1991; Kornberg, 1966).

^{13}C label entering the TCA cycle from acetyl-CoA is treated as a step function. Flux of ^{13}C label entering the acetyl-CoA compartment is only $V_{TCA} \cdot F_C$, since other endogenous substrates are oxidized via the TCA cycle. Therefore, the labeling kinetics for acetyl-CoA are represented by the following equation:

$$\frac{d}{dt}[^{13}\mathrm{C}-\mathrm{AC}] = V_{\mathrm{TCA}} \cdot F_{\mathrm{C}} - V_{\mathrm{TCA}} \cdot \frac{[^{13}\mathrm{C}-\mathrm{AC}]}{[\mathrm{AC}]}$$

where $[^{13}\mathrm{C}-\mathrm{AC}]$ is the concentration of [2-^{13}C]acetyl-CoA and [AC] is the total concentration of acetyl-CoA. If AC^* represents the ratio of ^{13}C-enriched acetyl-CoA to total acetyl-CoA, i.e.,

$$\mathrm{AC}^* = \frac{[^{13}\mathrm{C}-AC]}{[\mathrm{AC}]}$$

then

$$\frac{d}{dt}\mathrm{AC}^* = \frac{V_{\mathrm{TCA}}}{[\mathrm{AC}]} \cdot (F_{\mathrm{C}} - \mathrm{AC}^*)$$

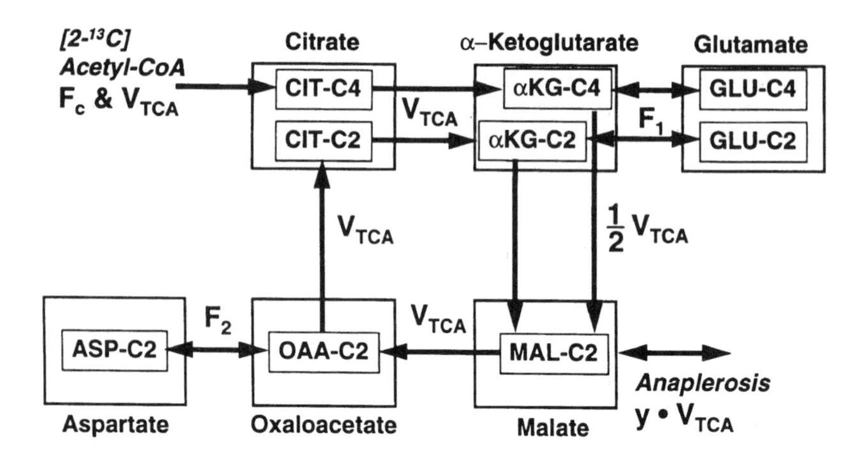

Figure 7. Metabolic compartment model used for kinetic analysis with elimination of non-rate-determining compartment. The concentration of each compartment and the influx and efflux of label through the compartment can be represented by a series of linear differential equations. Large boxes represent metabolite pools. Small boxes represent mathematical compartments of the model. Labeling of the 3-carbon sites is the same as that of the 2-carbon sites and is therefore eliminated. CIT, citrate; α-KG, α-ketoglutarate; GLU, glutamate; MAL, malate; OAA oxaloacetate; ASP, aspartate; V_{TCA}, TCA cycle flux; F_1, interconversion flux rate between α-ketoglutarate and glutamate; F_2, interconversion flux rate between aspartate and oxaloacetate; F_C, fractional enrichment of the 2-carbon position of acetyl-CoA entering the TCA cycle at citrate synthase. Reprinted with permission from X. Yu *et al.*, 1995, Kinetic analysis of dynamic ^{13}C NMR spectra: Metabolic flux, regulation, and compartmentation in hearts, *Biophys. J.* **69**:2090. Copyright 1995 Biophysical Society.

To simplify the problem, assume the natural abundance level of 1.1% ^{13}C is negligible, so that the enrichment of acetyl-CoA is not appreciably different from zero, i.e., $AC^*(0) \approx 0$. The labeling of acetyl-CoA then follows a simple monoexponential equation,

$$AC^*(t) = F_C \cdot \left[1 - \exp\left(-\frac{t}{[AC]/V_{TCA}}\right)\right]$$

Acetyl-CoA enrichment occurs as an exponential with the time constant being the ratio of the total acetyl-CoA concentration to TCA cycle flux. The physiological concentration for acetyl-CoA is approximately 0.2 μmoles/g dry weight only (96), so that the time constant for the enrichment of acetyl-CoA is very small. At TCA cycle flux rates ranging from 1.7 to 10.5 μmoles/min/g dry weight, the time constant for acetyl-CoA enrichment ranges from 0.12 down to 0.02, respectively. From these rates, the enrichment level of acetyl-CoA is 75% of the end point within 10 seconds, and is 98% within the first 30 seconds. Thus, the ^{13}C enrichment of acetyl-CoA can be assumed to be immediately enriched to steady-state levels, accounted for by F_C, in consideration of the time resolution of the NMR experiment.

Having examined the input of label at acetyl-CoA, let us set up the model for isotope flux through the TCA cycle. We will define TCA cycle flux as V_{TCA}, and the interconversion rate between α-ketoglutarate and glutamate, oxaloacetate and aspartate, as F_1 and F_2, respectively. With conservation of mass, the rate of change in ^{13}C enrichment in a compartment equals the difference between the amount of ^{13}C label entering and leaving that compartment. Flux of ^{13}C leaving the compartment is the total flux leaving that compartment multiplied by the enrichment level of that compartment. ^{13}C influx is the carbon flux entering that compartment multiplied by the enrichment level from the upstream compartment. Therefore, the concentration history of isotope in each compartment is then characterized by a set of differential equations that describe both the TCA cycle flux and the interconversion rates between metabolic compartments. The equations for each of the significant compartments are:

Citrate

$$\frac{d}{dt}[\mathrm{CIT-C4}] = V_{\mathrm{TCA}} \cdot F_{\mathrm{C}} - V_{\mathrm{TCA}} \cdot \frac{[\mathrm{CIT-C4}]}{[\mathrm{CIT}]} \tag{1}$$

$$\frac{d}{dt}[\mathrm{CIT-C2}] = V_{\mathrm{TCA}} \cdot \frac{[\mathrm{OAA-C2}]}{[\mathrm{OAA}]} - V_{\mathrm{TCA}} \cdot \frac{[\mathrm{CIT-C2}]}{[\mathrm{CIT}]} \tag{2}$$

α-Ketoglutarate

$$\frac{d}{dt}[\alpha\mathrm{KG-C4}] = V_{\mathrm{TCA}} \cdot \frac{[\mathrm{CIT-C4}]}{[\mathrm{CIT}]} - (V_{\mathrm{TCA}} + F_1) \cdot \frac{[\alpha\mathrm{KG-C4}]}{[\alpha\mathrm{KG}]} + F_1 \cdot \frac{[\mathrm{GLU-C4}]}{[\mathrm{GLU}]} \tag{3}$$

$$\frac{d}{dt}[\alpha\mathrm{KG-C2}] = V_{\mathrm{TCA}} \cdot \frac{[\mathrm{CIT-C2}]}{[\mathrm{CIT}]} - (V_{\mathrm{TCA}} + F_1) \cdot \frac{[\alpha\mathrm{KG-C2}]}{[\alpha\mathrm{KG}]} + F_1 \cdot \frac{[\mathrm{GLU-C2}]}{[\mathrm{GLU}]} \tag{4}$$

Glutamate

$$\frac{d}{dt}[\mathrm{GLU-C4}] = F_1 \cdot \frac{[\alpha\mathrm{KG-C4}]}{[\alpha\mathrm{KG}]} - F_1 \cdot \frac{[\mathrm{GLU-C4}]}{[\mathrm{GLU}]} \tag{5}$$

$$\frac{d}{dt}[\text{GLU}-\text{C2}] = F_1 \cdot \frac{[\alpha\text{KG}-\text{C2}]}{[\alpha\text{KG}]} - F_1 \cdot \frac{[\text{GLU}-\text{C2}]}{[\text{GLU}]} \tag{6}$$

Malate

$$\frac{d}{dt}[\text{MAL}-\text{C2}] = \frac{1}{2} V_{\text{TCA}} \cdot \left(\frac{[\alpha\text{KG}-\text{C4}]}{[\alpha\text{KG}]} + \frac{[\alpha\text{KG}-\text{C2}]}{[\alpha\text{KG}]}\right) - (1+y) \cdot V_{\text{TCA}} \cdot \frac{[\text{MAL}-\text{C2}]}{[\text{MAL}]} \tag{7}$$

Multiplying by 1/2 accounts for the equal distribution of label at the 2- and 3-carbons of malate. Unlabeled carbon entering the TCA cycle through anaplerosis is represented by y, while labeled TCA cycle intermediate leaves the cycle through cataplerotic flux. The anaplerotic flux and cataplerotic flux are equal in order to maintain constant TCA cycle intermediate pool sizes.

Oxaloacetate

$$\frac{d}{dt}[\text{OAA}-\text{C2}] = V_{\text{TCA}} \cdot \frac{[\text{MAL}-\text{C2}]}{[\text{MAL}]} - (V_{\text{TCA}} + F_2) \cdot \frac{[\text{OAA}-\text{C2}]}{[\text{OAA}]} + F_2 \cdot \frac{[\text{ASP}-C2]}{[\text{ASP}]} \tag{8}$$

Aspartate

$$\frac{d}{dt}[\text{ASP}-\text{C2}] = F_2 \cdot \frac{[\text{OAA}-\text{C2}]}{[\text{OAA}]} - F_2 \cdot \frac{[\text{ASP}-\text{C2}]}{[\text{ASP}]} \tag{9}$$

Since the total citrate is constant, Eq. (1) can be expressed in the following form:

$$[\text{CIT}] \cdot \frac{d}{dt}\frac{[\text{CIT}-\text{C4}]}{[\text{CIT}]} = V_{\text{TCA}} \cdot F_{\text{C}} - V_{\text{TCA}} \cdot \frac{[\text{CIT}-\text{C4}]}{[\text{CIT}]}$$

CIT4 represents the fraction of citrate labeled at the 4-carbon, i.e.,

$$\text{CIT4} = \frac{[\text{CIT}-\text{C4}]}{[\text{CIT}]}$$

So Eq. (1) becomes

$$\frac{d}{dt}\text{CIT4} = \frac{V_{\text{TCA}}}{[\text{CIT}]} \cdot (F_{\text{C}} - \text{CIT4})$$

Similarly, the equations can all be expressed as the corresponding fractional enrichment levels. Therefore, pre-steady-state ^{13}C enrichment of each compartment is described by the following equations:

$$\frac{d}{dt}\text{CIT4} = \frac{V_{\text{TCA}}}{[\text{CIT}]} \cdot (F_{\text{C}} - \text{CIT4}) \tag{10}$$

$$\frac{d}{dt}\alpha\text{KG4} = \frac{V_{\text{TCA}}}{[\alpha\text{KG}]} \cdot \text{CIT4} - \frac{V_{\text{TCA}} + F_1}{[\alpha\text{KG}]} \cdot \alpha\text{KG4} + \frac{F_1}{[\alpha\text{KG}]} \cdot \text{GLU4} \tag{11}$$

$$\frac{d}{dt}\text{GLU4} = \frac{F_1}{[\text{GLU}]} \cdot (\alpha\text{KG4} - \text{GLU4}) \tag{12}$$

$$\frac{d}{dt}\text{CIT2} = \frac{V_{\text{TCA}}}{[\text{CIT}]} \cdot (\text{OAA2} - \text{CIT2}) \tag{13}$$

$$\frac{d}{dt}\alpha\text{KG2} = \frac{V_{\text{TCA}}}{[\alpha\text{KG}]} \cdot \text{CIT2} - \frac{V_{\text{TCA}} + F_1}{[\alpha\text{KG}]} \cdot \alpha\text{KG2} + \frac{F_1}{[\alpha\text{KG}]} \cdot \text{GLU2} \tag{14}$$

$$\frac{d}{dt}\text{GLU2} = \frac{F_1}{[\text{GLU}]} \cdot (\alpha\text{KG2} - \text{GLU2}) \tag{15}$$

$$\frac{d}{dt}\text{MAL2} = \frac{V_{\text{TCA}}}{[\text{MAL}]} \cdot [\tfrac{1}{2} \cdot \alpha\text{KG2} + \tfrac{1}{2} \cdot \alpha\text{KG4} - (1 + y) \cdot \text{MAL2}] \tag{16}$$

$$\frac{d}{dt}\text{OAA2} = \frac{V_{\text{TCA}}}{[\text{OAA}]} \cdot \text{MAL2} - \frac{V_{\text{TCA}} + F_2}{[\text{OAA}]} \cdot \text{OAA2} + \frac{F_2}{[\text{OAA}]} \cdot \text{ASP2} \tag{17}$$

$$\frac{d}{dt}\text{ASP2} = \frac{F_2}{[\text{ASP}]} \cdot (\text{OAA2} - \text{ASP2}) \tag{18}$$

These nine linear differential equations describe pre-steady-state ^{13}C enrichment of both the 4- and 2-carbon positions of the TCA cycle intermediates. TCA cycle kinetics are characterized by three flux parameters (i.e., V_{TCA}, F_1, and F_2), the fractional enrichment of ^{13}C acetyl-CoA (F_{C}), relative anaplerotic flux (y), and the metabolite concentrations. A single 9×1 vector **q** can be used to represent the fractional enrichment of each compartment over time, i.e.,

$$\mathbf{q} = \begin{bmatrix} \text{CIT4} \\ \alpha\text{KG4} \\ \text{GLU4} \\ \text{CIT2} \\ \alpha\text{KG2} \\ \text{GLU2} \\ \text{MAL2} \\ \text{OAA2} \\ \text{ASP2} \end{bmatrix} = \begin{bmatrix} [\text{CIT} - \text{C4}]/[\text{CIT}] \\ [\alpha\text{KG} - \text{C4}]/[\alpha\text{KG}] \\ [\text{GLU} - \text{C4}]/[\text{GLU}] \\ [\text{CIT} - \text{C2}]/[\text{CIT}] \\ [\alpha\text{KG} - \text{C2}]/[\alpha\text{KG}] \\ [\text{GLU} - \text{C2}]/[\text{GLU}] \\ [\text{MAL} - \text{C2}]/[\text{MAL}] \\ [\text{OAA} - \text{C2}]/[\text{OAA}] \\ [\text{ASP} - \text{C2}]/[\text{ASP}] \end{bmatrix}$$

The model can then be described in matrix form as

$$\frac{d}{dt}\mathbf{q} = \mathbf{M}_{\mathrm{TCA}} \cdot \mathbf{q} + \mathbf{U}_{\mathrm{Acetyl\text{-}CoA}} \tag{19}$$

where $\mathbf{M}_{\mathrm{TCA}}$ is a 9×9 matrix characteristic of the TCA cycle. The input vector, $\mathbf{U}_{\mathrm{Acetyl\text{-}CoA}}$, is governed by the fraction of ^{13}C-enriched acetyl-CoA entering the TCA cycle through citrate synthase (F_C). The only nonzero element in $\mathbf{U}_{\mathrm{Acetyl\text{-}CoA}}$ is the labeling of the 4-carbon of citrate from acetyl-CoA. The evolution of $\mathbf{q}$ can then be viewed as a linear system in response to a stepwise stimulation.

By combining this analysis with NMR data of ^{13}C incorporation into glutamate pool, TCA cycle flux (V_{TCA}) and the interconversion rates between TCA cycle intermediates and amino acid pools (F_1 and F_2) can be determined by least-squares fitting of the kinetic model to ^{13}C-enrichment data from NMR spectra. For the purpose of these experiments using labeled acetate or butyrate, F_1 and F_2 are represented by F_1 alone. For optimization of data fitting from least-squares analysis, the Levenberg–Marquardt (Levenberg, 1944; Marquardt, 1963) method can then be used. This method performs an iterative evaluation of the coupled differential equations to achieve optimized values for the parameters V_{TCA} and F_1 until the best fit is achieved. The optimization can be run on either UNIX-based workstations or personal computer software (MATLAB, The MathWorks Inc., Natick, MA). The results of simulating enrichment rates by least-squares fitting of the model to experimental NMR data are shown in Fig. 8 and demonstrate very close agreement of the model to the experimental enrichment curves. As described above, the fitting can be performed using either the standard errors of the grouped mean values for all experiments in a set, or, more preferably, each individual experiment can be analyzed by utilizing the noise level of the NMR spectra from each experiment as the cost function in the least-squares analysis (Yu *et al.*, 1997).

While the optimal value for TCA cycle flux is obtained from least-squares fitting of the model to dynamic ^{13}C NMR observation, an optional penalty term that reflects the difference between the fitted TCA cycle flux and the TCA cycle flux calculated from oxygen consumption can be added to the cost function of the least-squares analysis, if desired (Yu *et al.*, 1997). As described in the previous section, this option allows the range of estimated TCA cycle flux values to be from measured oxygen consumption while taking into consideration the measurement error and the oxidation of other fuels. Under the experimental conditions shown above, for oxidation of ^{13}C acetate or butyrate, the oxygen-consumption constraint only accounts for a 10% difference in the output TCA cycle flux (Yu *et al.*, 1995).

The robustness of this model to errors in the measured concentrations of the key metabolites has been tested by sensitivity analysis (Yu *et al.*, 1997; Yu *et al.*, 1995). Simulations of ^{13}C turnover in glutamate were compared under perturbations of each pool size. The results show that the model is very robust against dramatic

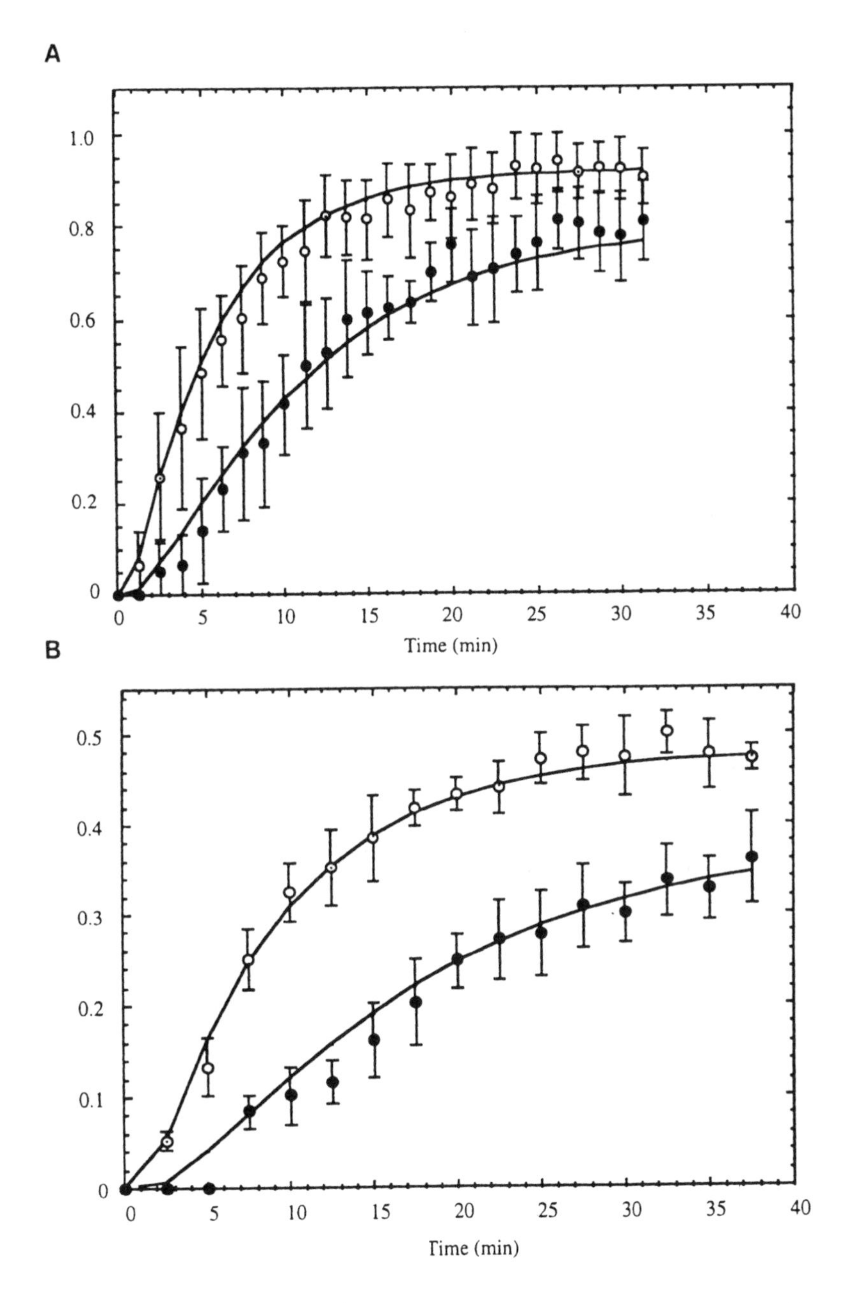

Figure 8. Time course of glutamate ^{13}C enrichment from both NMR measurements and kinetic analysis for hearts oxidizing (A) [2-^{13}C]acetate and (B) [2-^{13}C]butyrate. Signal intensities from dynamic ^{13}C NMR spectra are normalized to steady-state enrichment levels of the 4-carbon of glutamate. Open circles represent 4-carbon of glutamate. Closed circles represent 2-carbon of glutamate. Solid lines are modeled enrichment rates from least-squares fitting.

changes in all metabolite pools, with the exception of glutamate which introduced significant changes in labeling curves with changes in pool size of 20%.

This simple model is readily applicable to *in vivo* conditions, given some knowledge of metabolite pool sizes. As indicated by the study with [2,4-^{13}C]butyrate, the fractional enrichment of acetyl-CoA has no influence on the kinetics of glutamate labeling. Hence, ^{13}C NMR analysis of tissue extracts can be avoided while relative anaplerosis can be calculated from the steady-state enrichment levels of both the 4- and 2-carbons of glutamate. Furthermore, the labeling kinetics of glutamate are minimally influenced by the pool sizes of the TCA cycle intermediates and aspartate (Yu *et al.*, 1995). Therefore, only glutamate concentration needs to be determined accurately. An additional consideration, however, is that the current model assumes instant enrichment of acetyl-CoA pool, which is a valid approximation for isolated heart. However, since *in vivo* heart derives a significant amount of carbon source from endogenous fuels, this assumption may not longer be valid. In this case, it may be necessary to obtain the ^{13}C-enrichment curve of the plasma.

8. METABOLIC FLUX AND REGULATION FROM DYNAMIC ^{13}C NMR SPECTROSCOPY

Energy production in the normal myocardium is an overwhelmingly oxidative process with little contribution from the nonoxidative carbon flux through glycolysis in the normal heart (Lewandowski and Ingwall, 1993). The production of ATP via oxidative phosphorylation is coupled to the electron transport chain, consuming reducing equivalents, or electrons, produced by the progressive oxidation of carbon-based fuels within the mitochondrial matrix. In the myocardium, the principle energy yielding fuel are free fatty acids which are actively transported across the mitochondrial membrane and cleaved into two-carbon acetyl groups via β-oxidation, in the process generating reducing equivalents that are stored within the electron acceptor molecules, NADH and $FADH_2$. The two-carbon, acetyl groups are then activated to form acetyl-CoA for entry into the primary pathway of substrate oxidation, the tricarboxylic acid (TCA) cycle. The reducing equivalents generated by the TCA cycle are then used by the electron transport chain to set up an electrochemical proton gradient across the inner mitochondrial membrane. This proton gradient drives the F1-ATPase enzyme that is responsible for the phosphorylation of ADP to produce ATP. The electrons generated by the oxidative reactions of the TCA cycle are ultimately accepted by molecular oxygen to form water. Thus, the "front-line" of energy production in the heart occurs within the TCA cycle, where net flux must match the energy demands of physiological function.

The concerted activities of the enzymes of oxidative intermediary metabolism are well regulated via the redox state of the mitochondria (LaNoue *et al.*, 1970;

Randle *et al.*, 1970), intramitochondrial calcium ion levels (Strzelecki *et al.*, 1988; Denton and McCormick, 1981), substrate availability (Neely and Morgan, 1974; Bing, 1961; Shipp *et al.*, 1961; Williamson and Krebs, 1961), and, due now to our better understanding with ^{13}C NMR, the exchange of intermediates between the mitochondria and cytosol. This exchange can influence substrate-dependent flux through the rate-determining dehydrogenase enzymes of the TCA cycle (Yu *et al.*, 1996; Yu *et al.*, 1995; LaNoue *et al.*, 1973; Safer and Williamson, 1973; Randle *et al.*, 1970). The ability to monitor the responses of metabolic flux to these control factors have previously been limited to extracellular measurements of radioisotopes or experiments on isolated mitochondria. Thus, performing the experiments on isolated hearts described in this section determines a very important scientific direction for current ^{13}C NMR studies to measure on-line metabolic flux inside the cell.

From the rates of ^{13}C enrichment at the 4-carbon and 2- or 3-carbon positions of glutamate, the rate of TCA cycle flux can be quantified. Such flux measurements, performed in the heart, can be easily validated by comparison to measured oxygen consumption rates (Yu *et al.*, 1995). Of additional interest is the rate of interconversion between ^{13}C-enriched α-ketoglutarate and glutamate which is also a rate-determining process in glutamate labeling (Yu *et al.*, 1996; Yu *et al.*, 1995). This interconversion rate has been examined with ^{13}C NMR for comparison to more traditional biochemical measurements and calculations of the enzymatic interconversion between α-ketoglutarate and glutamate. Direct mathematical models of isotope flux through the TCA cycle intermediate pools and glutamate can be used to obtain these flux rates from dynamic ^{13}C NMR spectra. Approaches to kinetic models are discussed in the previous section. Having already examined experimental conditions that influence the observation of isotope enrichment rates in the glutamate pool with NMR, the physiological basis for observed TCA cycle flux and glutamate enrichment rates is now presented in this section.

A number of published studies demonstrate the abilities to monitor TCA cycle flux in intact or *in vivo* organs using ^{13}C NMR spectroscopy (Yu *et al.*, 1996; Yu *et al.*, 1995; Robitaille *et al.*, 1993a; Robitaille *et al.*, 1993b; Lewandowski, 1992a; Weiss *et al.*, 1992). Most notably from these studies, the ability to monitor mechanical function and oxygen consumption in the heart have validated these measurements. As expected, proportional changes in the TCA cycle flux occur to match the metabolic demands of different levels of mechanical work in the heart, as shown for hearts under normal workloads, stimulated workloads, and basal metabolic rates in the nonbeating, arrested heart. However, these experimental observations have been made under extreme differences in metabolic demand, while the observed ^{13}C kinetics have been shown to also be sensitive to more subtle changes in TCA cycle flux due to metabolic regulation. Specifically, differences in the rate of ^{13}C enrichment of glutamate have been observed in hearts oxidizing [2-^{13}C]acetate and hearts oxidizing the labeled short-chain fatty acid, [2-

^{13}C]butyrate, due to differences in the pathways of substrate oxidation with either fuel (Yu *et al.*, 1995; Lewandowski *et al.*, 1991b). The principle difference between fuels is that butyrate is oxidized first through β-oxidation prior to entry into the TCA cycle, while acetate is directly activated to acetyl-CoA for oxidation in the TCA cycle. Thus, more subtle physiological control of the TCA cycle beyond that observed at extreme difference in workload can be studied.

The process of β-oxidation includes two dehydrogenase reactions that generate reducing equivalents. For each acetyl-CoA molecule produced from the fatty acid chain, an $FADH_2$ and $NADH + H^+$ are also produced. These reducing equivalents are in addition to the four pairs of reducing equivalents produced in the TCA cycle for every acetyl-CoA molecule that enters the cycle. Thus, β-oxidation contributes 20% of the total reducing equivalents generated for each acetyl group cleaved from a fatty acid molecule. The additional reducing equivalents contribute to the total redox state of the mitochondria, to regulate flux through the TCA cycle (Williamson *et al.*, 1976).

Evidence for this simple level of metabolic regulation via redox potential ($NADH/NAD^+$) in the mitochondria of intact hearts has been provided by ^{13}C NMR spectroscopy. The dependence of the myocardium on β-oxidation for utilizing butyrate as an oxidative fuel has been established by ^{13}C-enrichment studies of glutamate in the presence of [2-^{13}C]butyrate and an inhibitor of β-oxidation, 4-bromocrotonic acid (Lewandowski *et al.*, 1991b). The reducing equivalents from both β-oxidation and the TCA cycle, which provide the currency for oxidative energy production, are ultimately accepted by molecular oxygen in the respiratory chain. Therefore, at the same oxygen consumption to support the energy demands of the normal heart, TCA cycle flux from butyrate oxidation is slightly slower compared to acetate oxidation, yet the energy yield is the same.

This redox state-dependent control of the TCA cycle in mitochondria is readily apparent from ^{13}C NMR spectroscopy of the heart which shows a slight reduction in the rate of glutamate enrichment from [2-^{13}C]butyrate than from [2-^{13}C]acetate (Yu *et al.*, 1995; Lewandowski *et al.*, 1991b). Figure 8 shows these differences in the glutamate enrichment rates of hearts oxidizing either of the two substrates. Flux measurements were determined from kinetic analysis using a simple compartment model of the TCA cycle (Yu *et al.*, 1995). The results of the kinetic analysis show that, for hearts closely matched in workload and oxygen consumption rates, the observed mean TCA cycle flux rates were 10.1 μmoles/min/g dry weight with acetate and 7.1 μmoles/min/g dry weight with butyrate. Considering the fact that hearts oxidizing butyrate produced 25% more reducing equivalents (2 pair vs. 4 pair), the approximate 30% reduction in mean TCA cycle flux observed with ^{13}C NMR during butyrate oxidation is very close to the expected response. These data demonstrate the potential for ^{13}C NMR spectroscopy to provide new insights into metabolic control and regulation in intact, functioning tissues.

9. METABOLITE COMPARTMENTATION EFFECTS ON ^{13}C KINETICS

As discussed in the preceding sections, the observation of ^{13}C enrichment of glutamate provides an indirect measure of the ^{13}C enrichment of the TCA cycle intermediates. The rate of labeling of the observed glutamate ^{13}C resonance signals is closely related to metabolic flux through the TCA cycle. These principles are based on the assumption of isotopic equilibrium between the TCA cycle and glutamate, which is due to the chemical exchange of label between α-ketoglutarate and glutamate through a transaminase reaction. In fact, early considerations of ^{13}C NMR spectra have drawn conclusions along the assumption that the TCA cycle intermediate, α-ketoglutarate, is in rapid exchange with glutamate to the extent that this exchange exerts no significant rate-determining step. However, the assumption that the interconversion rate between α-ketoglutarate and glutamate is not a rate-determining step had not been directly tested until a study by Yu *et al.* in 1995. The results of that study indicate that not only is the interconversion of labeled α-ketoglutarate and glutamate much slower than previously assumed, but that the rate-determining process is not the transaminase reaction but rather the physical separation of the glutamate pool from the TCA cycle.

While very rapid isotope exchange is a reasonable assumption based purely on the rate of the enzymatic reaction catalyzed by the glutamate–oxaloacetate transaminase (GOT), this assumption does not consider the physical separation of the TCA cycle within the mitochondria from the NMR observable glutamate pool in the cytosolic compartment. Since at least 90% of the glutamate is located in the cytosol, the labeling of glutamate also involves the transport of metabolites across the mitochondrial membrane through the carrier-mediated transporters of the malate–aspartate shuttle. Thus, while glutamate may be in isotopic equilibrium with α-ketoglutarate, the transaminase reaction does not alone account for the rate of isotope exchange between α-ketoglutarate and glutamate. Instead, physical transport of labeled metabolites across the mitochondrial membrane contributes a rate-determining step in the ^{13}C enrichment of glutamate (see Fig. 9).

This problem was addressed by determining the enzyme kinetics of the cardiac GOT in combination with dynamic ^{13}C NMR observations of glutamate enrichment at different metabolic rates in intact, isolated, perfused rabbit hearts. Since the GOT enzyme is not allosterically regulated and is characterized by simple double-displacement ("ping-pong") reaction kinetics, flux through GOT can be easily calculated from measured V_{max}, the K_m values for the substrates, and the concentrations of the reactants. Therefore, GOT flux can be compared to the NMR-observed, ^{13}C enrichment of glutamate. While GOT flux in the heart was indeed found to be fast in comparison to the TCA cycle flux rate, the reaction rate was found to be much too fast to account for the observed rates of glutamate labeling. The results of these experiments demonstrate that the rate of interconversion between α-ketoglutarate

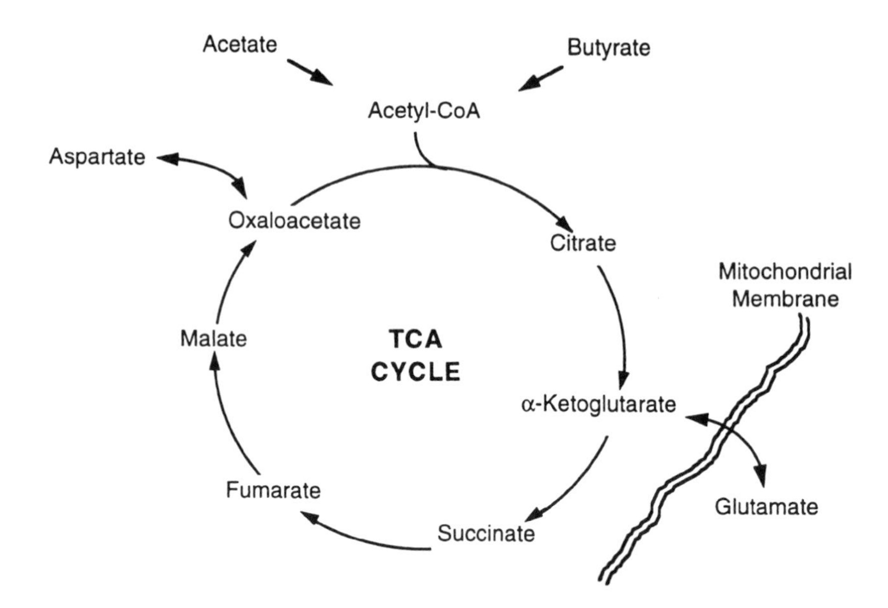

Figure 9. Compartmentation of a large glutamate pool in cytosol from the TCA cycle in the mitochondria. Physical separation of the metabolite pools indicates that metabolite transport contributes a potential rate-determining step in the exchange of ^{13}C between α-ketoglutarate and glutamate.

and glutamate is similar to TCA cycle flux rates and is actually much slower than the rate of GOT flux in the heart by at least twenty-fold. Thus, it is unlikely that glutamate enrichment is solely determined by the TCA cycle flux. Therefore, care must be taken in interpreting ^{13}C NMR spectra of glutamate as a direct index of the TCA cycle activity unless a comprehensive evaluation is given.

These values for TCA cycle flux, GOT flux, and the rate of interconversion (F_1) between α-ketoglutarate and glutamate in the isolated rabbit heart are displayed in Table 2. These values clearly demonstrate that the exchange of label between α-ketoglutarate and glutamate is indeed a rate-determining component of the isotopic enrichment of glutamate. If the exchange rate of ^{13}C between α-ketoglutarate and glutamate were determined by the transaminase rate alone, then labeling of the glutamate 4- and 2-carbon positions would be much faster than the observed rates. In Fig. 10, the simulated enrichment rates of the 4- and 2-carbon positions from a kinetic model using the values obtained for GOT flux are shown in comparison to the actual experimentally observed enrichment curves in a heart oxidizing [2-^{13}C]butyrate. From these data, the interconversion of α-ketoglutarate and glutamate demonstrates a rate-limiting component that cannot be accounted for by the enzymatic reaction alone. Rather, the physical transport of the metabolic intermediates across the mitochondrial membrane is a likely contribution to the

Table 2
Flux Rates, Kinetic Analysis, and GOT Measurements[a]

Substrate	V_{TCA}	F_1	F_{GOT}
Acetate	10.1 ± 0.2	9.3 ± 0.6	223
Butyrate	7.1 ± 0.2	6.4 ± 0.5	362
Acetate + KCl	3.1 ± 0.1	5.8 ± 0.6	181
Butyrate + KCl	1.8 ± 0.1	4.4 ± 1.0	158

[a] V_{TCA}, TCA cycle flux; F_1, flux of chemical exchange between glutamate and α-ketoglutarate; F_{GOT}, flux through cytosolic glutamate–oxaloacetate transaminase (GOT). All flux values are presented as μmoles/min/g dry tissue weight.

observed ^{13}C-enrichment rates. Therefore, the evolution of pre-steady-state ^{13}C NMR spectra is unlikely to be determined solely by TCA cycle activity and may, in fact, be dominated by metabolic communication between subcellular compartments. Recent experiments have directly addressed the potential for these isotope kinetics to be sensitive to the exchange of metabolites across the mitochondrial membrane. The evidence for NMR measurements of transport kinetics and the

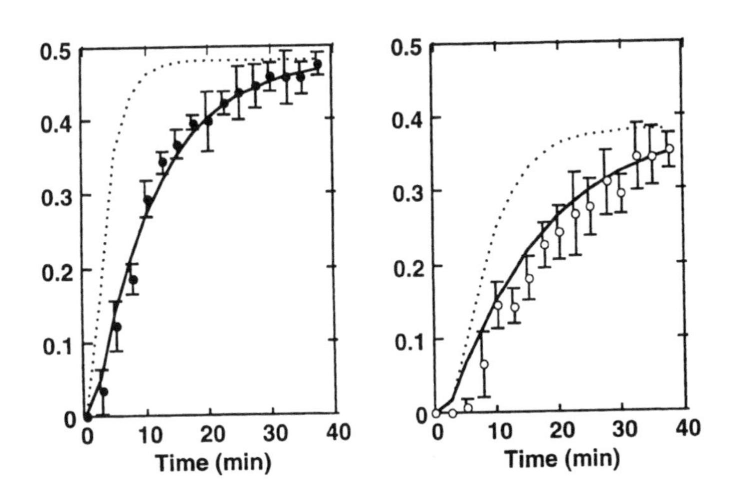

Figure 10. Comparison of NMR observed ^{13}C enrichment of 2- and 4-carbon of glutamate in hearts oxidizing [2-^{13}C]butyrate to the simulated ^{13}C-enrichment rate for both carbons of glutamate if the interconversion rate between α-ketoglutarate and glutamate is equal to flux through glutamate–oxaloacetate transaminase (GOT). The vertical axis represents fractional enrichment with ^{13}C. Experimental data are shown as circles with a solid line to represent least-squares fitting of the model. The dotted line represents the simulated time course of enrichment from GOT flux. Solid circles at left represent 4-carbon enrichment. Open circles at right represent 2-carbon enrichment.

implications of monitoring subcellular metabolite exchange within intact tissues is discussed in the following section.

10. ^{13}C NMR OF SUBCELLULAR TRANSPORT RATES

In the previous sections, the relationship of metabolic activity to isotope kinetics in the ^{13}C NMR observed glutamate pool have been described along with methods of analysis and demonstrations of the sensitivity of ^{13}C NMR to oxidative metabolic rates. These developments and applications of dynamic-mode NMR observations of ^{13}C enrichment of glutamate have lead to the consideration of the mechanisms of physical exchange between intermediates of oxidative metabolism in the mitochondria and the large pool of glutamate in the cytosol. Having documented the discrepancy between enzymatic exchange of carbon isotope to glutamate and the observed rate of exchange of isotopic enrichment of glutamate in the previous section, we must now consider the sensitivity of dynamic ^{13}C NMR to the exchange of metabolites across the mitochondrial membrane. This section describes the metabolite transport through the exchanger proteins of the malate–aspartate shuttle as a rate-determining process in the ^{13}C enrichment of glutamate.

The exchange of ^{13}C-enriched TCA cycle intermediates with the NMR-observed glutamate pool in the cytosol is achieved by the coordinated activity of two transporters on the mitochondrial membrane. One transporter protein is responsible for the reversible exchange of malate and α-ketoglutarate. The other transporter is unidirectional by virtue of being electrogenic, exchanging cytosolic glutamate for mitochondrial aspartate (Tischler *et al.*, 1976). Together with the transaminase enzymes in both the cytosol and mitochondria, these transporters maintain a balance in carbon mass between these cytosolic and mitochondrial intermediates and form the functional system known as the malate–aspartate shuttle. This shuttle is diagrammed in Fig. 11.

The malate–aspartate shuttle serves the functional purpose of transporting reducing equivalents in the form of NADH, from the cytosol to the mitochondrial matrix for the respiratory chain. The mitochondrial membrane is impermeable to NADH, but the reducing equivalent is carried by malate, which is then reconverted to oxaloacetate in the mitochondria. In the process, malate transfers the reducing equivalent to NAD^+. The net result is transfer of the NADH from the cytosol to NADH in the mitochondrion. At high cytosolic redox state ($NADH/NAD^+$), the malate–α-ketoglutarate exchanger is driven in the forward direction, with net shuttle activity increasing to accommodate the increased transport of reducing equivalents from the cytosol to the mitochondrial matrix.

The reversible α-ketoglutarate–malate transporter provides the mechanism by which ^{13}C-labeled α-ketoglutarate enters the cytosol for interconversion with the large cytosolic glutamate pool. Thus, the exchange of ^{13}C-enriched intermediates

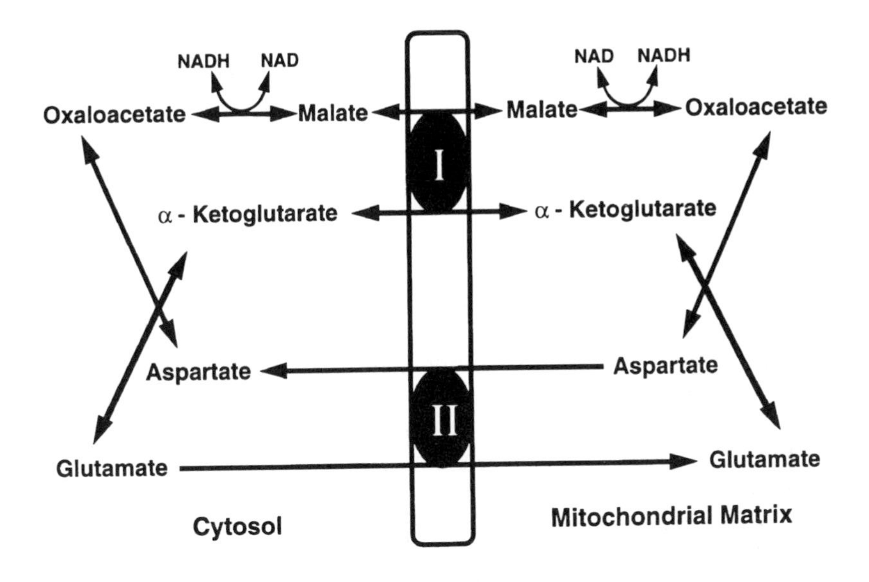

Figure 11. The malate–aspartate shuttle. Cytosolic malate is exchanged for mitochondrial α-ketoglutarate at the exchanger I. Net influx of malate into the mitochondria results in the transport of reducing equivalents which are transferred to NAD^+, to increase mitochondrial NADH. The unidirectional glutamate–aspartate antiport system (II) balances the intermediate pools on either side of the mitochondrial membrane. These transporters provide a mechanism of exchange between ^{13}C-enriched TCA cycle metabolites from the mitochondria with the glutamate pool in the cytosol.

between the TCA cycle in the mitochondrial matrix and the cytosolic space presents another potential rate-determining step in the NMR-observed, ^{13}C enrichment of glutamate. To test whether metabolite transport accounts for the discrepancy between the very fast transaminase rate and the much slower interconversion rate between α-ketoglutarate and glutamate, the effect of stimulated malate–aspartate shuttle on dynamic changes in ^{13}C NMR spectra was recently examined.

An important experimental consideration in elucidating rate-limiting steps of isotope enrichment is that the use of inhibitors is inappropriate. Any inhibition will cause the observed step in the kinetic process, even rapid processes, to become rate-limiting. Any fast component of the isotopic enrichment of glutamate can be partially inhibited, forcing it to become a new rate-limiting step. Instead, to identify a potential rate-limiting step, it must be sped up, releasing the limitation. Therefore, to test the potential for the malate–aspartate shuttle to be rate-limiting in the process of glutamate labeling, metabolite transport needed to be stimulated rather than inhibited.

To increase the rate of metabolite transport via net forward flux through the malate–aspartate shuttle, an experiment was designed to increase cytosolic redox state in isolated hearts with lactate (Yu *et al.*, 1996). Hearts were provided with the short-chain fatty acid [2-^{13}C]butyrate, which, from known effects of fatty acids on lactate or pyruvate oxidation (Johnston and Lewandowski, 1991; Weiss *et al.*, 1989; Dennis *et al.*, 1979), is oxidized in preference to the lactate. In fact, high-resolution ^{13}C spectra indicated that the lactate was not oxidized at all, but rather served to increase the amount of NADH in the cytosol by virtue of reverse flux through lactate dehydrogenase. The net result was that the ^{13}C-enriched butyrate was oxidized at a relatively normal rate in the presence of increased metabolite exchange across the mitochondrial membrane.

With the increase in the redox-driven malate–aspartate shuttle activity, the ^{13}C enrichment of glutamate was observed to occur at an increased rate. A series of spectra from individual hearts is shown in Fig. 12 to demonstrate the visually apparent increase in glutamate labeling that occurred at the increased transport rates. Figure 13 shows the enrichment curves for hearts perfused with the labeled butyrate at normal cytosolic redox state and for hearts perfused with labeled butyrate at elevated redox state. At comparable rates of oxygen consumption and TCA cycle flux, the interconversion rate between α-ketoglutarate and glutamate was slightly over 3 μmoles/min/g dry tissue weight in hearts at normal redox state vs. a significantly higher 14 μmoles/min/g dry tissue weight in hearts subjected to the high cytosolic redox state. The enrichment of glutamate was obviously stimulated by the increased malate–aspartate shuttle activity. These results clearly demonstrated that the transport process acts as a rate-determining step in the NMR-observed ^{13}C enrichment of glutamate. These results showed for the first time that dynamic ^{13}C NMR is sensitive not just to chemical exchange, but also to metabolite transport across the mitochondrial membrane. This recent finding indicates that future work holds potential for exciting insights into the metabolic communication of subcellular compartments in intact, functioning tissues.

However, the exchange of label between TCA cycle intermediates and the bulk of the NMR-observed glutamate pool in the cytosol need not require net forward flux through the malate–aspartate shuttle. Net forward flux through this shuttle system involves both the unidirectional glutamate–aspartate exchanger and the reversible α-ketoglutarate–malate exchanger (21,23). However, the efflux and influx of isotope-labeled α-ketoglutarate across the mitochondrial membrane for interconversion with the cytosolic glutamate pool can be accommodated by the reversible α-ketoglutarate-malate exchanger alone. Therefore, the transport of metabolites across the mitochondrial membrane may play a significant role in coordinating the TCA cycle flux with cytosolic metabolism even in the absence of significant activity through the malate–aspartate shuttle. This provides researchers the opportunity to examine the role of the metabolite transport proteins on the mitochondrial membrane and the recruitment of the malate–aspartate shuttle as mechanisms for the regulation of oxidation metabolism in the mitochondria and nonoxidative metabolism in the cytosol.

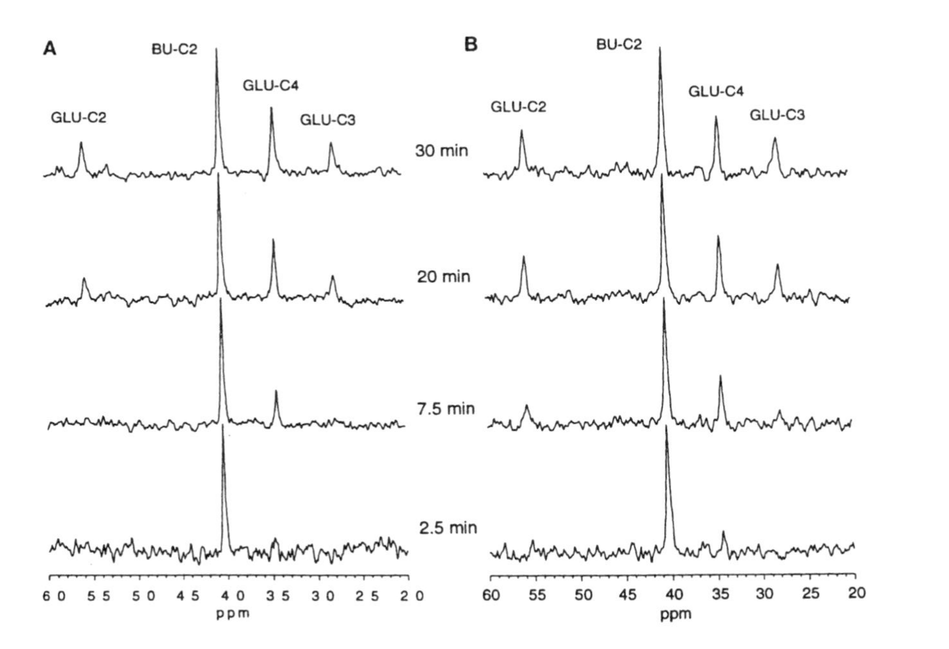

Figure 12. Effects of increased malate–aspartate shuttle activity on ^{13}C-enrichment rates of glutamate. (A) Series of spectra from isolated rabbit heart oxidizing [2-^{13}C]butyrate at normal cytosolic redox state. (B) Series of spectra from isolated rabbit heart oxidizing [2-^{13}C]butyrate in the presence of lactate to increase cytosolic redox state. Effect of high cytosolic redox state is to drive the malate–aspartate shuttle in the forward direction, increasing the rate of exchange between mitochondrial α-ketoglutarate and cytosolic glutamate. At high redox state, the enrichment rate of glutamate is increased due to increased malate–aspartate shuttle activity. Reprinted with permission from X. Yu *et al.*, 1996, Subcellular metabolite transport and carbon isotope kinetics in the intramyocardial glutamate pool, *Biochemistry* **35**:6963. Copyright 1996 American Chemical Society.

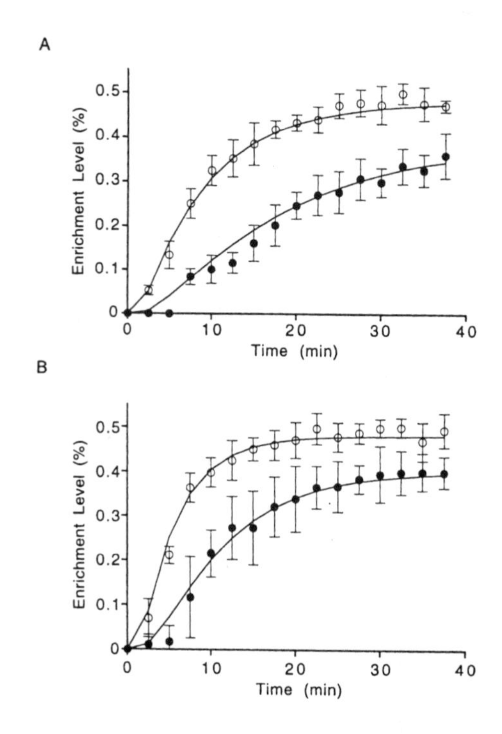

Figure 13. ^{13}C enrichment of glutamate at different cytosolic redox states. ^{13}C NMR signal intensities are normalized to steady-state enrichment levels. (A) ^{13}C enrichment from [2-^{13}C]butyrate at normal redox state. (B) ^{13}C enrichment from [2-^{13}C]butyrate at high redox state. Note increase in ^{13}C-enrichment rate at high cytosolic redox state. Open circles represent ^{13}C enrichment of 4-carbon of glutamate. Closed circles represent ^{13}C enrichment of 2-carbon of glutamate. Solid lines represent least-squares fit of the kinetic model to the NMR data. Reprinted with permission from X. Yu *et al.*, 1996, Subcellular metabolite transport and carbon isotope kinetics in the intramyocardial glutamate pool, *Biochemistry* **35**:6963. Copyright 1996 American Chemical Society

11. SUMMARY

Recent advances in ^{13}C NMR observations of metabolism in intact tissues have taken the applications of this method from static measurements of relative metabolic activity to quantification of metabolic flux rate and metabolite transport. An exciting application from this new understanding comes from the translation of metabolic conditions between the cytosol and mitochondria for physiological studies of metabolic regulation.

The compartmentation of metabolic processes in living cells represents an important regulatory mechanism in coordinating energy production to cell functions. Metabolic communication between subcellular compartments is achieved largely by selective permeability of membranes to different metabolites, or carrier-mediated transport, which allows for efficient control of reaction pathways. Through the adjustment of metabolite distribution between the mitochondria and cytosol, the metabolic demands of overall physiological function by the cell can be translated to the energetic machinery of the mitochondria.

In this manner, the malate–aspartate shuttle interacts directly with the TCA cycle through carrier-mediated transport of α-ketoglutarate, malate, aspartate, and glutamate across the mitochondrial membrane. However, the regulatory role of the individual metabolite transporter on the mitochondrial membrane has only been studied in isolated mitochondria with an artificial cytosolic environment and certainly no physiological function. The dynamic-mode ^{13}C NMR observations of intact tissues that have been discussed in this chapter can now contribute similar in-depth analysis of metabolic regulation in response to the physiological function of intact and *in vivo* organs.

REFERENCES

Bailey, I. A., Gadian, D. G., Matthews, P. M., Radda, G. K., and Seeley, P. J., 1981, *FEBS Lett.*, **123**:315.

Bing, R., 1961, *Am. J. Med.* **30**:679.

Buxton, D. B., Schwaiger, M., Nguyen, A., Phelps, M. E., and Schelbert, H. R., 1988, *Circ. Res.* **63**:628.

Chance, E. M., Seeholzer, S. H., Kobayashi, K., and Williamson, J. R., 1983, *J. Biol. Chem.* **258**:13785.

Chatham, J. C., Forder, J. R., Glickson, J. D., and Chance, E. M., 1995, *J. Biol. Chem.* **270**:7999.

Cohen, S. M., 1983, *J. Biol. Chem.* **258**:14294.

Cohen, S. M., 1987, *Biochemistry* **26**:563.

Cohen, S. M., Ogawa, S., and Shulman, R. G., 1979, *Proc. Natl. Acad. Sci. U.S.A.* **76**:1603.

Damico, L. A., White, L. T., Yu, X., and Lewandowski, E. D., 1996, *J. Mol. Cell. Cardiol.* **28**:989.

Dennis, S. C., Padma, A., DeBuysere, M. S., and Olson, M. S., 1979, *J. Biol. Chem.* **254**:1252.

Denton, R. M., and McCormick, J. G., 1981, *Clin. Sci. (London)* **61**:135.

Des Rosiers, C., Di Donato, L., Comte, B., Laplante, A., Marcoux, C., David, F., Fernandez, C. A., and Brunengraber, H., 1995, *J. Biol. Chem.* **270**:3731.

Eakin, R. T., Morgan, L. O., Gregg, C. T., and Matwiyoff, N. A., 1972, *FEBS Lett.* **28**:259.

Fitzpatrick, S. M., Hetherington, H. P., Behar, K. L., and Shulman, R. G., 1990, *J. Cereb. Blood Flow Metab.* **10**:170.

Johnston, D. L., and Lewandowski, E. D., 1991, *Circ. Res.* **68**:714.

Jones, J. G., Sherry, A. D., Jeffrey, F. M. H., Storey, C. J., and Malloy, C. R., 1993, *Biochemistry* **32**:12240.

Kelleher, J. A., 1985, *Am. J. Physiol.* **248**:E252.

Kornberg, H. L., 1966, Anaplerotic sequences and their role in metabolism, in *Essays in Biochemistry* (P. N. Campbell and G. D. Greville, eds.), Academic Press, New York, pp. 1–31.

Kupriyanov, V. V., Lakomkin, V. L., Steinschneider, A. Y., Severina, M. Y., Kapelko, V. I., Runge, E. K., and Saks, V. A., 1988, *J. Mol. Cell. Cardiol.* **20**:1151.

LaNoue, K., Nicklas, W. J., and Williamson, J. R., 1970, *J. Biol. Chem.* **245**:102.

LaNoue, K. F., Bryla, J., and Bassett, D. J. P., 1974, *J. Biol. Chem.* **249**:7514.

LaNoue, K. F., Walajtys, E. I., Williamson, J. R., 1973, *J. Biol. Chem.* **248**:7171.

Laughlin, M. R., Taylor, J., Chesnick, A. S., DeGroot, M., and Balaban, R. S., 1993, *Am. J. Physiol.* **264**:H2068.

Levenberg, K., 1944, *Quart. Appl. Math.* **2**:164.
Lewandowski, E. D., 1992a, *Circ. Res.* **70**:576.
Lewandowski, E. D., 1992b, *Biochemistry* **31**:8916.
Lewandowski, E. D., and Johnston, D. L., 1990, *Am. J. Physiol.* **258**:H1357.
Lewandowski, E. D., and Hulbert, C., 1991, *Magn. Reson. Med.* **19**:186.
Lewandowski, E. D., and Ingwall, J. S., 1994, The physiological chemistry of energy production in the heart, in *Hurst's the Heart: Arteries and Veins*, 8th ed. (J. W. Hurst, ed.), McGraw-Hill, New York, pp. 153–164.
Lewandowski, E. D., and White, L. T., 1995, *Circulation* **91**:2071.
Lewandowski, E. D., Johnston, D. L., and Roberts, R., 1991a, *Circ. Res.* **68**:578.
Lewandowski, E. D., Chari, M., Roberts, R., and Johnston, D. L., 1991b, *Am. J. Physiol.* **261**:H354.
Lewandowski, E. D., Damico, L. A., White, L. T., and Yu, X., 1995a, *Am. J. Physiol.* **269**:H160.
Lewandowski, E. D., Yu, X., White, L. T., and Damico, L. A., 1995b, *Circulation* **92**:I-387.
Lewandowski, E. D., Doumen, C., White, L. T., LaNoue, K. F., Damico, L. A., and Yu, X., 1996, *Magn. Reson. Med.* **35**:149.
Lopaschuk, G. D., Spafford, M. A., and Norman, J. D., 1990, *Circ. Res.* **66**:546.
Malloy, C. R., Sherry, A. D., and Jeffrey, F. M. H., 1988, *J. Biol. Chem.* **263**:6964.
Malloy, C. R., Thompson, J. R., Jeffrey, F. M. H., and Sherry, A. D., 1990, *Biochemistry* **29**:6756.
Marquardt, D., 1963, *SIAM J. Appl. Math.* **11**:431.
Mason, G. F., Rothman, D. L., Behar, K. L., and Shulman, R. G., 1992, *J. Cereb. Blood Flow Metab.* **12**:434.
Mason, G. F., Gruetter, R., Rothman, D. L., Behar, K. L., Shulman, R. G., and Novotny, E. J., 1995, *J. Cereb. Blood Flow Metab.* **15**:12.
Neely, J. R., and Morgan, H. E., 1974, *Ann. Rev. Physiol.* **36**:413.
Neely, J. R., Rovetto, M. J., Whitmer, J. T., and Morgan, H. E., 1973, *Am. J. Physiol.* **233**:651.
Neurohr, K. J., Barrett, E. J., and Shulman, R. G., 1983, *Proc. Natl. Acad. Sci. U.S.A.* **80**:1603.
O'Donnell, J. M., White, L. T., Doumen, C., Yu, X., LaNoue, K. F., and Lewandowski, E. D., 1996, *Circulation (Suppl.)* **94**:I-547.
Peuhkurinen, K. J., Takala, T. E., Nuutinen, E. M., and Hassisen, I. E., 1983, *Am. J. Physiol.* **244**:H281.
Randle, P. J., England, P. J., and Denton, R. M., 1970, *Biochem. J.* **117**:677.
Reimer, K. A., Jennings, R. B., and Tatum, A. H., 1983, *Am. J. Cardiol.* **52**:72A.
Renstrom, B., Nellis, S. H., and Leidtke, A. J., 1989, *Circ. Res.* **65**:1094.
Robitaille, P.-M. L., Rath, D. P., Abduljalil, A. M., O'Donnell, J. M., Jiang, Z., Zhang, J., and Hamlin, R. L., 1993a, *J. Biol. Chem.* **268**:26296.
Robitaille, P.-M. L., Rath, D. P., Skinner, T. E., and Abduljalil, A. M., 1993b, *Magn. Reson. Med.* **30**:262.
Russell, R. R., and Taegtmeyer, H., 1991, *J. Clin. Invest.* **87**:384.
Safer, B., 1975, *Circ. Res.* **37**:527.
Safer, B., and Williamson, J. R., 1973, *J. Biol. Chem.* **248**:2570.
Schneider, C. A., and Taegtmeyer, H., 1991, *Circ. Res.* **68**:1045.
Sherry, A. D., Nunnally, R. L., and Peshock, R. M., 1985, *J. Biol. Chem.* **260**:9272.
Sherry, A. D., Malloy, C. R., Roby, R. E., Rajagopal, A., and Jeffrey, F. M., 1988, *Biochem. J.* **254**:593.
Sherry, A. D., Sumegi, B., Miller, B., Cottam, G. L., Gavva, S., Jones, J. G., and Malloy, C. R., 1994, *Biochemistry* **33**:6268.
Shipp, J. C. S., Opie, L. H., and Challoner, D., 1961, *Nature* **189**:1018.
Strisower, E. H., Kohler, G. D., and Chaikoff, I. L., 1952, *J. Biol. Chem.* **198**:115.
Strzelecki, T., Strzelecki, D., Koch, C. D., La Noue, K. F., 1988, *Arch. Biochem. Biophys.* **264:**310.
Taegtmeyer, H., Hems, R., and Krebs, H. A., 1980, *Biochem. J.* **186**:701.
Taegtmeyer, H., Roberts, A. F., and Raine, A. E., 1985, *J. Am. Coll. Cardiol.* **6**:864.
Tischler, M., Pachence, J., Williamson, J. R., and LaNoue, K. F., 1976, *Arch. Biochem. Biophys.* **173**:448.
Weinman, E. O., Strisower, E. H., and Chaikoff, I. L., 1957, *Physiol. Rev.* **37**:252.

Weiss, R. G., Chacko, V. P., and Gerstenblith, G., 1989, *J. Mol. Cell. Cardiol.* **21**:469.
Weiss, R. G., Gloth, S. T., Kalil-Filho, R., Chacko, V. P., Stern, M. D., and Gerstenblith, G., 1992, *Circ. Res.* **70**:392.
Weiss, R. G., Kalil-Filho, R., Herskowitz, A., Chacko, V. P., Litt, M., 1993, *Circulation* **87**:270.
Williamson, J. R., and Krebs, H. A., 1961, *Biochem. J.* **80**:540.
Williamson, J. R., Ford, C., Illingworth, J., and Safer, B., 1976, *Circ. Res.* **38**:39.
Yu, X., White, L. T., Doumen, C., Damico, L. A., LaNoue, K. F., Alpert, N. M., and Lewandowski, E. D., 1995, *Biophys. J.* **69**:2090.
Yu, X., White, L. T., Alpert, N. M., and Lewandowski, E. D., 1996 *Biochemistry* **35**:6963.
Yu, X., Alpert, N. M., and Lewandowski, E. D., 1997, *Am. J. Physiol.: Cell Physiology (Modeling in Physiology)*, in press.
Zimmer, S. D., Michurski, S. P., Mohanakrishnan, P., Ulstad, V. K., Thoma, W. J., Ugurbil, K., 1990, *Biochemistry* **29**:3731.

5

Assessing Cardiac Metabolic Rates during Pathologic Conditions with Dynamic ^{13}C NMR Spectra

Robert G. Weiss and Gary Gerstenblith

1. INTRODUCTION

Prior chapters have explored some basic elements of myocardial metabolism and dynamic ^{13}C NMR methods for quantifying metabolic flux. This chapter aims to review applications of specific dynamic ^{13}C NMR methods for the understanding of myocardial pathophysiology. It will focus on insights provided by dynamic ^{13}C NMR spectroscopy into two timely and clinically relevant questions of interest to our laboratory, namely, paradigms of ischemic dysfunction and ischemic preconditioning.

Robert G. Weiss and Gary Gerstenblith • Carnegie 584, The Johns Hopkins Hospital, 600 N. Wolfe Street, Baltimore, Maryland 21287-6568.

Biological Magnetic Resonance, Volume 15: In Vivo Carbon-13 NMR, edited by L. J. Berliner and P.-M. L. Robitaille. Kluwer Academic / Plenum Publishers, New York, 1998.

2. DYNAMIC ^{13}C NMR SPECTROSCOPY IN PARADIGMS OF MYOCARDIAL DYSFUNCTION: "STUNNED" AND "HIBERNATING" MYOCARDIUM

2.1. "Stunned" Myocardium

Myocardial contractile function is closely linked to the rates of ATP production. The tricarboxylic acid cycle provides more than 90% of the ATP produced by the aerobic heart (Neely and Kobayashi, 1979; Neely and Morgan, 1974) and thus quantitative measures of total TCA cycle carbon oxidation are essential for assessing ATP production and understanding the relationship between metabolism and normal function. Myocardial contractile function is directly related to measures of TCA cycle flux, such as myocardial oxygen consumption. Although contractile dysfunction can result from nonmetabolic causes, such as changes in calcium handling, primary alterations in TCA cycle oxidation can reduce cardiac contractile function. During total severe ischemia, oxidative metabolism is inhibited and contractile function rapidly ceases. The role of altered metabolism in contractile dysfunction during modest reductions in coronary flow and during reperfusion after severe ischemia are less well determined.

Several terms have been developed to describe myocardial dysfunction in the clinical setting. The term "stunned myocardium" was coined several years ago to describe the phenomenon of transient contractile dysfunction occurring after a brief but severe episode of ischemia (Heyndrickx *et al.*, 1975; Braunwald and Kloner, 1982). Experimental studies demonstrate that stunned myocardium is not necrotic, has normal coronary blood flow, and can respond to inotropic stimulation (Kloner *et al.*, 1974; Stahl *et al.*, 1988). Normal contractile function will spontaneously return after minutes to days, depending on the experimental model or the duration or severity of the ischemic insult in the clinical setting. The precise pathophysiology underlying the dysfunction of stunned myocardium has not been fully elucidated although several factors such as altered calcium handling, cellular calcium overload, and free-radical generation are thought to contribute (Bolli, 1992). An important possibility that has been evaluated by dynamic ^{13}C spectra is the issue of whether oxidative metabolism is persistently altered after ischemia and contributes to reduced contractile function.

Two studies used ^{13}C NMR kinetics to evaluate TCA cycle activity in post-ischemic "stunned" myocardium and both were conducted in isolated, perfused hearts. Lewandowski *et al.* studied glutamate labeling kinetics immediately after a 10-minute period of total ischemia in perfused rabbit hearts (Lewandowski and Johnston, 1990). Contractile function, abolished during ischemia, recovered to about 50% of pre-ischemic values during reperfusion. [2-^{13}C]acetate was administered at the time of reperfusion and the kinetics of TCA cycle activity were indexed by serial ^{13}C NMR assessments of relative glutamate C2 and C4 enrichment. The glutamate C2/C4 ratio prior to attainment of isotopic steady state of reperfused stunned myocardium was significantly less than that in control hearts never exposed to ischemia and functioning at a roughly twofold higher, or normal, rate–pressure product. It was concluded that TCA cycle activity is reduced, at least at the onset of reperfusion when the ^{13}C tracer was administered, in "stunned" myocardium.

In another study, TCA cycle activity was studied by ^{13}C NMR spectroscopy after both brief and prolonged periods of ischemia in isolated rat hearts (Weiss *et al.*, 1993). During reperfusion after the prolonged 45 minutes of normothermic ischemia, there was ultrastructural evidence of irreversible injury by electron microscopy, no contractile activity, absent high-energy phosphates by ^{31}P NMR, and no detectable TCA cycle activity by ^{13}C NMR (see Fig. 1). These data were the first to demonstrate that irreversibly injured reperfused myocardium lack ^{13}C NMR detectable TCA cycle activity. Reperfusion after more brief ischemia (17–20 min) resulted in myocardial "stunning" as contractile recovery was depressed by about 25–30% from pre-ischemic values, creatine phosphate recovered fully, and there was no ultrastructural evidence of irreversible injury. [2-^{13}C]acetate was administered at seven minutes of reperfusion when contractile recovery had reached a steady but depressed level and when the major TCA cycle-linked metabolite pools, glutamate and aspartate, were not changing. The steady-state levels of ^{13}C glutamate enrichment were lower than those of control hearts, but the rate of glutamate enrichment was not slower but more rapid (Fig. 2). TCA cycle activity was independently assessed by empirical and mathematical modeling analyses of the ^{13}C NMR data in which both methods account for the lower pool sizes (Weiss *et al.*, 1992). Both analyses indicated that mean TCA cycle flux was higher by 60–100% in reperfused stunned hearts than in workload-matched hearts not exposed to ischemia and reperfusion. The empirical method, based on the time difference between ^{13}C labeling of different positions in the glutamate molecule enriched in subsequent "turns" of the TCA cycle, will be discussed below. The mathematical modeling approach, a simplified version of that initially reported by Chance *et al.* (1983), also indicated a significant relative increase in TCA cycle flux and, in addition, suggested a modest reduction in amino-transaminase activity in reperfused stunned hearts. These findings of an increase in TCA cycle activity in stunned hearts relative to that accompanying a similar workload in hearts not exposed to prior ischemia are consistent with multiple studies in different species showing a relative increase in myocardial oxygen consumption in stunned hearts (Sako *et al.*, 1988; Laster *et al.*, 1989; Neubauer *et al.*, 1988; Stahl *et al.*, 1988). The increase in TCA cycle activity in stunned hearts cannot be attributed to mitochondrial uncoupling (Sako *et al.*, 1988) but is likely related, at least in part, to nonsystolic energy demands, such as energy-consuming diastolic calcium cycling (Weiss *et al.*, 1990b), which are increased in stunned myocardium. Taking the available evidence from these two studies of TCA cycle kinetics in reperfused "stunned" myocardium (Lewandowski and Johnston, 1990; Weiss *et al.*, 1993), one

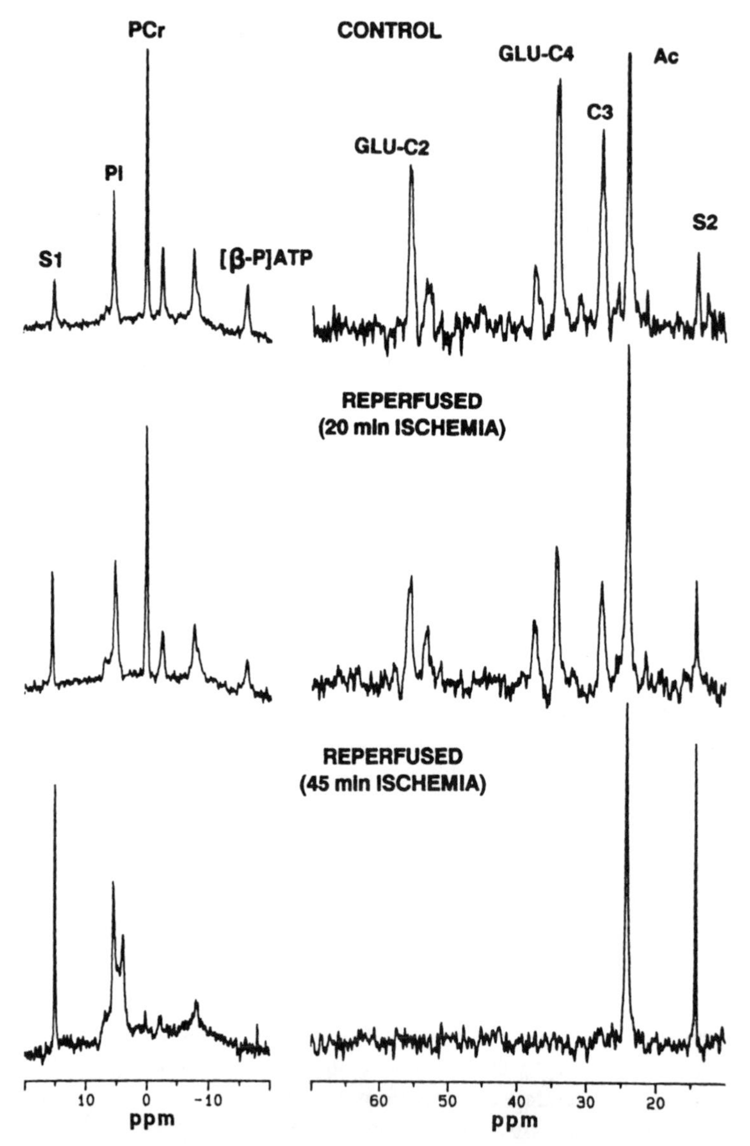

Figure 1. Representative ^{31}P (left panel) and ^{13}C (right panel) NMR spectra obtained in control (top), reperfused "stunned" (middle), and reperfused nonviable (bottom) rat hearts. Peak assignments are Pi, inorganic phosphate (intracellular and buffer); PCr, creatine phosphate; [*β*-P]ATP, beta-phosphate of ATP; GLU-C2, glutamate C2; GLU-C4, glutamate C4; C3, glutamate C3; Ac, acetate C2; S1, phenylphosphonic acid standard; S2, hexanoate C6 standard. Stunned hearts (middle) had nearly complete recovery of PCr but reduced ATP and, at isotopic steady state, lower levels of glutamate. After prolonged ischemia, irreversibly injured hearts had barely detectable PCr and ATP and no detectable TCA cycle activity by ^{13}C NMR. Reproduced from Weiss *et al.* (1993).

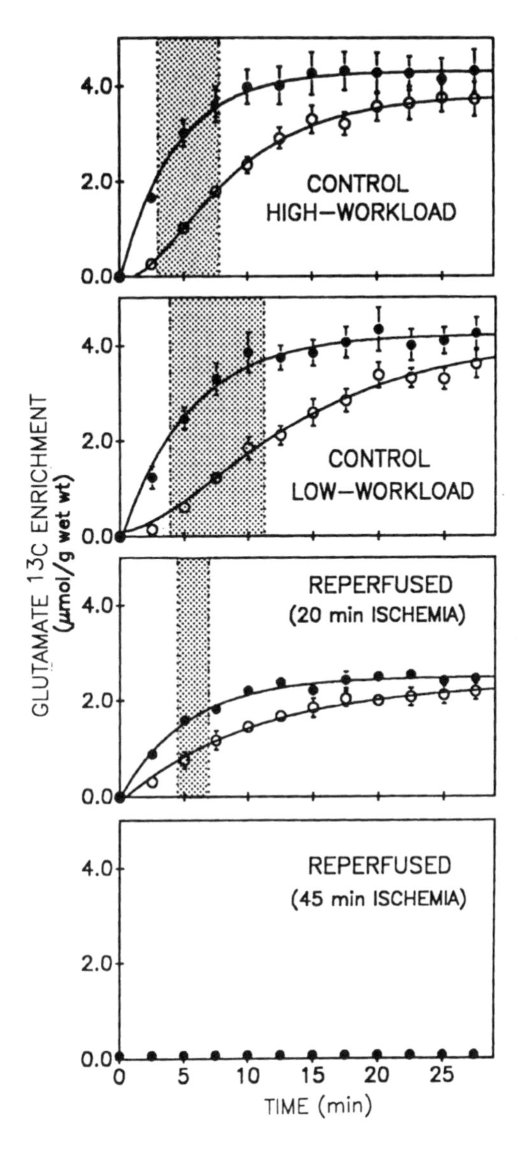

Figure 2. Plots of dynamic ^{13}C NMR data for glutamate C4 (filled circles) and glutamate C2 (open circles) in control (top two panels), stunned (third panel), and irreversibly injured (bottom panel) rat hearts. The x-axis represents the time after administration of [2-^{13}C]acetate. The time to half-maximal enrichment for each isotopomer is demarcated by a dotted line, and the difference between these times, the glutamate Δt_{50}, in indicated by the projection of the cross-hatched region onto the time axis. Stunned hearts have lower steady-state glutamate levels and shorter glutamate Δt_{50} as compared with matched control hearts (second panel); the latter suggests more rapid TCA cycle activity. Glutamate enrichment by the TCA cycle was not observed in irreversibly injured hearts (bottom panel). Reproduced from Weiss *et al.* (1993).

could conclude that TCA cycle activity is limited or slowed only during the early minutes of reperfusion but quickly attains levels at or above those of similarly functioning control hearts. Therefore, dynamic ^{13}C NMR kinetics demonstrate that reduced TCA cycle activity does not exist beyond the initial reflow period and cannot, therefore, account for the depressed function of reperfused stunned myocardium.

2.2. "Hibernating" Myocardium

Another paradigm of myocardial dysfunction was initially described as "hibernating" myocardium by Rahimtoola (1985). The concept is that modest chronic reductions in coronary blood flow (supply) could be matched by chronic reductions in systolic function (demand) such that a new, down-regulated or "hibernating" state would be obtained. The theoretical possibility that this stable state of depressed contractility with reduced but matched supply and demand could be reversed with restoration of normal flow was supported by clinical observations of rapidly improved regional function immediately after coronary artery bypass (Topol *et al.*, 1984). Initial experimental studies demonstrated the attainment of normal high-energy phosphate levels in models of hibernating myocardium (Zhang *et al.*, 1993), indicating that classic ischemia with an imbalance of oxygen supply and demand and high-energy phosphate depletion was not chronically present. Dynamic ^{13}C NMR spectra have also been obtained in an isolated rat heart model with modest reductions in coronary flow (Weiss *et al.*, 1989), similar to those which result in acute myocardial hibernation. Although severe (~80%) reductions in coronary flow in such crystalloid perfused hearts resulted in classic ischemia with depletion of high-energy phosphates and incomplete post-ischemic recovery, modest reductions in coronary flow (~60%) resulted in unchanged high-energy phosphate levels and a stable 50% depression in left ventricular developed pressure for one hour which fully recovered with restoration of baseline flow. The latter characteristics are those anticipated for hibernating myocardium. During perfusion with [1-^{13}C]glucose, modest reductions in coronary flow resulted in a twofold increase in [3-^{13}C]lactate levels and depressed and slowed glutamate enrichment by the TCA cycle as compared with those of normal hearts. The delayed TCA cycle kinetics were not due to reduced [1-^{13}C]glucose delivery during reduced coronary flow as reduced [1-^{13}C]glucose delivery in the absence of reduced flow did not reduce TCA cycle kinetics (Weiss *et al.*, 1989). The reduced TCA cycle kinetics were mimicked by reducing cardiac workload in nonischemic hearts, but that maneuver did not increase lactate levels as in hearts with reduced coronary flow. A relative increase in the contribution of anaplerotic to acetyl-CoA-derived carbon entering the TCA cycle, a potential source of nonoxidative, substrate-level phosphorylation energy production, was suggested in hearts with modest flow reductions from ^{13}C NMR spectra obtained well after attainment of ^{13}C glutamate isotopic steady state. These observations from dynamic ^{13}C NMR spectra demonstrate that acute reductions in coronary flow which mimic the paradigm of "hibernating myocardium" result in augmentation of anaerobic metabolism with down-regulated TCA cycle oxidation.

The findings of these dynamic ^{13}C NMR studies suggest two additional insights which go beyond myocardial stunning and hibernation in these models. First, dynamic ^{13}C NMR spectra can provide unique and valuable information on metabolic flux through specific pathways that cannot be reliably gained from measures of tissue uptake alone, like that from radio-tracer studies. For example, in the studies of myocardium reperfused after a short period of ischemia, total tissue ^{13}C uptake was less than that of normal tissues (see Fig. 2). However, it is incorrect to assume that reduced tissue uptake means reduced TCA cycle flux, since total glutamate pools and amino-transaminase flux are reduced while TCA cycle flux is actually increased in stunned hearts. Dynamic ^{13}C NMR spectra alone provide sufficient nondestructive information to correctly identify increased TCA cycle flux in the setting of reduced pool sizes and altered related metabolic pathways. Second, all of these studies suggest that dynamic ^{13}C NMR spectra may provide a unique method for metabolically characterizing and distinguishing "hibernating," "stunned," and infarcted dysfunctional myocardium in more clinically relevant settings.

2.3. Implications for Distinguishing Types of Dysfunctional Myocardium in the Clinical Setting: *In Vivo* ^{13}C NMR

Myocardial dysfunction in ischemic heart disease is a major cause of morbidity and mortality. Although many conventional noninvasive techniques (e.g., echocardiography, gated blood pool nuclear techniques, and magnetic resonance imaging) as well as invasive techniques (e.g., ventriculography at cardiac catheterization) provide reliable measures of the extent and severity of myocardial dysfunction, they cannot identify the underlying paradigm causing dysfunction or whether the noncontracting tissue is viable. In addition, distinguishing the cause of dysfunction is important clinically since revascularization strategies (e.g., coronary artery bypass or balloon angioplasty) cannot improve the contractile function of nonviable infarcted myocardium, do improve dysfunction in "hibernating" myocardium, and are not required in "stunned" regions whose function would improve spontaneously. Unfortunately, there is no routine method for prospectively distinguishing among these paradigms. Preserved metabolic uptake, detected by PET methods (Tillisch *et al.*, 1986), is generally accepted as one of the best research methods for identifying viable myocardium and distinguishing it from nonviable tissue. Unfortunately, PET technology is available in relatively few centers and is not widely used. We propose, based on the above studies, that *in vivo* ^{13}C NMR has the potential to uniquely distinguish viable from nonviable myocardium as well as viable "stunned" from viable "hibernating" myocardium.

The theoretical framework and the rationale for the idea that dynamic ^{13}C NMR may one day be useful for distinguishing different kinds of dysfunctional myocardium is outlined in Table 1. In this simplified framework, severely dysfunctional or noncontracting myocardium can be classified as viable "stunned," viable "hibernating," or nonviable infarcted myocardium. These cannot be distinguished based on contractile function which can be equally depressed in all. Revascularization would improve function in "hibernating" myocardium but not in nonviable, in-

Table 1
Observations Relating Metabolic Profiles to Myocardial Pathophysiology

Myocardium	Function	^{31}P NMR: HEP's	^{13}C NMR: TCA	Rx
"Stunned"	depressed	normal	increased TCA flux	wait
"Hibernating"	depressed	normal	slowed TCA flux increased lactate	revascularize
Infarcted	depressed	reduced levels	absent TCA flux	no revasc.

farcted tissues. Function would spontaneously improve in stunned myocardium without intervention. Based on animal studies (Rehr *et al.*, 1989; Deboer *et al.*, 1983; Weiss *et al.*, 1993) and a single study in patients (Yabe *et al.*, 1995), myocardial high-energy phosphate levels would be reduced or absent in infarcted myocardium. Myocardial high-energy phosphate ratios and levels would be expected to be normal in both "hibernating" and "stunned" cardiac tissues (Sako *et al.*, 1988; Zhang *et al.*, 1993; Weiss *et al.*, 1989; Weiss *et al.*, 1993). Dynamic ^{13}C NMR spectra could be useful for prospectively distinguishing all three, in that infarcted tissues would lack evidence of metabolic uptake and TCA cycle activity (Weiss *et al.*, 1993), "stunned" myocardium would exhibit normal or increased TCA cycle activity (Weiss *et al.*, 1993), while "hibernating" myocardium would evidence decreased TCA cycle activity but increased lactate formation (Weiss *et al.*, 1989). Therefore, based on the available theoretical experimental evidence, dynamic ^{13}C NMR spectroscopy has the potential to metabolically distinguish these three states and could be useful in prospectively predicting the functional benefit of revascularization procedures to dysfunctional regions based on the metabolic characterization of the underlying pathophysiology.

It is important to first emphasize that this is a theoretical framework and has not been clinically demonstrated. The metabolic ^{13}C NMR profiles of stunned and hibernating myocardium have not been tested in more physiologic settings than isolated hearts and ^{13}C NMR is not routinely implemented in the clinical setting. Nevertheless, spatially localized cardiac ^{31}P NMR spectroscopy has been used for several years in clinical studies to identify stress-induced changes in cardiac high-energy phosphate ratios in subjects with ischemia in viable regions (Weiss *et al.*, 1990a; Yabe *et al.*, 1994), to identify reductions in metabolite ratios at rest in subjects with ischemic and nonischemic cardiomyopathy (Hardy *et al.*, 1991; Neubauer *et al.*, 1992), as well as to quantify reduced high-energy phosphate levels or concentrations in infarcted, nonviable tissues (Yabe *et al.*, 1995). *In vivo* human cardiac ^{13}C NMR spectroscopy is far less developed than ^{31}P NMR, in part due to the need for proton decoupling as well as the administration of expensive ^{13}C-labeled substrates. Spatially localized, proton-decoupled cardiac ^{13}C NMR spectra have been obtained of the naturally abundant carbons of the human heart (Bottomley *et al.*, 1989) although cardiac acquisitions during infusion of ^{13}C-enriched substrates have not yet been published, as in the human brain. *In vivo* ^{13}C studies are primarily limited by the relatively high cost of ^{13}C isotopes. The cost of stable ^{13}C-enriched substrates has fallen in recent years as production has increased and will likely to continue to fall to the range of radioisotopes used in nuclear imaging if produced in increasing quantities. Despite the limitations, this theoretical framework is logical and based on available experimental evidence. It also provides a useful rationale to further test the developing technology of *in vivo* cardiac ^{13}C spectroscopy. In addition, clinical enthusiasm is especially high because high-field magnetic resonance systems suitable for ^{13}C NMR are available at most hospitals, because ^{13}C-labeled substrates are stable and do not require elaborate on-site synthesis, and since identification of myocardial viability would significantly and economically affect the care of many patients.

2.4. A Strategy for Measuring TCA Cycle Flux with ^{13}C NMR

If *in vivo* ^{13}C NMR spectroscopy could be performed routinely on the human heart in the future, how would the spectra be analyzed to estimate TCA cycle flux? As earlier chapters indicate, there are both empirical and mathematical modeling means for analyzing dynamic ^{13}C NMR spectra during infusion and washout of ^{13}C-labeled isotopes which can be used for calculating TCA cycle flux. We think all methods have strengths and weaknesses, and that the hypothesis tested and the experimental model should primarily dictate which strategy is used to measure TCA cycle flux from a given set of ^{13}C NMR data.

We will review and critique one method for indexing TCA cycle flux that we initially described since it has some unique strengths, especially for *in vivo* applications. Prior (Chance *et al.*, 1983) and subsequent (Weiss *et al.*, 1992; Robitaille *et al.*, 1993) mathematical modeling approaches for estimating TCA cycle flux require, in addition to ^{13}C NMR data, information about the sizes of various metabolite pools (e.g., glutamate) that can typically be obtained only by biochemical measures from tissue extracts, precluding human studies. Several years ago we sought to develop an empirical method for quantifying TCA cycle flux based solely on ^{13}C NMR data. The underlying basis for the approach exploited a novel strength of ^{13}C NMR, to quantify labeling of distinct carbons of the same molecule enriched during different "turns" of the TCA cycle (Weiss *et al.*, 1992). The central idea (see Fig. 3) is that dynamic ^{13}C NMR acquisitions can index the time for carbon to make one "turn" of the cycle from the time difference in enrichment of glutamate carbons enriched in subsequent TCA turns. In other words, during administration of a ^{13}C-enriched substrate which generates [2-^{13}C]acetyl-CoA (e.g., [1-^{13}C]glucose, [3-^{13}C]pyruvate, and so forth), the time for ^{13}C to appear and enrich glutamate C4 includes the time for isotope administration, delivery to the heart, cellular uptake, metabolism through pathways producing acetyl-CoA (glycolysis for glu-

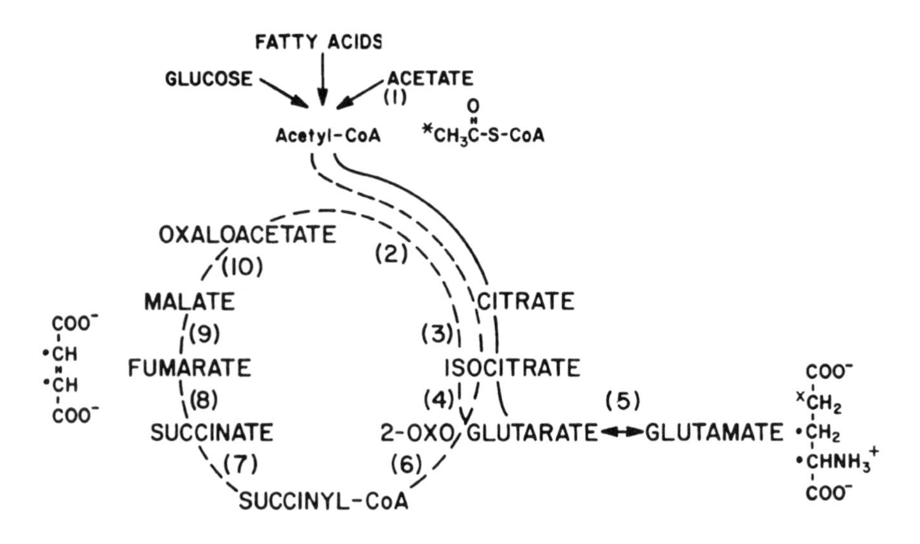

Figure 3. Schematic of TCA cycle reactions and the path of carbon acetate and other substrates to the glutamate isotopomers. The solid line denotes the path for reactions influencing the time for the metabolism of [2-^{13}C]acetate to glutamate C4 (×) and the dotted line the reactions influencing the time course for the metabolism of [2-^{13}C]acetate to glutamate C2 or C3 (●). The time difference between glutamate C4 and C2 (or C3) enrichment is directly proportional to the time for carbon movement through the citric acid cycle and inversely proportional to cycle flux. Reproduced from Weiss *et al.* (1992).

cose or beta-oxidation for fatty acids), and flux through the first reactions of the cycle (Fig. 3, solid line) and through the amino transaminase reaction to glutamate C4. The time for enrichment of glutamate C3 (or C2) includes all of the same delays, but also includes flux through one turn of the TCA cycle including randomization at succinate (Fig. 3, dashed line). Therefore, the *time difference* (Δt_{50}) between glutamate C4 and C3 enrichment is directly related to the time for movement through the cycle, indirectly related to TCA cycle flux, and essentially independent of many other factors including time of delivery, uptake, glycolysis, and beta-oxidation. If one takes into account the sizes of metabolite pools detected by ^{13}C NMR and the fractional enrichment of acetyl-CoA entering the cycle, Δt_{50} can be used to calculate K_t, a measure directly proportional to TCA cycle flux (Weiss *et al.*, 1992).

There are at least six strengths to this approach for indexing TCA cycle flux. First, the method was experimentally validated by comparison with measures of myocardial oxygen consumption (MVO_2), which is proportional to TCA cycle flux, across a range of contractile activity (Weiss *et al.*, 1992). In fact, it is the only method of TCA cycle flux quantification by ^{13}C NMR to date which has been shown to closely correlate with MVO_2 across a wide range in more than a single experimental setting. Second, the consequence of relatively large changes in ^{13}C NMR detectable and smaller pool sizes was extensively evaluated with mathematical modeling and shown to generate only small variations within the noise of the experiment (Weiss *et al.*, 1992). Third, a careful potential flaw analysis, including variations in isotope delivery, was also conducted and shown to induce relatively little change in these measures of TCA cycle activity (Weiss *et al.*, 1993). Fourth, the measures can be obtained from *in vivo* or clinical data, since they arise entirely from ^{13}C NMR measures and do not require destructive measures or assumptions regarding pool sizes for estimating TCA cycle flux. Fifth, the approach can be used with both "wash in" and "wash out" ^{13}C methods, although the latter may be clinically problematic with physiologic substrates (i.e., it is easier to give a labeled substrate than it is to rapidly remove it from the blood *in vivo*). Sixth, the method does not require the assumption that amino-transaminase flux is much more rapid than TCA cycle flux, as do several other approaches (Fitzpatrick *et al.*, 1990; Robitaille *et al.*, 1993).

This latter consideration was recently explored experimentally in isolated rat hearts with the use of an amino-transaminase inhibitor. Administration of 0.1 mM aminooxyacetic acid inhibited *in vitro* aspartate aminotransferase activity by 95%, slowed ^{13}C labeling of glutamate C4 and C3, reduced calculated aminotransferase activity in the intact heart by 50%, but did not alter indices of total TCA flux (Weiss *et al.*, 1995). In this setting K_t was altered by only 1%, while some other measures of TCA cycle flux were altered by as much as 30% (Weiss *et al.*, 1995). Therefore in settings where amino-transaminase activity is reduced or unknown, this approach to estimating TCA cycle flux based on the time difference in glutamate enrichment may have some unique advantages.

The limitations of the approach are, first, that it does not provide an absolute measure of TCA cycle activity, only a measure which is directly proportional to TCA cycle flux. Second, since the approach is relatively independent of amino-transaminase activity, it does not provide a measure of amino-transaminase activity, as do several modeling methods. Third, the method does not provide an analytic solution to the problem of calculating flux, although it has a form strikingly similar to a recently reported analytic solution that also is directly related to TCA metabolite mass divided by the difference in dominant rate constants of glutamate C4 and C2 (Cohen and Bergman, 1995).

Once again, our bias is that the hypothesis to be tested and the experimental model should dictate the dynamic ^{13}C NMR data analysis method used to estimate TCA cycle flux. If one wants to quantify the absolute rate of TCA cycle turnover or wishes to measure amino-transaminase activity, then the K_t approach, which is dependent on the time difference between glutamate carbon labeling (Δt_{50}), will not provide the necessary information and should not be used. If, on the other hand, one needs a relative measure of TCA cycle flux, for example, for comparison with the same measure under other conditions or in different regions of the heart, which only requires ^{13}C NMR data and is relatively independent of the time of isotope delivery and amino-transaminase activity, then K_t should be considered.

Whether or not dynamic *in vivo* ^{13}C NMR spectroscopy will have clinical utility for metabolically characterizing and distinguishing viable dysfunctional myocardium in the clinical setting is still far from being demonstrated. Nevertheless, the experimental studies and methods reviewed here, as well as the work of many other groups, provide an experimental foundation and logical approach for future investigations into this important question in more physiological, clinically-relevant settings.

3. DYNAMIC ^{13}C NMR SPECTROSCOPY OF GLYCOLYSIS IN ISCHEMIC PRECONDITIONING

3.1. Metabolic Changes in Ischemic Preconditioned Hearts

"Ischemic preconditioning" was the term used by Murry, Jennings, and Reimer (Murry *et al.*, 1986) in their initial description of the phenomenon whereby short periods of ischemia significantly increase myocardial tolerance to a subsequent prolonged ischemic insult. From a clinical perspective, ischemic preconditioning is of tremendous investigative interest both because it affords significantly greater reductions in myocardial infarct size than all prior "anti-ischemic" therapies and because it occurs in all species studied, including humans (Deutsch *et al.*, 1990). Many mechanisms have been identified which contribute to the protection of ischemic preconditioning including adenosine receptor stimulation (Liu *et al.*, 1991), α-1-adrenergic stimulation (Bankwala *et al.*, 1994), ATP-dependent K^+ channel activation (Gross and Auchampach, 1992), and activation and translocation of protein kinase C (Mitchell *et al.*, 1995; Speechly-Dick *et al.*, 1994). Changes in intermediary metabolism also contribute to the protection of ischemic preconditioning and studies using NMR spectroscopy have detailed some of these effects.

In an early description of ischemic preconditioning, Murry *et al.* (1990) recognized that preconditioning critically reduced myocardial energy demand during the subsequent prolonged ischemic period, but they could not distinguish whether preserved ATP levels or reduced cellular load of catabolites (H^+ and lactate) was primarily responsible for delaying ischemic cell death. Using ^{31}P NMR spectroscopy in intact pigs, Kida *et al.* (1991) observed both preserved high-energy phosphates and cellular pH in preconditioned hearts during ischemia and suggested that the pH effect may be more important because it persisted longer during ischemia. Although preserved high-energy phosphates during ischemia are only occasionally observed in preconditioned hearts, attenuated acidosis during ischemia is observed in nearly all preconditioning models (Kida *et al.*, 1991; Steenbergen *et al.*, 1993; Chen *et al.*, 1995; Schaefer *et al.*, 1995; Wolfe *et al.*, 1993).

In studies where the attenuation of ischemic acidosis in preconditioned hearts is blocked, the metabolic and contractile benefits of preconditioning are also blocked (de Albuquerque *et al.*, 1994). It is likely that the attenuation of acidosis contributes to lessened alterations in intracellular Na^+ and Ca^{++} during ischemia and subsequent reperfusion (Steenbergen *et al.*, 1993) and thereby confers protection. Since attenuation of ischemic acidosis contributes to the protection afforded by preconditioning, what is the mechanism accounting for the attenuation of acidosis in preconditioned hearts?

3.2. Attenuated Ischemic Acidosis in Preconditioned Hearts

Reduced acidosis during ischemia in preconditioned hearts could be due to decreased H^+ production or increased H^+ buffering during ischemia. In studies using dynamic ^{13}C NMR techniques, no increase in tissue buffering capacity is observed in ischemia preconditioned rat hearts (see Fig. 4) (de Albuquerque *et al.*, 1995). However, there is considerable evidence suggesting decreased proton production. The early studies of Murry *et al.* reported reduced lactate levels in preconditioned dog hearts during ischemia (Murry *et al.*, 1990) and this has been reproduced in other experimental models (Finegan *et al.*, 1995). The studies of Wolfe *et al.* in intact rats suggest that glycogen depletion may account for the attenuation of lactate and proton accumulation during ischemia in preconditioned hearts and contributes to improved recovery (Wolfe *et al.*, 1993). In experiments where glycogen repletion is allowed to occur after preconditioning episodes, ischemic acidosis is no longer attenuated and the improvement in post-ischemic recovery is not observed as in conventionally preconditioned hearts (Wolfe *et al.*, 1993). Although attenuation of ischemic acidosis appears important for at least some of the protection afforded by preconditioning in nearly all models, the mechanisms resulting in reduced proton production and the role of glycogen depletion are unclear. If glycogen depletion alone accounts for the attenuation of ischemic acidosis in preconditioning, then glycolytic and proton production rates should be similar in control and preconditioned hearts during early ischemia but differ later after glycogen stores are exhausted. Conversely, if attenuated acidosis results from slowed glycolysis prior to exhaustion of glycogen stores, then glycolytic and proton production rates would be less in preconditioned hearts throughout ischemia. Distinguishing between these two mechanisms is important for designing strategies to mimic some of the protective effects of preconditioning by attenuating ischemic acidosis.

We recently used dynamic ^{13}C NMR to address the mechanisms accounting for reduced H^+ production during ischemia in preconditioned hearts since ^{13}C NMR alone can provide simultaneous measures of intracellular pH (Chacko and Weiss, 1993), lactate appearance (Hoekenga *et al.*, 1988; Weiss *et al.*, 1989; Lewandowski *et al.*, 1991), and glycogen degradation (Hoekenga *et al.*, 1988; Brainard *et al.*,

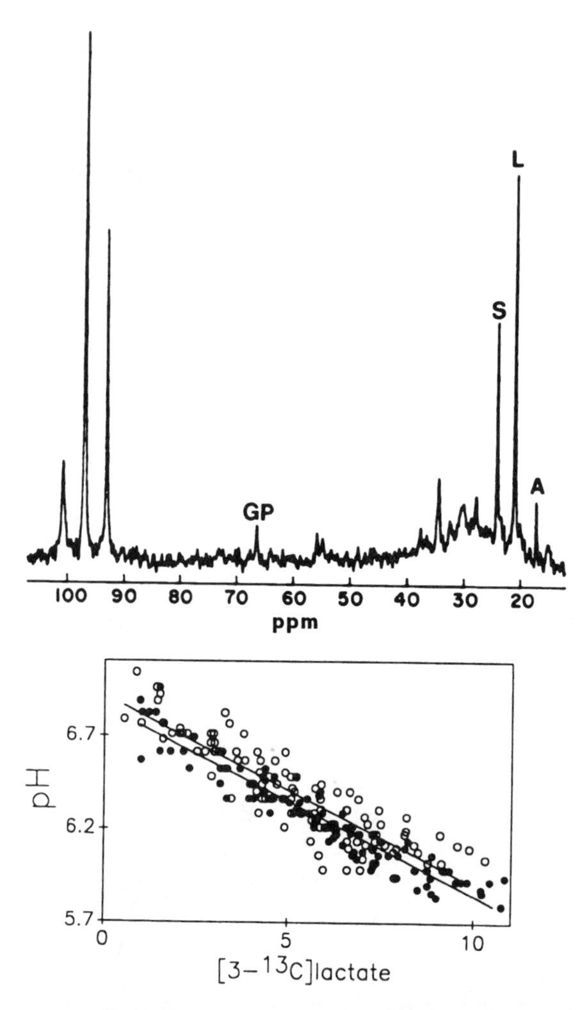

Figure 4. Representative ^{13}C NMR spectrum (upper panel) acquired over 1.25 min at 11.75 T in an isolated rat heart after [1-^{13}C]glucose perfusion and 10 minutes of total ischemia is shown. The peaks include [3-^{13}C]glycerol phosphate (GP, used for pH determination), [2-^{13}C]acetate standard contained in the intraventricular balloon (S), [3-^{13}C]lactate (L), and [3-^{13}C]alanine (A). The relationship between intracellular pH and [3-^{13}C]lactate concentrations (μmol/g ww) during global, no-flow ischemia in control and preconditioned hearts ($n = 9$ each) is presented in the lower panel. There is no difference in the buffering capacity between control and preconditioned hearts under these conditions, as indicated by (1/slope) of this relationship. Adapted from de Albuquerque *et al.* (1995).

1989; Kalil-Filho *et al.*, 1991), even during total ischemia. We proposed that the changing rate of glycolysis could be determined during total ischemia by quantifying the rate of appearance of glycolytic end-products (lactate and alanine) from serial ^{13}C NMR acquisitions in [1-^{13}C]glucose perfused and [^{13}C]glycogen-loaded hearts (Weiss *et al.*, 1996). Likewise, the rates of [^{13}C]glycogenolysis and [^{13}C]glucose utilization during total ischemia were also calculated from serial ^{13}C NMR spectra by the reductions in each of these peak areas over time. Figure 5 presents examples of serial ^{13}C NMR spectra showing changes in glycolytic products and end-products over time and some calculated metabolic rates. Under the conditions studied in isolated rat hearts, the approach was validated by observations that ^{13}C NMR measures of absolute lactate concentrations agreed closely with biochemical

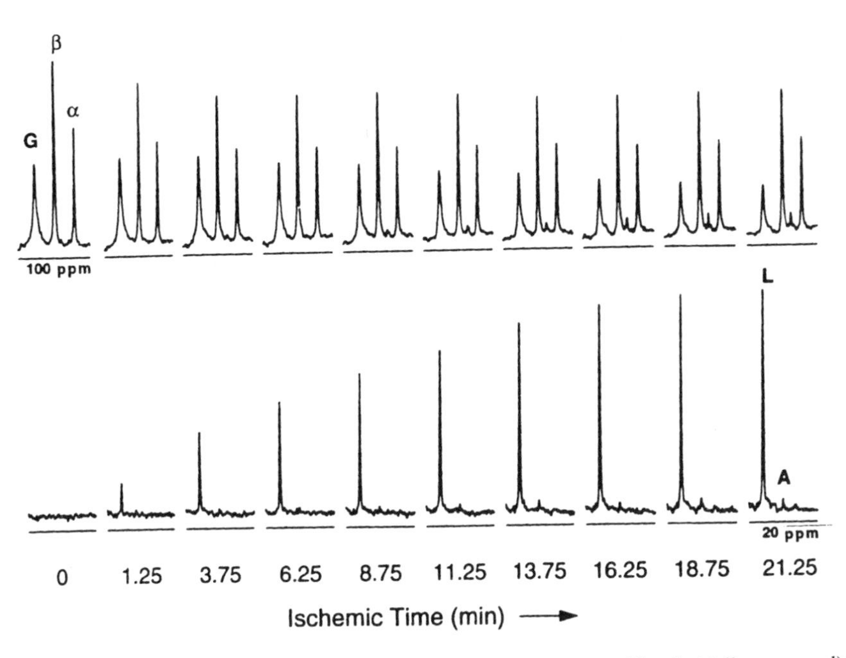

Figure 5. Expanded portions of the glycogen (G), and glucose resonances (β and α) (all upper panel) and of the lactate (L) and alanine (A) regions (lower panel) of ^{13}C NMR spectra acquired every 2.5 minutes in the same control heart. The numbers below each spectrum indicate the time in minutes during total ischemia when the spectrum was acquired with time 0 indicating the pre-ischemic spectra. The increase in peak areas between sequential spectra of the lower panel reflect the ^{13}C glycolytic flux occurring during those acquisitions. The progressively smaller differences in lactate and alanine peaks from one acquisition to the next during later ischemia demonstrate the rate at which glycolytic flux slows during total ischemia. The upper panel displays the time course of the sources of that glycolytic flux and demonstrates that nearly all glucose utilization occurs during the first minutes of ischemia and that glycogen utilization begins after one minute of ischemia. Reproduced from Weiss *et al.* (1996).

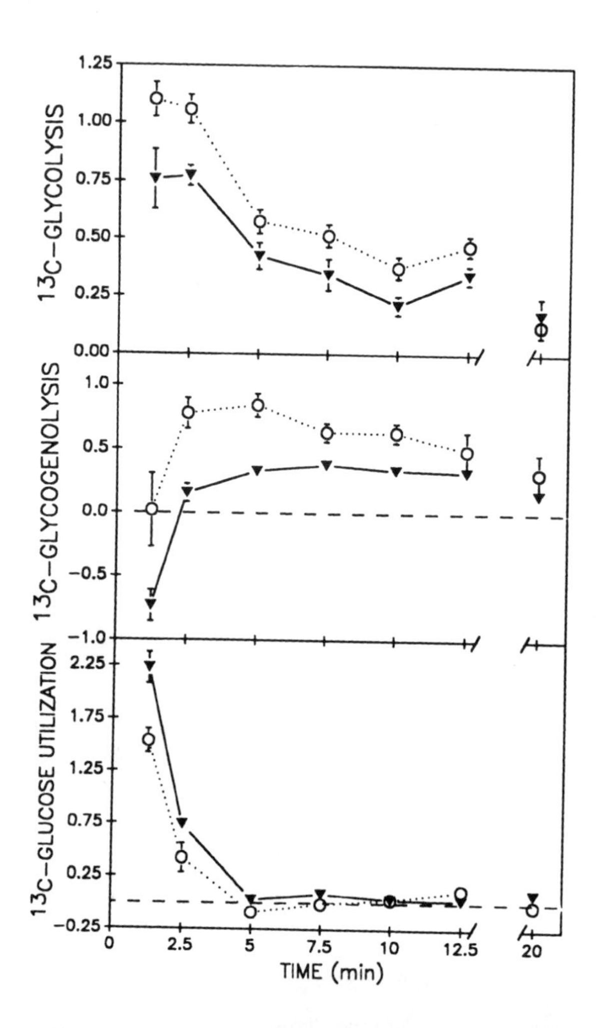

Figure 6. Rates of ^{13}C-glycolysis, ^{13}C-glycogenolysis, and ^{13}C-glucose utilization during ischemia in control (open circles, dotted line) and ischemia preconditioned hearts (filled triangles, solid line) expressed as μmol/min/g ww. In control hearts, glycolytic rates fall after several minutes of ischemia as glucose utilization rates are declining rapidly and glycogenolytic rates are rising, but persistent glycolysis is readily detectable for at least 20 minutes of total ischemia. Glycolytic rates are significantly lower throughout ischemia in ischemia preconditioned hearts than in controls. Higher glucose utilization is observed at 1.25 min in preconditioned hearts, which coincides with brief glycogen synthesis. The rate of glycogenolysis is significantly lower in ischemic preconditioned than in control hearts throughout ischemia. Reproduced from Weiss *et al.* (1996).

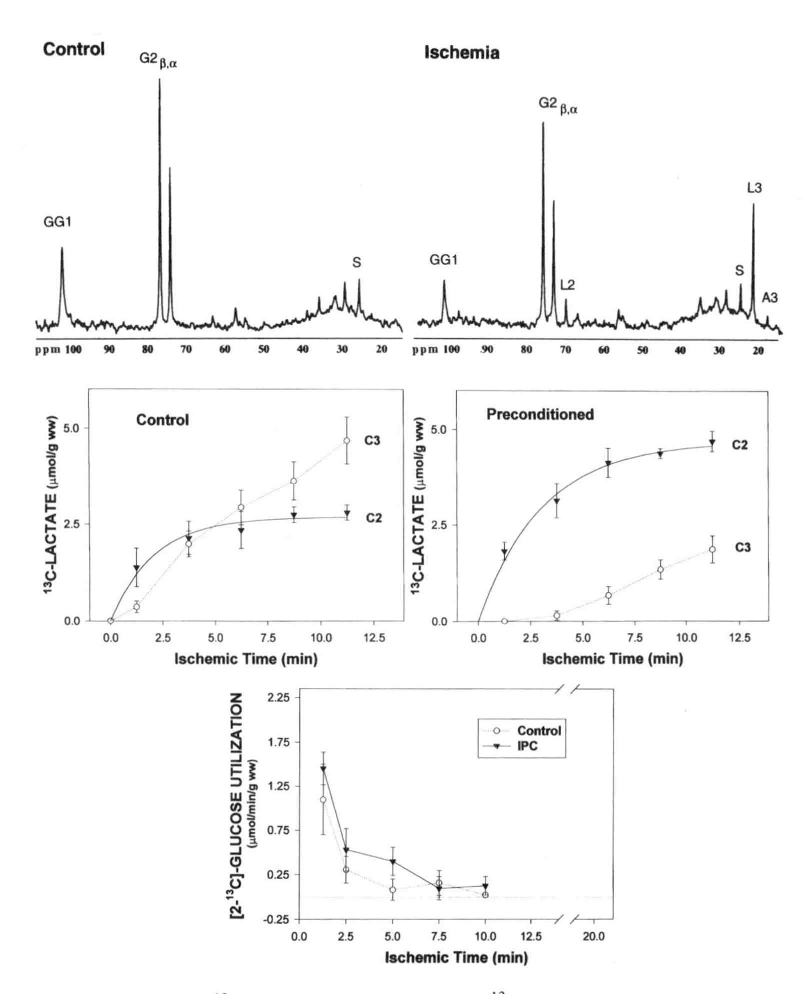

Figure 7. Representative ^{13}C NMR spectra (upper panel), ^{13}C-lactate levels (middle panel, μmol/g ww), and ^{13}C-glucose utilization rates (bottom panel, μmol/min/g ww) acquired from control and preconditioned hearts (n = 4, each) with glycogen synthesized from [1-^{13}C]glucose and extra- and intracellular glucose pools subsequently replaced with [2-^{13}C]glucose prior to total ischemia. The peaks denote [1-^{13}C]glycogen (GG1), the β- and α-anomers of [2-^{13}C]glucose ($G2_{\beta\alpha}$), the intraventricular balloon [2-^{13}C]acetate standard (S), [2-^{13}C]lactate (L2), [3-^{13}C]lactate (L3), and [3-^{13}C]alanine (A3). During ischemia (middle panel), [2-^{13}C]lactate (filled triangles, solid line), derived from [2-^{13}C]glucose, appears rapidly and achieves maximal levels within about 3 minutes. [3-^{13}C]lactate (open circles, dotted line), derived from [1-^{13}C]glycogen, appears more gradually but accounts for nearly all of the lactate formed after the first minute or so of ischemia in control hearts. In preconditioned hearts (middle panel, right), depressed glycogen utilization (L3) and increased glucose utilization (L2) are observed during early ischemia, in agreement with the earlier experiments (Fig. 6 above). ^{13}C-glucose utilization rates (bottom panel) derived directly from the appearance of [2-^{13}C]lactate in these experiments trended higher in preconditioned hearts and also agree well, although with more variability, than similar measures derived from the disappearance of the [1-^{13}C]glucose peaks (open circles, dotted line) described before. These dynamic ^{13}C NMR observations are also consistent with a primary reduction in glycogenolysis accounting for attenuated glycolysis throughout ischemia in preconditioned hearts. Reproduced from Weiss *et al.* (1996).

measures in the same tissues, that the carbons from glycolysis do not enter the TCA cycle shortly after the onset of ischemia, and that the production of glycolytic end-products stoichiometrically agreed with the loss of glycolytic products, with the small excess in the latter likely appearing as glycolytic intermediates (Weiss *et al.*, 1996).

These studies demonstrated directly for the first time that glycolysis is inhibited during the first few minutes of total ischemia and that measurable glycolysis and glycolytic ATP production rates are observable for up to 20 minutes of ischemia. In addition, glucose provides the primary glycolytic source during the first minute of ischemia, but that glycogenolysis is quickly activated and, as expected, provides the primary fuel during most of ischemia. In terms of ischemic preconditioning, glycolytic rates were slowed in preconditioned hearts from the onset of ischemia and long before exhaustion of glycogen stores. Glycogenolysis was attenuated from the onset of ischemia in preconditioned hearts (Fig. 6) and this was out of proportion to the reduction in glycolysis and occurred during a modest increase in glucose utilization. Separate dual isotope studies with [1-^{13}C]glycogen and [2-^{13}C]glucose confirmed the primary reduction in glycogenolysis, increased glucose utilization in preconditioned hearts, and the validity of the basic approach (Fig. 7) (Weiss *et al.*, 1996). Thus dynamic ^{13}C NMR directly demonstrated for the first time that glycogenolysis is attenuated throughout ischemia and is the cause of attenuated proton production in preconditioned hearts, not simply depletion of glycogen stores. This probably explains, at least in part, why glycogen depleting interventions alone do not reproduce entirely the beneficial effects of ischemic preconditioning (Schaefer *et al.*, 1995).

4. SUMMARY

Approaches for assessing both glycolytic and TCA cycle metabolic activity from dynamic ^{13}C NMR spectra have been reviewed. In each case, specific examples are provided into ways in which ^{13}C NMR measures provide unique pathophysiologic insights into clinically important questions regarding myocardial stunning and hibernation as well as ischemic preconditioning. ^{13}C NMR studies in isolated hearts suggest that the TCA cycle may be transiently slowed immediately after ischemia but that the persistent dysfunction of "stunned" myocardium cannot be attributed to reduced TCA cycle activity since it is relatively increased as compared with that in normal myocardium with similar contractile activity. "Hibernating" myocardium is associated with reduced TCA cycle activity and increased anaerobic metabolism. ^{13}C NMR may provide a unique means to distinguish these causes of reversible myocardial dysfunction. The attenuated ischemic acidosis, which contributes to protection in ischemic preconditioning, has been shown by ^{13}C NMR to be attributable to a primary reduction in glycogenolysis throughout ischemia and not simply due to an exhaustion of glycogen stores by the preconditioning episodes. It is anticipated that these and other observations will help guide future work as ^{13}C NMR studies are implemented in more physiologic and ultimately clinical settings.

REFERENCES

Bankwala, A., Hale, S. L., and Kloner, R. A., 1994, *Circulation* **90**:1023.

Bolli, R., 1992, *Circulation* **86**:1671.

Bottomley, P. A., Hardy, C. J., Roemer, P. B., and Mueller, O. M., 1989, *Magn. Reson. Med.* **12**:348.

Brainard, J. R., Hutson, J. Y., Hoekenga, D. E., and Lenhoff, R., 1989, *Biochemistry* **28**:9766.

Braunwald, E., and Kloner, R. A., 1982, *Circulation* **66**:1146.

Chacko, V. P., and Weiss, R. G., 1993, *Am. J. Physiol.* **264**:C755.

Chance, E. M., Seeholzer, S. H., Kobayashi, K., and Williamson, J. R., 1983, *J. Biol. Chem.* **258**:13785.

Chen, W., Gabel, S., Steenbergen, C., and Murphy, E., 1995, *Circ. Res.* **77**:424.

Cohen, D. M., and Bergman, R. N., 1995, *Am. J. Physiol.* **268**:E397.

de Albuquerque, C. P., Gerstenblith, G., and Weiss, R. G., 1994, *Circ. Res.* **74**:139.

de Albuquerque, C. P., Gerstenblith, G., and Weiss, R. G., 1995, *J. Mol. Cell. Cardiol.* **27**:777.

Deboer, L. W. V., Rude, R. E., Kloner, R. A., Ingwall, J. S., Maroko, P. R., Davis, M. A., and Braunwald, E., 1983, *Proc. Natl. Acad. Sci. U.S.A.* **80**:5784.

Deutsch, E., Berger, M., Kussmaul, W. G., Hirshfeld, J. W., Herrmann, H. C., and Laskey, W. R., 1990, *Circulation* **82**:2044.

Finegan, B. A., Lopaschuk, G. D., Gandhi, M., and Clanachan, A. S., 1995, *Am. J. Physiol.* **269**:H1767.

Fitzpatrick, S. M., Hetherington, H. P., Behar, K. L., and Shulman, R. G., 1990, *J. Cereb. Blood Flow. Metab.* **10**:170.

Gross, G. J., and Auchampach, J. A., 1992, *Circ. Res.* **70**:223.

Hardy, C. J., Weiss, R. G., Bottomley, P. A., and Gerstenblith, G., 1991, *Am. Heart J.* **122**:795.

Heyndrickx, G. R., Millard, R. W., McRitchie, R. J., Maroko, P. R., and Vatner, S. F., 1975, *J. Clin. Invest.* **56**:978.

Hoekenga, D. E., Brainard, J. R., and Hutson, J. Y., 1988, *Circ. Res.* **62**:1065.

Kalil-Filho, R., Gerstenblith, G., Hansford, R. G., Chacko, V. P., Vandegaer, K. M., and Weiss, R. G., 1991, *J. Mol. Cell. Cardiol.* **23**:1467.

Kida, M., Fujiwara, H., Ishida, M., Kawai, C., Ohura, M., Miura, I., and Yabuuchi, Y., 1991, *Circulation* **84**:2495.

Kloner, R. A., Ganote, C. E., Walen, D. A., and Jennings, R. B., 1974, *Am. J. Pathol.* **74**:399.

Laster, S. B., Becker, L. C., Ambrosio, G., and Jacobus, W. E., 1989, *J. Mol. Cell. Cardiol.* **21**:419.

Lewandowski, E. D., and Johnston, D. L., 1990, *Am. J. Physiol.* **258**:H1357.

Lewandowski, E. D., Johnston, D. L., and Roberts, R., 1991, *Circ. Res.* **68**:578.

Liu, G. S., Thornton, J., Van Winkle, D. M., Stanley, A. W. H., Olsson, R. A., and Downey, J. M., 1991, *Circulation* **84**:350.

Mitchell, M. B., Meng, X., Ao, L., Brown, J. M., Harken, A. H., and Banerjee, A., 1995, *Circ. Res.* **76**:73.

Murry, C. E., Jennings, R. B., and Reimer, K. A., 1986, *Circulation* **74**:1124.

Murry, C. E., Richard, V. J., Reimer, K. A., and Jennings, R. B., 1990, *Circ. Res.* **66**:913.

Neely, J. R., and Kobayashi, K., 1979, *Circ. Res.* **44**:166.

Neely, J. R., and Morgan, H. E., 1974, *Annu. Rev. Physiol.* **36**:413.

Neubauer, S., Hamman, B. L., Perry, S. B., Bittl, J. A., and Ingwall, J. S., 1988, *Circ. Res.* **63**:1.

Neubauer, S., Krahe, T., Schindler, R., Horn, M., Hillenbrand, H., Entzeroth, C., Mader, H., Kromer, E. P., Riegger, G. A. J., Lackner, K., *et al.* 1992, *Circulation* **86**:1810.

Rahimtoola, S. H., 1985, *Circulation* **72**:V123.

Rehr, R. B., Tatum, J. L., Hirsch, J. I., Quint, R., and Clarke, G., 1989, *Radiology* **172**:53.

Robitaille, P. M. L., Rath, D. P., Skinner, T. E., Abduljalil, A. M., and Hamlin, R. L., 1993, *Magn. Reson. Med.* **30**:262.

Sako, E. Y., Kingsley-Hickman, P. B., From, A. H. L., Foker, J. E., and Ugurbil, K., 1988, *J. Biol. Chem.* **263**:10600.

Schaefer, S., Carr, L. J., Prussel, E., and Ramasamy, R., 1995, *Am. J. Physiol.* **268**:H935.

Speechly-Dick, M. E., Mocanu, M. M., and Yellon, D. M., 1994, *Circ. Res.* **75**:586.

Stahl, L. D., Weiss, H. R., and Becker, L. C., 1988, *Circulation* **77**:865.

Steenbergen, C., Perlman, M. E., London, R. E., and Murphy, E., 1993, *Circ. Res.* **72**:112.

Tillisch, J. H., Brunken, R., Marshall, R. C., Schwaiger, M., Mandelkern, M., Phelps, M. E., and Schelbert, H. R., 1986, *N. Engl. J. Med.* **314**:884.

Topol, E. J., Weiss, J. L., Guzman, P. A., Dorsey-Lima, S., Blanck, T. J. J., Humphrey, L. S., Baumgartner, W. A., Flaherty, J. T., and Reitz, B. A., 1984, *JACC* **4**:1123.

Weiss, R. G., Chacko, V. P., Glickson, J. D., and Gerstenblith, G., 1989, *Proc. Natl. Acad. Sci. U.S.A.* **86**:6426.

Weiss, R. G., Bottomley, P. A., Hardy, C. J., and Gerstenblith, G., 1990a, *N. Engl. J. Med.* **323**:1593.

Weiss, R. G., Gerstenblith, G., and Lakatta, E. G., 1990b, *J. Clin. Inv.* **85**:757.

Weiss, R. G., Gloth, S. T., Kalil-Filho, R., Chacko, V. P., Stern, M. D., and Gerstenblith, G., 1992, *Circ. Res.* **70**:392.

Weiss, R. G., Kalil-Filho, R., Herskowitz, A., Chacko, V. P., Litt, M., Stern, M. D., and Gerstenblith, G., 1993, *Circulation* **87**:270.

Weiss, R. G., Stern, M. D., Vandegaer, K. M., Chacko, V. P., de Albuquerque, C. P., and Gerstenblith, G., 1995, *Biochim. Biophys. Acta* **1243**:543.

Weiss, R. G., de Albuquerque, C. P., Vandegaer, K. M., Chacko, V. P., and Gerstenblith, G., 1996, *Circ. Res.* **79**:435.

Wolfe, C. L., Sievers, R. E., Visseren, F. L. J., and Donnelly, T. J., 1993, *Circulation* **87**:881.

Yabe, T., Mitsunami, K., Okada, M., Morikawa, S., Inubushi, T., and Kinoshita, M., 1994, *Circulation* **89**:1709.

Yabe, T., Mitsunami, K., Inubushi, T., and Kinoshita, M., 1995, *Circulation* **92**:15.

Zhang, J., Path, G., Chepuri, V., Xu, Y., Yoshiyama, M., Bache, R. J., From, A. H. L., and Ugurbil, K., 1993, *Magn. Reson. Med.* **30**:28.

6

Applications of ^{13}C Labeling to Studies of Human Brain Metabolism *In Vivo*

Graeme F. Mason

1. INTRODUCTION

The carbon isotope ^{13}C has proven useful for studies of metabolism in systems that range from cellular preparations (Shulman *et al.*, 1979; Sillerud *et al.*, 1981; Sillerud and Shulman, 1983; den Hollander *et al.*, 1986; Reibstein *et al.*, 1986; Sonnewald *et al.*, 1993; Westergaard *et al.*, 1994) to the human brain *in vivo* (Beckmann *et al.*, 1991; Gruetter *et al.*, 1994; Mason *et al.*, 1995). The usefulness of NMR detection of ^{13}C in the brain stems from a variety of sources. NMR spectroscopy permits the separate detection not only of specific chemicals, but of isotopic labeling and patterns of labeling among specific carbon positions in those chemicals. The detection is nondestructive and can penetrate the brain to any depth. The detection of the ^{13}C can be combined with the spectroscopic detection of total (unlabeled and ^{13}C-labeled) levels of metabolites like lactate, choline, the neuronal marker *N*-acetylaspartate, and other compounds. If an NMR spectrometer is equipped with gradients, anatomic imaging and in many cases functional imaging

Graeme F. Mason • Magnetic Resonance Center, Yale University, School of Medicine, 333 Cedar Street, P.O. Box 208043, New Haven, Connecticut 06520-8043.

Biological Magnetic Resonance, Volume 15: In Vivo Carbon-13 NMR, edited by L. J. Berliner and P.-M. L. Robitaille. Kluwer Academic / Plenum Publishers, New York, 1998.

can be performed in the same experimental session, permitting unambiguous correlation of results with anatomy.

^{13}C detection by NMR spectroscopy has many applications in the brain, including measurements of metabolic rates, relative rates of substrate utilization, and metabolic concentrations in health, in disease, and during pharmacological intervention. Since the early studies of cellular preparations (Shulman *et al.*, 1979; Sillerud *et al.*, 1981; Sillerud and Shulman, 1983; den Hollander *et al.*, 1986; Reibstein *et al.*, 1986) and perfused organs (Cohen *et al.*, 1980; Cohen and Shulman, 1980; Chance *et al.*, 1983; Bendall *et al.*, 1985), the methods have been applied to a variety of neurological applications. ^{13}C-labeling studies have been performed in cellular and subcellular neuronal and glial preparations (Petroff *et al.*, 1991; Portais *et al.*, 1991; Sonnewald *et al.*, 1991), brain slices (Badar-Goffer *et al.*, 1990, 1992), extracts of brain (Brainard *et al.*, 1989; Cerdán *et al.*, 1990; Shank *et al.*, 1993; Chapa *et al.*, 1995), animal brain *in vivo* (Behar *et al.*, 1986; Fitzpatrick *et al.*, 1990; Cerdán *et al.*, 1990; Mason *et al.*, 1992a), and the human brain *in vivo* (Beckmann *et al.*, 1991; Rothman *et al.*, 1992; Gruetter *et al.*, 1992, 1994; Chen *et al.*, 1994). The kinetics of blood-brain transport of glucose have been determined *in vivo* in rats (Mason *et al.*, 1992a) and humans in health (Gruetter *et al.*, 1992) and diabetes (Novotny *et al.*, 1993). The rates of the Krebs cycle, exchange between α-ketoglutarate and glutamate, and the cycling of glutamate and glutamine have been determined in rats (Fitzpatrick *et al.*, 1990; Mason *et al.*, 1992b; Martin *et al.*, 1995) and humans (Rothman *et al.*, 1992; Mason *et al.*, 1995). Studies of the metabolism of GABA have been performed during pharmacological treatment with anti-epileptic agents (Manor *et al.*, 1996) and thyroid-replacement therapy (Chapa *et al.*, 1995). Based on the success, variety, and progressive improvement of ^{13}C-labeling studies, one may project that the applications of ^{13}C-labeling and NMR spectroscopy to studies of brain metabolism *in vivo* will increase in the numbers of pathways, the metabolic perturbations studied, and the depth of detail that is examined in each experiment.

The majority of studies of brain *in vivo* have addressed the activity of the TCA cycle by measurements of labeling of glutamate, and a summary of the known sensitivities to measured and known parameters would be useful to investigators interested in pursuing such measurements in either humans or animals. The issues increase in importance as studies are performed to understand the regulation of brain metabolism in states relevant to human consciousness or disease, such as sensory stimulation (Hyder *et al.*, 1996a) and hyperammonemia (Sibson *et al.*, 1996). For this reason, this chapter addresses primarily the measurement of the TCA cycle rate in the brain.

The following section of this chapter describes how the TCA cycle rate is measured *in vivo*. There follows a detailed description of the use of mathematical modeling of ^{13}C-labeling time course data for the determination of metabolic flow rates. An explanation of the derivation of secondary parameters like oxygen consumption then precedes a discussion of sensitivity analyses of various assumptions and parameters in the model. A discussion of the effects of metabolic compartmentation on the interpretation of the labeling data follows. A description of the probable future of ^{13}C-labeling studies of brain *in vivo* ends this chapter.

2. MEASUREMENT OF THE TCA CYCLE RATE IN THE BRAIN

This section consists primarily of a general description of ^{13}C-labeling time course studies *in vivo* and the application of mathematical modeling analysis to interpret those studies. A brief discussion of ^{13}C detection methods currently applied *in vivo* for measurement of the TCA cycle rate follows.

2.1. A General Description

In studies *in vivo*, ^{13}C-labeled substrates are infused into the bloodstream, and the label appears in a variety of products at various positions in the carbon chains (Fig. 1). The appearance of the substrate and its products in the brain is detected using an NMR probe, two types of which are shown in Fig. 2. Surface coils are

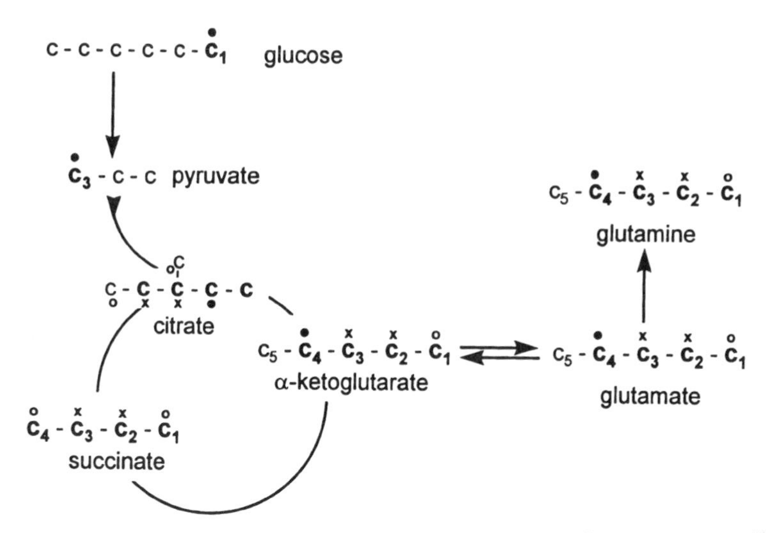

Figure 1. Schematic of ^{13}C labeling of glutamate from [1-^{13}C]glucose. ^{13}C label flows from ^{13}C-labeled glucose through glycolysis and into the TCA cycle, where it labels the C4 of α-ketoglutarate, glutamate, and glutamine (filled circles). The ^{13}C label continues through the cycle, labeling the symmetric molecule succinate at C2 and C3, and the C2 and C3 of α-ketoglutarate, glutamate, and glutamine on subsequent turns of the cycle (indicated by X). After the second turn of the cycle, the label also appears in the C1 of α-ketoglutarate, glutamate, and glutamine.

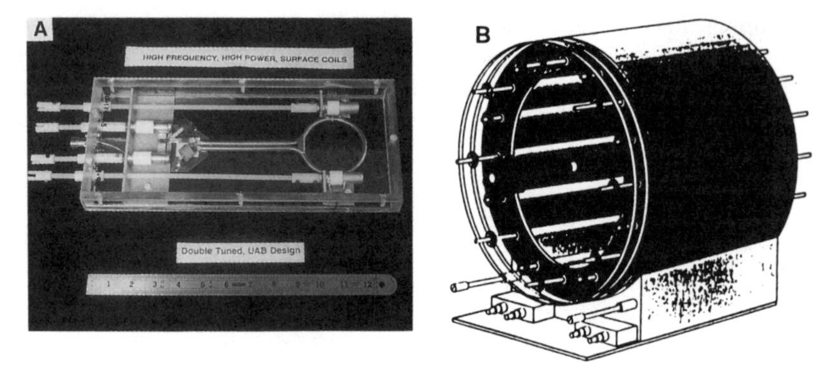

Figure 2. Two of types of radiofrequency coils (probes) used for NMR experiments; (A) surface coil, usually placed on the surface of the head or another body part; (B) resonant cavity coil (Vaughan *et al.*, 1994), inside which the sample (e.g., a head) is placed. Reproduced from Mason *et al.*, (1996a).

circuits in the form of flat loops that are placed against the head. Surface coils are sensitive, but limited to the detection of metabolites not much deeper than one radius of the loop. Because surface coils for studies of the brain are commonly 7–10 cm in diameter, the depth of detection is approximately 3.5–5 cm. In contrast,

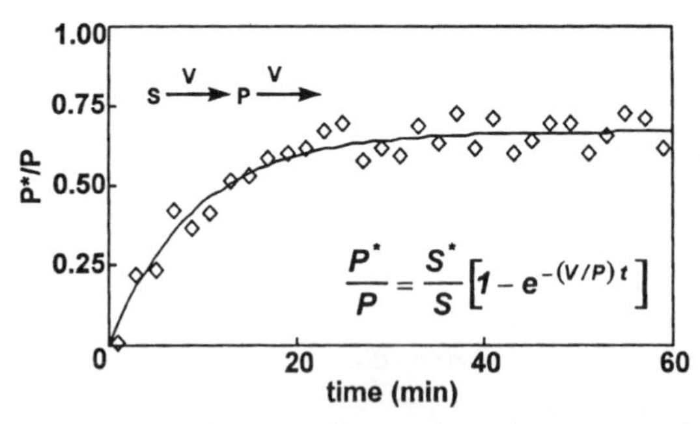

Figure 3. A demonstration of the behavior of ^{13}C flow from a substrate S to a product P. *S* and *P* represent the total (labeled + unlabeled) concentrations of the c substrate and product compounds, and S^* and P^* represent the labeled concentrations. The equations to describe the simple schematic shown in the figure are $dP/dt = V - V = 0$ and $dP^*/dt = (S^*/S)V - (P^*/P)V$. In this example , the ratio S^*/S was a step function that rose instantaneously from an initial value of 0 to a final value of 67%, and the concentration *P* was constant. An analytical expression for the fractional enrichment of the product *P* then has the form of the exponential equation in the figure. The equation can be fitted to the data by adjusting the exponential time constant *V*/*P*. If the concentration of *P* is known, then *V* can be determined. In reality, models may certainly be more complex, but the same principles of analysis of isotopic flow hold true.

volume coils surround the head and are used to study volumes anywhere in the brain. However, the sensitivity of volume coils is less than that of surface coils. In summary, the appropriate choice of detection device—surface or volume coil—depends on the required location and sensitivity of the measurement.

The NMR probe is used to detect the appearance of the ^{13}C label, and kinetic constants of transport and/or metabolism are derived. Mass and isotope balance equations are used to express the ^{13}C flow through the TCA cycle, using the principles shown in Fig. 3. In most ^{13}C-labeling studies of the brain, metabolites are in a metabolic (as opposed to *isotopic*) steady state, and most of the mass balance equations are therefore equal to zero. In the case of time-varying metabolite pools, one must know the time course of concentration in order to fit the ^{13}C-labeling data. Given an accurate set of model equations, the level of detail that can be obtained *in vivo* depends upon several factors:

Concentrations of Compounds. The NMR signal amplitude is directly proportional to the quantities of observed compounds in the sample. Successive NMR spectroscopic measurements are typically added to increase the signal-to-noise ratio, thereby allowing ^{13}C labeling to be measured by signal averaging. In studies of extracts, which yield stable solutions for studies, measurements over periods of many hours are feasible. However, experiments conducted *in vivo* have more stringent limitations imposed by pool sizes, rates of metabolism, or the length of time that an experimental preparation can be maintained in a certain physiologic state.

Spectral Resolution. Peaks in an NMR spectrum have widths that are a function of several parameters, among which is inhomogeneity of the static magnetic field. Inhomogeneity broadens spectral resonances and, in many cases, the widths of the peaks become comparable to the frequency separation between the peaks and prevent the separate detection of resonances of some compounds. Static field inhomogeneities are usually larger *in vivo* than in tissue extracts or other ex vivo preparations. Some nuclei, like ^{13}C, possess a large frequency dispersion compared to ^{1}H and yield greater detail despite lower signal-to-noise ratios.

Time for the System to Reach an Isotopic Steady State. Determination of the rates of metabolic pathways involves measurements that must be made kinetically, measured before the system reaches an isotopic steady state. For example, if ^{13}C is administered and its uptake into glutamate leads to an isotopic steady state after some time t, then each measurement in the time course must be acquired in a time significantly less than t. An analogous case for radioisotopic labeling studies would be one in which more time is required to obtain a tissue sample than is needed for complete accumulation of the isotope. In such cases, the rate cannot be determined beyond the determination of a lower limit.

2.2. Theoretical Basis of the Kinetic Modeling: Mass and Isotope Balance

The kinetic modeling sometimes appears daunting to those not regularly involved with mathematics. However, despite the presence of multiple simultaneous equations and descriptions of numerical solutions, the most important points for an investigator to understand are the very simple basis for the model, and these points are the concepts of mass and isotope balance (Fig. 3).

2.2.1. Mass Balance

Any active metabolic pool has carbon atoms flowing into and out of it, depositing and removing mass, and the relationship between the inflow and outflow can be described by a mass balance equation. A good example is the mass balance of brain pyruvate. For example, the brain pyruvate pool has carbon influx from glycolysis at a rate CMR_{gl} via the enzyme phosphoenol pyruvate dehydrogenase and from lactate at a rate V_{ldh} via lactate dehydrogenase. Because the lactate dehydrogenase reaction is readily reversible, it also provides a path for carbon efflux, as do pyruvate dehydrogenase and pyruvate carboxylase at the rates V_{pdh} and V_{pc}, respectively. Therefore, the rate of change in pyruvate concentration is

$$\frac{d[\text{Pyruvate}]}{dt} = \text{Influx} - \text{Efflux}$$

or

$$\frac{d[\text{Pyruvate}]}{dt} = (\text{CMR}_{gl} + V_{ldh}) - (V_{ldh} + V_{pdh} + V_{pc})$$

If the concentration of pyruvate does not change during a study, then

$$\frac{d[\text{Pyruvate}]}{dt} = 0.$$

However, there is no requirement that the concentration of any metabolite be constant for a metabolic labeling study to be interpretable by metabolic modeling; one need only characterize the way in which the compound changes during the study so that the mass balance portion of the model can be interpreted correctly. In fact, such is the case with the brain glucose pool in the first few minutes of a typical $[1\text{-}^{13}C]$glucose infusion.

2.2.2. Isotope balance

Because ^{13}C lacks significant isotope effects and is metabolized almost identically to ^{12}C (Attwood *et al.*, 1986; Melzer and Schmidt, 1987; Tipton and Cleland, 1988), during a ^{13}C-labeling study of the brain the mass flow just discussed continues as it would in the absence of ^{13}C label. Therefore, the mass balance equations are unaffected in the presence of ^{13}C. However, when label is supplied, a time-dependent fraction of the carbon mass flow is ^{13}C, and equations must be written to balance the fraction of the isotopically labeled flow. The rate of change in the concentration of ^{13}C-labeled pyruvate is

$$\frac{d[^{13}\text{C-Pyruvate}]}{dt} = [^{13}\text{C Influx}] - [^{13}\text{C Efflux}]$$

The rate of ^{13}C influx (first bracketed term) is composed of each individual rate of mass inflow multiplied by the fraction of the source that is labeled with ^{13}C, and the rate of ^{13}C efflux (second bracketed term) is composed of the sum of the rates of mass efflux multiplied by the fraction of the pyruvate that is labeled with ^{13}C:

$$\frac{d[^{13}\text{C-Pyruvate}]}{dt} = \left[\text{CMR}_{\text{gl}}\left(\frac{[^{13}\text{C-Pyruvate}]}{[\text{Pyruvate}]}\right) + V_{\text{ldh}}\left(\frac{[^{13}\text{C-Lactate}]}{[\text{Lactate}]}\right)\right] - \left[\left(\frac{[^{13}\text{C-Pyruvate}]}{[\text{Pyruvate}]}\right)(V_{\text{ldh}} + V_{\text{pdh}} + V_{\text{pc}})\right]$$

In cases of perfused organs or incubated cell cultures, investigators are able to supply an instantaneous dose of ^{13}C-labeled substrates. In many such cases, the differential equations can be solved as analytic solutions. Some of these ideal models are simple enough that time courses of labeling can be calculated using pocket calculators. However, in the brain *in vivo*, the fractional ^{13}C enrichment of substrates cannot be changed instantaneously, due to blood-brain substrate transport and finite mixing times of substrate pools in the blood. A variety of solutions is available for solving systems of differential equations, and investigators may either program their own or use software packages with such as MATLAB (The MathWorks, Inc., Natick, Massachusetts). What is most important is that an investigator be able to describe the relevant biochemical pathways in a way that can be translated into mass and isotopic balance equations.

2.3. Overview of Interpretation of the Data by Mathematical Modeling

Time courses of labeling data obtained *in vivo* are fitted with mathematical models in order to derive metabolic rates. To date, the parameters measured in the brain *in vivo* have been the TCA cycle rate (V_{tca}), the glutamate/glutamine carbon flow rate (V_{gln}), and the rate of exchange between α-ketoglutarate and glutamate (V_{x}) (Fig. 4). Briefly, the value of V_{gln} is determined by comparison of the time courses of glutamate and glutamine C4 labeling, the value of V_{x} is determined by comparison of the time courses of glutamate C4 and C3 labeling, and the value of V_{tca} is determined from the glutamate C4 labeling and the value of V_{x}.

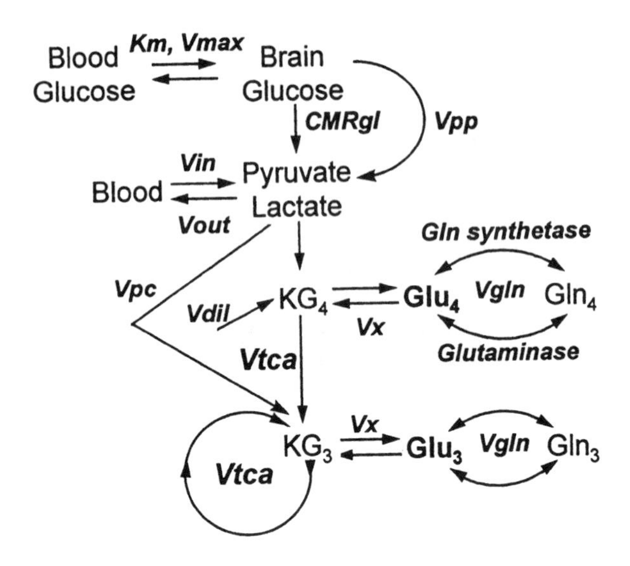

Figure 4. Schematic representation of the model. Glucose in the plasma (G_o) and the brain (G_i) exchange via the Michaelis–Menten kinetic parameters K_m (4.9 mM) and V_{max} ($3.6 \times CMR_{gl}$). Carbon flows at the rate $2CMR_{gl}$ (μmol min^{-1} g^{-1}) through the glycolytic intermediates, assumed to have negligible concentrations, and arrives at pyruvate; pyruvate and lactate are assumed to be in isotopic equilibrium and are therefore added to from a single pool [$L = 0.6$ μmol g^{-1} (Hanstock *et al.*, 1988)]. There is an efflux of lactate V_{out} [0.12 μmol min^{-1} g^{-1} (calculated from Knudsen *et al.*, 1991; Juhlin-Dannfelt, 1977)]. Unlabeled carbon enters the acetyl-CoA pool to be utilized at the rate V_{dil}. ^{13}C also enters the TCA cycle at this point and labels α-ketoglutarate and glutamate [9.1 μmol g^{-1} (Gruetter *et al.*, 1994)] at carbon 4, and the ^{13}C is exchanged between the mitochondrial and the cytosolic pools of each; these exchanges are reduced to a single exchange rate (V_x) between a single combined pool of cytosolic and mitochondrial glutamate (Glu_4) and one grouped pool of cytosolic and mitochondrial α-ketoglutarate [α-$KG_4 = 0.0183 \times Glu_4$ (Hawkins and Mans, 1983)]. Glutamine [4.1 μmol g^{-1} (Gruetter *et al.*, 1994)] is synthesized from glutamate at a rate V_{gln} (μmol min^{-1} g^{-1}). Aspartate is labeled from oxaloacetate C2 and C3 at a rate V_{xa}. The ^{13}C label continues through the cycle to KG_3 and Glu_3 and, on subsequent turns of the TCA cycle, one-half of the label returns to KG_3 and Glu_3.

The quantity V_{gln} is an important parameter, because the concentration of glutamine is significant compared to that of glutamate, and carbon flow between glutamate and glutamine can alter the kinetics of labeling of glutamate significantly. Such an alteration requires that the kinetic effects be known in order to derive an accurate value of the TCA cycle rate. In the conscious human, the rate of glutamine C4 labeling has been shown to be nearly as rapid as that of glutamate labeling (Mason *et al.*, 1995), while in the α-chloralose-anesthetized rat the rate of glutamate C4 labeling is slower (Martin *et al.*, 1995).

The quantity V_x is an important parameter, because the exchange between cytosolic and mitochondrial α-ketoglutarate and glutamate carries label from the

TCA cycle in the mitochondria to cytosolic glutamate, which forms the majority of glutamate. Because the cytosolic glutamate is more abundant than α-ketoglutarate or mitochondrial glutamate, the NMR experiments detect primarily the labeling of cytosolic glutamate. In order to carry the ^{13}C from mitochondrial α-ketoglutarate to cytosolic glutamate, one step is the movement across the mitochondrial membrane (Fig. 5). Another step is the movement of ^{13}C label from α-ketoglutarate to glutamate, because α-ketoglutarate is too dilute to be detected in the human brain *in vivo*. Investigators must rely on the multistep exchange of ^{13}C label from the mitochondria, where α-ketoglutarate is produced in the TCA cycle, to cytosolic glutamate, which forms the majority of the brain glutamate pool. If V_x is fast, the fractional labeling of glutamate will reflect the TCA cycle rate. If V_x is slow, glutamate labeling will not reflect V_{tca} as much as V_x.

When four pools exchange in the patter of mitochondrial and cytosolic α-ketoglutarate and glutamate (Fig. 5), and the method of measuring isotopic labeling does not distinguish between glutamate pools, the exchanges may be expressed as occurring through a single exchange reaction (see the exchange between KG_4 and Glu_4 in Fig. 4). The pools cannot be distinguished in the brain, because the concentration of α-ketoglutarate is too low to detect by NMR, and mitochondrial and cytosolic glutamate have the same chemical shift. Therefore, a ^{13}C-labeling experiment cannot be used to distinguish between fast mitochondrial/cytosolic exchange and fast α-ketoglutarate/glutamate exchange, and the four exchange reactions must be combined into one effective exchange whose rate is V_x (Mason *et al.*, 1992b). In the rat and in conscious, resting humans, the value of V_x has been shown to be much faster than the TCA cycle rate.

Once the value of V_x has been determined, it is possible to determine the value of V_{tca} from the glutamate C4 labeling time course. In studies of the brain reported thus far, the rate of V_x has been measured to be 50–150 times greater than the TCA cycle rate (Mason *et al.*, 1992b, 1995), validating the assumption that the glutamate C4 labeling time course can be used to determine V_{tca} under resting conditions.

The model shown in Fig. 4 is clearly simplified from the case of complex metabolic flows that exist in the brain. Therefore, sensitivity analyses have been performed to test the validity of experimental results and to direct future experiments to answer questions to which studies demonstrate high sensitivities. For

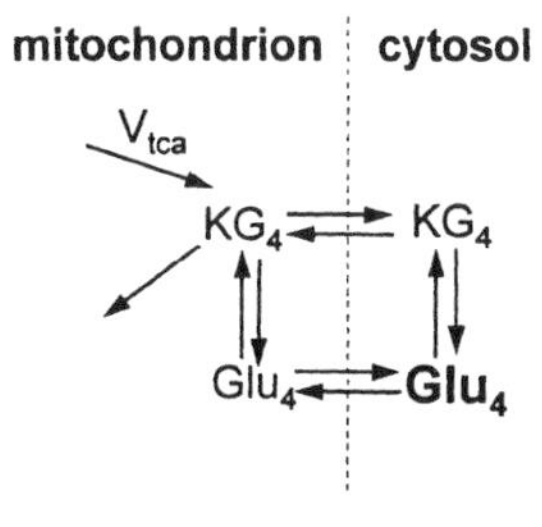

Figure 5. The four-way exchanges among cytosolic and mitochondrial glutamate and α-ketoglutarate. By simplification of the exchanges to the single exchange step shown between α-ketoglutarate and glutamate in Fig. 4, one can determine the value of an apparent rate-limiting step. The apparent limiting rate may be one of those shown in the four-way exchange, or it may result from the combined effects of more than one step.

example, one assumption that has been tested is the assertion that the rate of exchange between pyruvate and lactate is much faster than the rate of glycolysis, and another is that glycolytic intermediates are low enough in concentration so as to be neglected. Because the data analyses were based on the assumption of rapid exchange between lactate and pyruvate, and included also the assumption that glycolytic intermediates were too low in concentration to influence the results, sensitivity testing of those assumptions is a necessary part of the experimental analysis.

The fitting of experimentally obtained time courses and the testing of assumptions will be discussed in detail for [1-^{13}C]glucose measurements. Because ^{13}C-labeled acetate is also of interest for labeling studies in the brain and heart, measurements with acetate will be discussed briefly regarding the sensitivity of the acetate studies to the rates of the TCA cycle rate and exchange between α-ketoglutarate and glutamate.

2.4. Methods of Detection Currently in Use for Metabolic Studies of Brain *In Vivo*

Each technique of ^{13}C detection has limitations and advantages that are best considered in the planning stages of an experiment. Some issues to consider for the evaluation of each technique are the following: (1) the time allowed for the measurement of each time point, (2) the technical feasibility of a type of measurement in the laboratory, given the limitations of equipment, and (3) the information yielded by each particular technique. For example, the signal-to-noise ratio that can be obtained in a given time determines the time required for signal averaging for each time point, because each time point must have an adequate signal-to-noise ratio. In turn, the signal-to-noise ratio and the maximum rate of data collection place limits on the type of kinetic data that can be acquired with a given pulse sequence. The NMR spectroscopic techniques are categorized as direct or indirect, depending upon whether the observed signal is from ^{13}C or from ^{1}H that is chemically bonded to ^{13}C.

2.4.1. Indirect ^{13}C Detection

Methods have been developed to detect ^{1}H nuclei that are coupled to ^{13}C. ^{1}H detection of the ^{13}C nucleus is more sensitive than direct ^{13}C detection, allowing more rapid acquisition of spectra or acquisition of data from smaller volumes. The sensitivity increase arises from many factors. ^{1}H NMR is 16 times more sensitive, per atom, than ^{13}C NMR. The sensitivity also increases linearly with the number of hydrogen atoms bonded to the observed ^{13}C nucleus, and more than ^{1}H atom is often bonded to each ^{13}C. These and other factors such as T_1 and T_2 relaxation rates determine the precise improvement in sensitivity obtained by indirect detection of

^{13}C (Rothman *et al.*, 1987; Novotny *et al.*, 1990). An added advantage is that ^{1}H detection methods are easily combined with volume-localization techniques (Ordidge *et al.*, 1986).

Several methods of indirect detection are discussed briefly here. The first, proton-observed, or ^{1}H-observed/^{13}C-edited (POCE) NMR spectroscopy (Rothman *et al.*, 1985) requires pairs of acquisitions. One of each pair yields the total metabolite signal, while the difference between the pair yields only that fraction of metabolites that are labeled with ^{13}C. The method has the advantage of permitting the measurement of fractional enrichments and has been used to detect labeling of glutamate in rats (Fitzpatrick *et al.*, 1990) and humans (Rothman *et al.*, 1992; Chen *et al.*, 1994; Pan *et al.*, 1996; Mason *et al.*, 1996b). A second method is spectral editing using a single acquisition with the BISEP pulse (Garwood and Merkle, 1991). The BISEP approach has been used to observe the formation of [3-^{13}C]lactate in gliomas implanted in rats and holds the potential to measure glutamate labeling in the brain *in vivo*. Another method of indirect detection in a single acquisition is gradient-enhanced heteronuclear multiple-quantum coherence (HMQC), which has been used to detect uptake of ^{13}C-labeled glucose in the cat brain *in vivo* (van Zijl *et al.*, 1993). Due to excellent water suppression, the HMQC technique is ideal for the detection of the ^{1}H resonance of ^{13}C-labeled glucose whose resonance lies near that of water.

Despite the high sensitivity, ^{1}H NMR detection of ^{13}C has some disadvantages. Although ^{1}H NMR has greater sensitivity than ^{13}C detection, the sensitivity observed *in vivo* is often less than ideal due to broad ^{1}H line widths. In addition, ^{1}H NMR of metabolites almost always requires water suppression and, for experiments performed *in vivo*, lipid suppression is usually needed as well. The suppression of water and lipids carries stringent requirements of magnetic field homogeneity, and the requirements are difficult to meet in moving samples or systems with irregular geometry (e.g., beating heart). In contrast to the heart, the more stationary and nearly spherical geometry of the brain makes it an ideal organ for application of indirect detection techniques, for two reasons: (1) motion can be minimized by holding the head still, and (2) NMR-visible lipids are primarily of extracerebral origin and can be suppressed by a surface spoiler (Chen and Ackerman, 1989) and other localization techniques. In summary, while indirect detection of ^{13}C sacrifices spectral resolution, it yields a greater amount of signal per atom of ^{13}C and thereby permits much higher spatial resolution and measurements of the TCA cycle rate in small volumes such as the human visual cortex (Chen *et al.*, 1994) or the somato-sensory cortex of the rat (Hyder *et al.*, 1996a).

2.4.2. Direct ^{13}C Detection

Direct detection of ^{13}C takes advantage of the wide chemical shift dispersion of ^{13}C (~200 ppm) to obtain high spectral resolution. In conjunction with ^{1}H

decoupling to collapse the ^{13}C–^{1}H spin–spin splitting, ^{13}C NMR allows the resolution of many compounds whose resonances are difficult to distinguish in the ^{1}H spectrum. The high spectral resolution also permits the separate detection of ^{13}C isotopomers that are identified by their ^{13}C–^{13}C homonuclear *J*-coupling. Direct ^{13}C methods require no water suppression and so are generally much simpler than ^{1}H NMR experiments, especially those performed *in vivo*. However, ^{13}C NMR is much less difficult than is ^{1}H NMR, and the large chemical shift dispersion makes localization more difficult than is the case for ^{1}H NMR. Despite the success of time course measurements with direct ^{13}C detection, significant improvements over ^{13}C in sensitivity and temporal resolution are achievable (Rothman, 1987) and have been obtained with ^{1}H NMR methods of indirect detection, as discussed previously. One way to improve the sensitivity of direct ^{13}C detection is the transfer of polarization from the ^{1}H to ^{13}C, prior to detecting the ^{13}C (Davis *et al.*, 1995); the application of this technique to concentrated solutions of glucose shows promise for use in the brain (Gruetter *et al.*, 1996).

3. MATHEMATICAL MODELING: A DETAILED DISCUSSION

3.1. Description of Metabolic Flow

A detailed discussion of the model of Fig. 4 is now in order. [1-^{13}C]glucose is infused into the bloodstream using a shaped infusion protocol (DeFronzo *et al.*, 1979). Current applications normally use a time-varying infusion that raises the plasma glucose concentration and [1-^{13}C]enrichment within 5 minutes to 200 mg/dl and 60–67%, respectively, and maintains those levels for one or more hours, depending on the length of the experiment. The plasma glucose (G_o) exchanges with the brain glucose (G_i), and within a few minutes the fractional [1-^{13}C] enrichments of the two pools are at equilibrium (i.e., $G_o^*/G_o = G_i^*/G_i$). The glucose is phosphorylated by hexokinase to form glucose-6-phosphate. At equilibrium, the enrichment of G6P is the same as that of glucose. A small fraction (3–5%) of the G6P is passed to the pentose phosphate shunt at the rate V_{pp}, removing C1 and therefore any labeled carbons in the shunt. The rest of the G6P is metabolized further, yielding triose units. Because only the C1 of glucose was labeled, only half of the triose products are labeled, and the fractional enrichment at this stage is reduced by 50%. Due to rapid equilibrium of glycolytic intermediates, the several small pools behave kinetically as one larger pool (Mason *et al.*, 1992b), here called glyceraldehyde phosphate (GAP).

The triose units are converted to pyruvate, which exchanges via lactate dehydrogenase with lactate. The exchange is so rapid that the two pools behave kinetically as one which is equal to the sum of the two pools (L). At isotopic steady state, the fractional enrichment of pyruvate/lactate is equal to (G_i^*/G_i) (1–

V_{pp}/CMR_{gl}). About 10% of the pyruvate/lactate pool is channeled through pyruvate carboxylase to generate oxaloacetate (Berl *et al.*, 1962). Due to rapid scrambling of label in the TCA cycle, labeled oxaloacetate generated due to the pyruvate carboxylase flux will be labeled at C2 and C3. However, most of the pyruvate/lactate carbon enters the TCA cycle to form citrate, which is further oxidized to form C4-labeled α-ketoglutarate in the first pass of the ^{13}C through the TCA cycle at carbon 4. Through exchange, the C4-labeled α-ketoglutarate is converted to C4-labeled glutamate, which in turn is converted to C4-labeled glutamine. Meanwhile, the carbon from α-ketoglutarate C4 continues through the TCA cycle, labeling oxaloacetate at carbons 2 and 3 on the second and subsequent passes. Oxaloacetate is in exchange with aspartate (V_{xa}), analogously to α-ketoglutarate/glutamate exchange. Oxaloacetate combines with acetyl-CoA and forms citrate which is labeled such that its products are α-ketoglutarate, glutamate, and glutamine labeled at carbons 2 and 3. To summarize the labeling of glutamate following metabolism of [1-^{13}C]glucose, the C4 of glutamate is labeled on the first pass of ^{13}C through the TCA cycle, and the C3 of glutamate is labeled on subsequent passes, with a small exception made for pyruvate carboxylase activity.

3.2. Equations and Procedures Used to Determine Model Parameters

The mathematical interpretation is given by the mass and isotopic balance equations in Table 1. Now, the procedures used to derive the metabolic rates from ^{13}C-labeling data obtained *in vivo* are explained. The NMR-observed enrichment of glutamate C4 is analyzed to determine the rate of ^{13}C labeling of glutamate C4.

Table 1
Differential Equations for the Model in Fig. 4[a]

$$dG_i/dt = V_{max}G_o/(K_m + G_o) - V_{max}G_i/(K_m + G_i) - CMR_{gl}$$
$$dL/dt = dKG/dt = dGlu/dt = dG\ln/dt = 0$$
$$dG_i/dt = V_{max}G_o^*/(K_m + G_o) - V_{max}G_i^*/(K_m + G_i) - CMR_{gl}(G_i^*/G_i)$$
$$dL^*/dt = CMR_{gl}(G_i^*/G_i) - (2CMR_{gl} + V_{out})(L^*/L)$$
$$dKG_4^*/dt = (2CMR_{gl} - V_{out})(L^*/L) - (V_{tca} + V_x)(KG_4^*/KG) + V_x(Glu_4^*/Glu)$$
$$dGlu_4^*/dt = V_x(KG_4^*/KG) - (V_x + V_{g\ln})(Glu_4^*/Glu) + V_{g\ln}(G\ln_4^*/G\ln)$$
$$dG\ln_4^*/dt = V_{g\ln}(Glu_4^*/Glu) - (V_{g\ln}(G\ln_4^*/G\ln)$$
$$dOAA^*/dt = \tfrac{1}{2}V_{tca}(KG_4^*/KG) + \tfrac{1}{2}V_{tca}(KG_3^*/KG) - (V_{tca} + V_{xa})(OAA^*/OAA) + V_{xa}(Asp^*/Asp)$$
$$dAsp^*/dt = V_{xa}(OAA^*/OAA) - (V_{xa}(Asp^*/Asp)$$
$$dkG_3^*/dt = V_{tca}(OAA^*/OAA) - (V_{tca} + V_x(KG_3^*/KG) + V_x(Glu_3^*/Glu)$$
$$dGlu_3^*/dt = (V_x(KG_3^*/KG) - (V_x + V_{gln}(Glu_3^*/Glu) + V_{g\ln}(G\ln_3^*/G\ln)$$
$$dG\ln_3^*/dt = (V_{g\ln}(Glu_3^*/Glu) - (V_{g\ln}(G\ln_3^*/G\ln)$$

[a] An asterisk (*) denotes ^{13}C labeling. *L* represents lactate and pyruvate. G_i and G_o represent brain and blood glucose. Subscript numbers are the carbon position that is labeled with ^{13}C.

This rate is called the glutamate turnover rate, or V_{gt}. The relationship between V_{gt} and the metabolic rates V_x and V_{tca} is determined by the divergence of flow at α-ketoglutarate. The total flow of carbon out of the α-ketoglutarate pool is V_{tca} + V_x, and the fraction that flows from α-ketoglutarate to glutamate is described by $V_x/(V_{tca} + V_x)$. Hence, the rate of isotopic flow into glutamate is $V_{gt} = V_{tca}V_x/(V_{tca} + V_x)$, which can be expressed as

$$V_{tca} = V_{gt}\left(\frac{1 + V_x/V_{tca}}{V_x/V_{tca}}\right) \tag{1}$$

First, the value of V_{gt} is measured by comparison of the plasma glucose1 and glutamate C4 enrichment time courses. Then, the values of V_{tca} and V_x/V_{tca} from V_{gt}, by iteration of the ratio V_x/V_{tca} in order to fit the model to the glutamate C3 labeling data, using Eq. (1) to calculate a new value of V_{tca} for each value of V_{gt}, thereby maintaining the measured rate of ^{13}C-label flow to glutamate C4.

3.3. Measurement of Glutamate Turnover Rate (V_{gt})

Data obtained in the human brain *in vivo* (Gruetter *et al.*, 1994; Mason *et al.*, 1995) will be examined next. First, the rate V_{gt} is determined by modification of the model of Fig. 4 to the form shown in Fig. 6. The portion of the model farther down the pathway from glutamate C4 has no effect on the calculation of V_{gt}. When a least-squares fit of the metabolic model depicted in Fig. 6 is made to the glutamate C4 labeling data, a rate of 0.73 μmol min^{-1} g^{-1} results for V_{gt} in the human brain. The fit from one subject is shown in Fig. 7A.

3.4. Determination of V_x/V_{tca} and V_{tca} from V_{gt} and C4/C3 Labeling Time Courses

The value of V_x/V_{tca} is determined next, by fitting the model of Fig. 4 to the glutamate C3 data. In order for the ^{13}C to flow from glucose C1 to glutamate C3, the label must first pass through α-ketoglutarate C4 (with a small exception of pyruvate carboxylase activity, which will be addressed later), where the label is diverted to glutamate C4. The rest of the ^{13}C continues through the TCA cycle, labeling α-ketoglutarate C2 and C3. As the label arrives at α-ketoglutarate C3, the fraction of label $(V_x/V_{tca})/(1 + V_x/V_{tca})$ flows into glutamate C3. In order to illustrate the effects of V_x on the labeling time courses, two extreme cases are discussed with regard to their effects on the labeling relationship between glutamate C4 and C3. The two extremes are $V_x/V_{tca} \ll 1$ and $V_x/V_{tca} \gg 1$.

When $V_x/V_{tca} \gg 1$, nearly all of the ^{13}C passes to glutamate C4 and is trapped there for some time before it continues through the TCA cycle and is incorporated into α-ketoglutarate C3. The large pool of glutamate (~9 μmol g^{-1} in the human

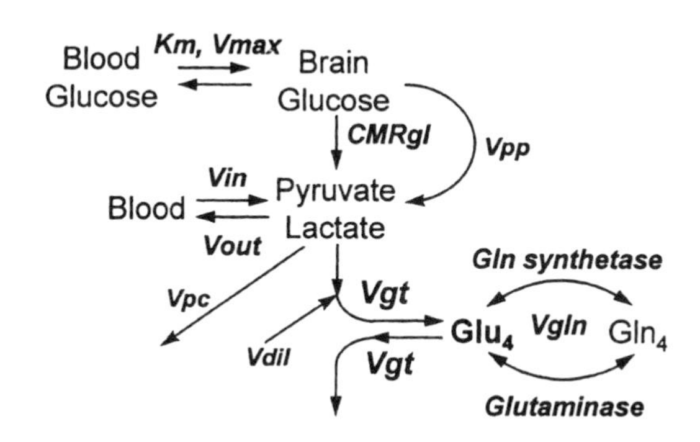

Figure 6. In order to determine the rate of carbon flow through Glu (V_{gt}), the model of Fig. 4 was simplified to assume flow directly into and out of Glu C4. At the stage of the analysis in which V_{gt} was determined, no assumptions were made regarding V_{tca} and V_x. The purpose was solely to obtain an estimate of the value of V_{gt}.

brain; Perry *et al.*, 1987; Gruetter *et al.*, 1994) significantly delays the appearance of ^{13}C label in α-ketoglutarate C3 and glutamate C3. As a result, in the case of $V_x/V_{tca} >> 1$, the fractional enrichment of the carbon flow into glutamate C3 will lag the flow into glutamate C4 considerably. In contrast, when $V_x/V_{tca} << 1$, very little of the ^{13}C will enter glutamate C4 at first, most of it continuing through the TCA cycle to label α-ketoglutarate at C3. Therefore, in the case of $V_x/V_{tca} << 1$, the time courses of α-ketoglutarate C4 and C3 positions will reach the same steady-state fractional enrichment almost simultaneously, and the observed time courses of glutamate C4 and C3 labeling will be similar.

Figure 7B shows a fitted time course from the same study shown in Fig. 7A. Least-squares fitting for the human brain yielded a value of $V_x/V_{tca} = 77$. In the present case of $V_x/V_{tca} >> 1$, Eq. (1) shows that $V_{gt} \approx V_{tca}$. Using the value of V_x/V_{tca} and Eq. (1), it was possible to use V_{gt} to determine the value of V_{tca}. Therefore, for the human brain, V_{tca} was 0.73 μmol min^{-1} g^{-1}.

3.5. Determination of Glutamate/Glutamine Carbon Flow (V_{gln})

If, in addition to the time course of glutamate C4, the time course of labeling of glutamine C4 is observed, it is possible to determine the rate of glutamate/glutamine carbon flow (V_{gln}). Because glutamine is synthesized from glutamate, when a single-compartment model is assumed, the glutamate C4 labeling time course can be used as an input function to model the glutamine C4 labeling time course, deriving the value of V_{gln}. In the human brain, glutamine C4 was resolved from glutamate C4, and the value of V_{gln} was 0.47 μmol min^{-1} g^{-1} (Fig. 7C).

While the value of V_{gln} was determined assuming a single-compartment model, a body of evidence exists to support a two-compartment model of brain glutamine metabolism. The evidence and the important implications for modeling that are raised by compartmentation are discussed later in this chapter.

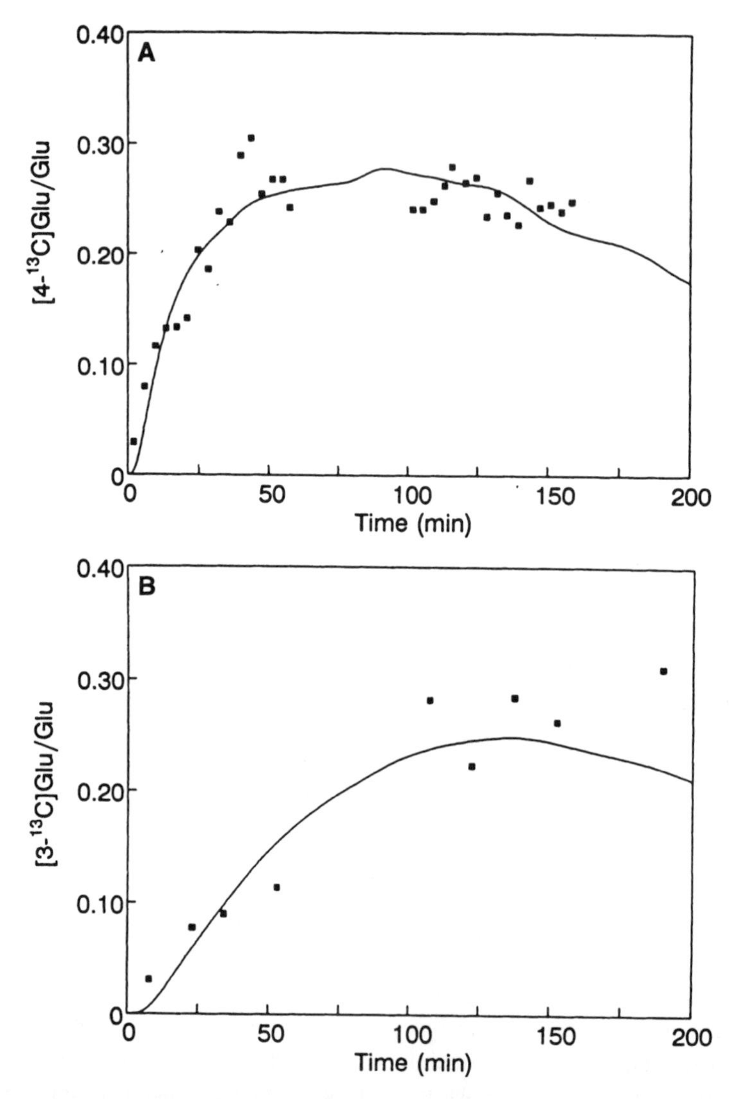

Figure 7. Fits of the model to ^{13}C-labeling time courses. The time courses of (A) glutamate C4 and (B) glutamate C3 are smoother than (C), the fit to glutamine C4. The fit shows bumps and waves which reflect the behavior of the glutamate C4 time course that was used as input. The fits to glutamate C4 and C3 are smoother than the fit to glutamine C4, because bumps and waves in the plasma glucose enrichment used as the input function for glutamate C4 and C3 were smoothed by intervening metabolite pools; the smoothing was a consequence of the damping nature of the first-order kinetic equations in Table 1 (from Mason *et al.*, 1995).

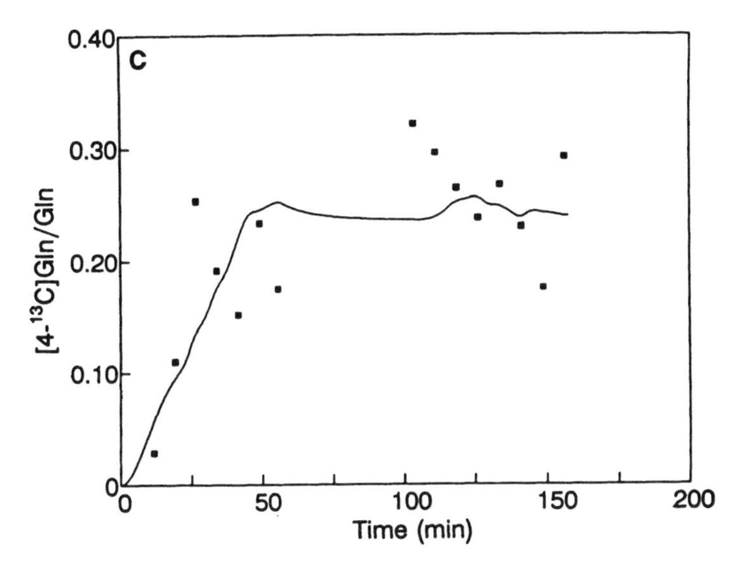

Figure 7. Continued

4. DERIVATION OF SECONDARY PARAMETERS

The ^{13}C-labeling studies described in this chapter yield as their primary parameters the rates of the TCA cycle (V_{tca}), exchange between α-ketoglutarate and glutamate (V_x), and carbon flow between glutamate and glutamine (V_{gln}). The quantities V_{tca}, V_x, and V_{gln} have just been discussed. In addition, the rates of brain glucose consumption (CMR_{gl}) and oxygen consumption ($CMRO_2$) are also derived as secondary parameters.

The rates V_{tca}, CMR_{gl}, and $CMRO_2$ are linked stoichiometrically. Through glycolysis, each hexose unit of glucose creates two triose units that can be metabolized through pyruvate dehydrogenase and enter the TCA cycle. If glucose and only glucose were oxidized, and if the oxidation were complete, the relationships of the rates would be $CMR_{gl} = 1/2\ V_{tca}$ and $CMRO_2 = 3V_{tca}$. However, because there is a dilution of labeled glutamate, it is apparent that there is a nonglucose contribution to the TCA cycle rate, and to determine the appropriate relationships of mass balance it is important to estimate the sources of the dilution.

In humans after a 12-h fast, a dilution of 14% was observed (Gruetter *et al.*, 1994). The diluting flux is probably due to a combination of the pentose phosphate shunt activity, lactate influx from the blood, and utilization of ketone bodies. The rate of protein degradation is extremely slow in rats [0.3% of V_{tca}, calculated from Berl *et al.* (1961)] and is probably slower in humans. In order to calculate CMR_{gl}, it is necessary to quantify carbon sources for the TCA cycle, so as to determine the appropriate linkage between CMR_{gl} and V_{tca}.

4.1. Estimation of Carbon Sources for the TCA Cycle

4.1.1. Pentose Phosphate Activity

The pentose phosphate shunt cleaves the labeled C1 from glucose-6-phosphate and releases it as $^{13}CO_2$. The activity in the rat brain is 1–3% of CMR_{gl} (Gaitonde *et al.*, 1983; Hawkins *et al.*, 1985) and forms a small, constant source of dilution throughout infusions of [1-^{13}C]glucose. Because the dilution is both low and constant, it has insignificant impact on the CMR_{gl}–V_{tca} linkage and the determination of the TCA cycle rate.

4.1.2. Transport of Unlabeled Lactate and Pyruvate Across the Blood–Brain Barrier

During infusions of [1-^{13}C]glucose, the labeling of lactate at C2 or C3 in the blood could lead to labeling of glutamate C4 in the brain. However, blood lactate has been observed to remain insignificantly labeled throughout the infusion, and lactate influx from the blood therefore acts as a diluting influence. In the human brain, the rate of unidirectional influx of lactate has been reported to be 0.08 μmol min^{-1} g^{-1} (Knudsen *et al.*, 1991). No information about pyruvate influx has been reported, but it is assumed that the human is similar to the rat in that pyruvate influx is much less than the rate of lactate influx (Pardridge and Oldendorf, 1977) and can be neglected. However, the diluting contribution from lactate is substantial.

A net efflux of lactate of ~10% of CMR_{gl} occurs in both rats [calculated from Pardridge and Oldendorf (1977) and Hawkins and Mans (1983)] and humans (Gottstein *et al.*, 1963; Juhlin-Dannfelt, 1977). Because the efflux component of lactate removes both labeled and unlabeled carbon from the brain, it does not dilute the fractional enrichment of lactate. However, the presence of a net efflux does alter slightly the mass balance equation for brain lactate, and its value should be included if it is known.

4.1.3. Utilization of Ketone Bodies

The utilization of ketone bodies in humans, calculated from the arterial-difference methods of Juhlin-Dannfelt (1977), was 0.86 and 4.00% of CMR_{gl}, respectively. Because each unit of ketone body produces 2 units of acetyl-CoA, the total rate of isotopic dilution for the dilution due to ketone bodies is 9.8% of CMR_{gl} in the human brain, contributing to a dilution that enters the system at the step of the incorporation of acetyl-CoA into citrate.

4.2. Determination of CMR_{gl}

CMR_{gl} was calculated using mass balance at lactate and acetyl-CoA, including carbon efflux to the TCA cycle flux, influx from glycolysis, pentose phosphate activity, and net influx of lactate from the blood. In the visual cortex of conscious humans, CMR_{gl} is 0.37 μmol min^{-1} g^{-1}, which is similar to results observed with PET in human grey matter (Heiss *et al.*, 1984).

4.3. Brain Oxygen Consumption

Because most of the TCA cycle carbon is supplied by glucose and lactate, a quick estimate of brain oxygen consumption can be made by multiplying the TCA cycle rate by 3. However, a more accurate determination should be made when possible, including the contributions of ketone bodies to the carbon in the cycle.

The rate of carbon flow through the TCA cycle is $V_{tca} = V_g + V_\beta + V_{ac}$, where V_g is the rate of flow glucose-derived carbon, also interpreted as any carbon that flows through pyruvate, V_β is the rate of utilization of β-hydroxybutyrate, and V_{ac} is the rate of utilization of acetoacetate. Due to differences in the reducing equivalents generated by each species, a more accurate estimate of oxygen utilization is not $3V_{tca}$, but $CMRO_2 = 3V_g + 2.25V_\beta + 2V_{ac}$. In the conscious human, a value of 2.14 μmol min^{-1} g^{-1} (5.07 ml 100 g^{-1} min^{-1}) was found, and the value is similar to the value of 5.41 ml 100g^{-1} min^{-1} observed with PET (Lebrun-Grandié *et al.*, 1983).

5. QUANTITATIVE ANALYSES OF SENSITIVITIES

The sensitivities of the fitted parameters V_x and V_{tca} to a variety of parameters whose values were assumed or neglected in the model were investigated numerically. It was shown that the effects of the assumed values on the calculated values of V_{tca} and V_x were small and did not significantly affect the rates derived by the model. The following parameters were evaluated: the time required for glucose isotopic enrichment, the rate of exchange between pyruvate and lactate, the labeling kinetics of metabolic intermediates in glycolysis and the TCA cycle, the rates of influx and efflux of lactate between the brain and blood, the pyruvate carboxylase rate, and ^{13}C recycling. The influence of the flow of carbon between glutamate and glutamine will be discussed separately, due to its special importance in health and disease and the ramifications of metabolic compartmentation of glutamate and glutamine metabolism in the brain.

5.1. Glucose Turnover Time

A short period of time is required for the brain glucose pool to reach isotopic equilibrium with the plasma glucose, and this equilibration time is characterized by the turnover half-time $\Delta t_{1/2}$, which is the difference in the times required by the brain and blood to reach half of their isotopic steady states. The plasma glucose half-time is controlled primarily by the infusion protocol for [1-^{13}C]glucose, while the brain glucose half-time is controlled by a combination of the brain glucose pool size, the blood–brain glucose transport parameters, and the rate of glucose utilization. For humans fasted 12–18 hours, the plasma glucose was raised from 4.5 mM to 12 mM in five minutes. The rapid enrichment and rise in concentration of the plasma glucose caused the brain glucose pool to be enriched with a $\Delta t_{1/2}$ of only 1.5 min. The sensitivity of the TCA cycle to $\Delta t_{1/2}$ arises due to a small but finite effect on the labeling kinetics of glutamate C4. Because the brain glucose pool has a concentration of 1–2 μmol g^{-1} wet weight, it is one of two pools that exerts the largest influence over the time of arrival of ^{13}C label at glutamate C4. Lactate is the other major pool, and its kinetics will be discussed later. If brain glucose were labeled instantaneously by infinitely fast exchange with the plasma glucose pool, then glucose would not cause any delay in the arrival of ^{13}C at glutamate C4. In contrast, if brain glucose were transported slowly across the blood–brain barrier, the 1–2 μmol g^{-1} pool would be slow to label and would slow down the rate of labeling of glutamate C4, causing the TCA cycle to appear slower than it actually is. In reality, blood–brain glucose transport is rapid but finite, having been reported to maintain a unidirectional rate of twofold or more of the rate of brain glucose utilization in the rat and human brain (Cremer *et al.*, 1981; Cremer *et al.*, 1983; Hawkins *et al.*, 1983; Otsuka *et al.*, 1991; Mason *et al.*, 1992a; Gruetter *et al.*, 1992).

To test the sensitivity of the accuracy of the assumed glucose transport kinetics on the labeling of glutamate C4, time courses of glutamate C4 labeling were

Table 2
Errors Induced in the Value of V_{tca} by Uncertainty in the Time Required to Enrich the Intracerebral Glucose Pool[a]

$t_{1/2}$ (min)	Error (%)
0.5	8.8
1.0	5.0
1.4	**0.0**
1.5	−0.6
2.0	−4.5
2.3	−6.6

[a]Glutamate C4 ^{13}C-enrichment time courses were simulated using the expected values of V_{tca} for the human brain (0.73 μmol min^{-1} g^{-1}) with ranges of $t_{1/2}$ values. The simulated time courses were used as input for the model's fitting procedure to determine the error in V_{tca}. The boldfaced value is the value of $\Delta t_{1/2}$ used in the model.

simulated using values of $\Delta t_{1/2}$ between 0 and 2 min. The simulated time courses were then used as input for the model's fitting procedure, and values of V_{tca} were calculated (Table 2). The effect of intracerebral glucose turnover on the calculated value of V_{tca} was small, with underestimates of less than 5% for deviations of ±30 s in the human brain.

5.2. Exchange of Lactate and Pyruvate

Because lactate is not in the direct path of glycolysis and because, like glucose, its concentration (0.6 μmol g^{-1}) and potential impact on glutamate labeling is large compared to the intermediates of glycolysis, the effects of uncertainties in its labeling kinetics are discussed here. Lactate and pyruvate are represented as a single pool in the metabolic model, due to the rapid rate of exchange through lactate dehydrogenase (V_{ldh}). If V_{ldh} is slow, then little of the label in pyruvate will flow to lactate at any given instant, and most of the label will pass out of the very small pyruvate pool ($\leq$0.1 μmol g^{-1}) and into the TCA cycle without significant delay. In contrast, if V_{ldh} is much faster than the rate of glycolysis, then most of the label that enters pyruvate will be forced to flow to the relatively large pool of lactate and be delayed there before returning to pyruvate and flowing into the TCA cycle, slowing the rate of labeling of glutamate C4.

In order to test the sensitivity of the calculation of V_{tca} to the value of V_{ldh}, time courses of glutamate C4 labeling were simulated for values of V_{ldh}/V_{tca} between 0 and ∞. As shown in Table 3, when $V_{ldh}/V_{tca} > 2$, the derived value of V_{tca} deviates from the assumed value of ∞ by less than 0.01%; V_{ldh}/V_{tca} has been reported to be ~100 in the rat brain (Lowry and Passonneau, 1964; Balázs, 1970). If a similar exchange rate can be assumed for the human brain, then the exchange between pyruvate and lactate is fast enough that the two pools can be considered to form a single kinetic pool, for the purposes of deriving the TCA cycle rate.

5.3. Effects of Metabolic Intermediates and Aspartate

There was uncertainty in the concentrations of the small pools of metabolic intermediates, as well as in the effective pool size of aspartate. Because the metabolic intermediates affect primarily the glutamate C4 labeling, while the presence of aspartate affects the glutamate C3 labeling, these two questions were addressed separately.

In addressing the issue of metabolic intermediates, a simplification of the model was to neglect small metabolic pools like glyceraldehyde 3-phosphate and citrate. To evaluate the effect of neglecting the low concentrations of such intermediates on the calculation of V_{tca}, glutamate C4 labeling time courses were simulated assuming a range of concentrations for the total intermediate pool size. Because most if not all metabolites have higher concentrations in the rat brain than in the

Table 3
Errors Induced in the Value of V_{tca} by Slowness in Lactate Dehydrogenase Activity[a]

V_{ldh}/V_{tca}	Error (%)
0.5	3.2
1.0	1.2
2.0	0.0
∞	**0.0**

[a]Glutamate C4 ^{13}C-enrichment time courses were simulated using the expected values of V_{tca} for the human brain (0.73 μmol min^{-1} g^{-1}) with ranges of values for V_{ldh}/V_{tca}. The simulated time courses were used as input for the model's fitting procedure to determine the error in V_{tca}. The boldfaced value is the value used in the modeling.

human brain, the sum of the glycolytic intermediates in the rat brain (0.3 μmol g^{-1}; Hawkins and Mans, 1983) was chosen as the upper limit. The simulated labeling time courses were then used as input for the model's fitting procedure. As shown in Table 4, the value of V_{tca} that was derived from the simulated time courses without intermediates was overestimated by less than 2.5%. This test indicates that the neglect of the intermediates has a negligible effect on the determination of the TCA cycle rate.

Having determined the sensitivity of the model to uncertainties in metabolic intermediates, the effect of the aspartate pool size on the glutamate C3 labeling time course was evaluated in order to test the sensitivity of V_x, which was the rate of

Table 4
Errors Induced in the Value of V_{tca} by Variations in the Concentrations of Metabolic Intermediates[a]

[intermediates]	Error (%)
0.0	0.0
0.1	−0.9
0.2	−1.8
0.3	−2.5

[a]Glutamate C4 ^{13}C-enrichment time courses were simulated using the expected values of V_{tca} for the human brain (0.73 μmol min^{-1} g^{-1}) with ranges of values for the total concentration of metabolic intermediates like citrate, glyceraldehyde 3-phosphate, etc. The simulated time courses were used as input for the model's fitting procedure to determine the error in V_{tca}.. The boldfaced value was the value used for the modeling analysis.

exchange between α-ketoglutarate and glutamate. The sensitivity of V_x to aspartate was tested by fitting the actual glutamate C3 labeling data, looking at the two extremes of complete inclusion and total neglect of aspartate. When aspartate was included, there was a difference of about 50% in V_x. However, because $V_x/V_{tca} >> 1$, the 50% change in V_x caused only a 1% change in V_{tca}.

5.4. Rates of Influx and Efflux of Lactate and Pyruvate

Influx and efflux of pyruvate can affect V_{tca} either by altering the fractional [1-^{13}C]enrichment of the carbon flowing into the TCA cycle or by changing the coupling between brain glycolysis and the TCA cycle. Assuming that the human resembles the data in rats, lactate and pyruvate are labeled insignificantly during infusions of [1-^{13}C]glucose (Fitzpatrick *et al.*, 1990). Therefore, lactate and pyruvate remain completely unlabeled throughout each study, and the fractional ^{13}C enrichments of cerebral pyruvate and glutamate are diluted by the same constant fraction, resulting in no effect of unidirectional influx and efflux of lactate and pyruvate on the calculation of V_{tca}.

While the unidirectional fluxes individually do not affect the calculation of V_{tca}, there is a small *net* efflux of lactate and pyruvate from the brain due to a concentration gradient between the blood and the brain, mediated by passive transport (Gottstein *et al.*, 1963; Pardridge and Oldendorf, 1977; Juhlin-Dannfelt, 1977). However, the net efflux from the human brain is only about 10% of the TCA cycle rate and exerts a negligible effect on the determination of V_{tca}.

5.5. Exchange Between α-Ketoglutarate and Glutamate

The rate of carbon flow through glutamate (V_{gt}) is a function of both the TCA cycle rate and the rate of exchange (V_x) between α-ketoglutarate and glutamate. That is, V_{gt} results from both flow into α-ketoglutarate according to TCA cycle activity and flow via exchange between α-ketoglutarate and glutamate. Therefore, in order to determine V_{tca} from V_{gt}, one must account for the sensitivity of V_{tca} to uncertainties in V_x. Examining Eq. (1), the uncertainty depends a great deal on the value of V_x/V_{tca}. For $V_x/V_{tca} = 100$, a 20% uncertainty in V_x/V_{tca} represents an uncertainty in V_{tca} of only ±0.2%, while for $V_x/V_{tca} = 1$, a 20% uncertainty in V_x/V_{tca} represents an uncertainty of –17% to 25%. Because V_x/V_{tca} was shown to be 77 in the human brain, V_{tca} is relatively insensitive to large variations in V_x/V_{tca}.

5.6. Pyruvate Carboxylase

The flux through pyruvate carboxylase (V_{pc}) affects V_{tca}, CMR$_{gl}$, and V_x in different ways. First of all the impact of V_{pc} on V_x will be discussed. The pyruvate carboxylase reaction carries label from C3-labeled pyruvate to the C3 of oxaloactate. By reversible reactions through malate and fumarate, at least some of the label is also placed in the C2 position, by scrambling. The ^{13}C from the C2 and C3 of oxaloacetate is delivered to the C2 and C3 of α-ketoglutarate and glutamate, thereby increasing the rate of C3 labeling of glutamate. Because the time course of C3 labeling of glutamate relative to that of C4 is used to determine the value of V_x, uncertainty in V_{pc} may influence the calculated value of V_x.

Using an infusion of [1-^{13}C]glucose in the brain increases the rate of glutamate C3 labeling. Remembering the qualitative arguments that increased lag between glutamate C4 and C3 signified faster values of V_{tca}, one can examine qualitatively the consequences were the model to include incorrect values of V_{pc}. When fitting V_x, the assumed value of V_{pc} is used, and V_x is adjusted to make the calculated lag conform to the observed data. If V_{pc} is overestimated, then for a given value of V_x, the time course of glutamate C3 labeling will rise too quickly in the calculation, and the model must increase V_x in order to reduce the calculated time course of glutamate C3. Therefore, with infusions of [1-^{13}C]glucose, overestimates in V_{pc} lead to overestimates of V_x. In contrast, if V_{pc} is underestimated, then for a given value of V_x, the calculated glutamate C3 labeling will increase too slowly, and the model must decrease V_x in order to increase the calculated time course of glutamate C3 labeling. Therefore, with infusions of [1-^{13}C]glucose, underestimates in V_{pc} lead to underestimates in V_x.

Simulations were performed for ^{13}C-labeling results from [1-^{13}C]glucose infusion studies of the human brain. In the analysis, V_{pc} was neglected, and the value of V_x/V_{tca} was reported to be 77. By the reasoning just described, the neglect of V_{pc} means that the already large value of V_x/V_{tca} was actually an *underestimate*. In summary, when V_{pc} is included in the analysis of the labeling in the human brain the already fast V_x is shown to be even faster.

In contrast, experiments performed with [2-^{13}C]acetate show *dilution* of glutamate C3 and C2 by pyruvate carboxylase. In the case of dilution by pyruvate carboxylase flux, one finds that underestimation of the pyruvate carboxylase flux leads to an overestimation of V_x, which is the opposite effect as that seen with [1-^{13}C]glucose. It also follows that overestimation of V_{pc} yields an underestimation of V_x when [2-^{13}C]acetate is used. Therefore, it is necessary to make an accurate assessment of the contribution of pyruvate carboxylase to the TCA cycle in order to determine V_x. If such an assessment is made by isotopomer analysis (Malloy *et al.*, 1990), it is necessary to continue the study until an isotopic steady state is achieved, so as not to underestimate the flux through pyruvate carboxylase.

An additional, smaller effect of uncertainty in V_{pc} is the alteration of the coupling between glycolysis and the flux through pyruvate dehydrogenase, which in the present model is equal to the TCA cycle flux V_{tca}. Because triose units for pyruvate carboxylase activity are supplied by metabolism of additional glucose units, the inclusion of V_{pc} forces the rate of glycolysis to be increased in order to maintain mass balance at pyruvate/lactate. Increasing the rate of glycolysis de-

Table 5
Errors Induced in the Value of V_{tca} by Variations in the Value of V_{pc}/V_{tca}[a]

V_{pc}/V_{tca}	Error (%)
0.0	0.0
0.1	2.0
0.2	3.6
0.3	5.1
0.4	6.3
0.5	7.5

[a]Glutamate C4 ^{13}C-enrichment time courses were simulated using the expected values of V_{tca} for the human brain (0.73 μmol min^{-1} g^{-1}) with ranges of values for the flux through pyruvate carboxylase (V_{pc}). The simulated time courses were used as input for the model's fitting procedure to determine the error in V_{tca}. The boldfaced value was the value used for the modeling analysis.

creases the isotopic equilibration time of brain glucose, lactate, and other metabolic intermediates, which in turn speeds the arrival of ^{13}C at glutamate C4. Table 5 shows errors induced in V_{tca} by underestimation of V_{pc} during an infusion of [1-^{13}C]glucose. The value of V_{pc} has a small but negligible influence on the rate of appearance of label at glutamate C4, and therefore practically no influence on the calculated V_{tca}.

5.7. Pentose Phosphate Shunt

In the basal condition, the pentose phosphate shunt has been reported to be 1–3% of the glycolytic flux in the adult rat brain (Gaitonde *et al.*, 1983; Hawkins *et al.*, 1985). The shunt removes intracerebral glucose from the glycolytic pathway and the [1-^{13}C] label is lost as $^{13}CO_2$. Therefore, the pentose phosphate pathway dilutes the glutamate ^{13}C enrichment by a constant fraction and does not affect the determination of V_{tca}. Assuming that the pentose phosphate shunt has similarly low activity in the human brain, the dilution is constant and small compared to the flux through the TCA cycle. Thus, uncertainty in the rate of the shunt exerts a negligible effect on the determination of the TCA cycle rate.

5.8. Glucose Label Scrambling

The recycling of isotopically labeled carbon atoms in glucose due to systemic metabolism and resynthesis results in the redistribution of the ^{13}C in the glucose molecule (Katz and Rognstad, 1976). Like the C1 of glucose, the C6 of glucose passes through the C3 of pyruvate, so recycling of ^{13}C from C1 to C6 of glucose contributes to the labeling of glutamate C4 on the first pass of the label through the TCA cycle. In contrast, recycling to the C2 and C5 of glucose adds to the labeling of glutamate C3 in the first pass of the cycle.

The effects of recycling are addressed in two ways. The first approach is to minimize the actual impact of recycling on the experimental results, by careful experimental procedure, while the second approach is to quantify the recycling that actually occurred.

The infusion protocol was designed to raise the plasma glucose concentration rapidly and maintain it elevated, which increased insulin levels and reduced production of glucose by the liver. The subjects also fasted 12–18 hours in order to lower the total available pool of endogenous glucose, which further increased liver uptake of glucose instead of production during the infusion.

While the protocol minimized recycling, prudence dictates that any residual recycling be measured. The total label recycled to C2–C6 in the human was found to be less than 0.5% in all subjects, even after 31/2 hours of infusion (Mason *et al.*, 1995). Even if all of that 0.5% were shunted to the C6 of glucose, there would be no detectable effect on the calculation of V_{tca}.

Recycling of label into the C2 and C5 of glucose has no effect on the time course of labeling of glutamate C4 and therefore does not alter the determination of V_{gt}. However, the C2 and C5 of glucose do appear at the C3 of α-ketoglutarate, glutamate, and glutamine. Therefore, recycling of ^{13}C to the C2 and C5 of glucose would increase the rate of labeling of glutamate C3, reducing the lag between the C4 and C3 of glutamate and thereby causing V_x to appear *slower* than was determined without consideration of recycling. For the case of the human study, in which all of the label in C2–C6 was less than 0.5%, the extreme case in which all of the 0.5% could have appeared in the C2 of glucose would mean that the value of V_x was underestimated by <1%.

5.9. Glutamate/Glutamine Carbon Flow

The rate of carbon flow between glutamate and glutamine exerts an important influence on the labeling kinetics of glutamate C4. The conversion of glutamate to glutamine via glutamine synthetase does not affect the kinetics of glutamate C4 labeling. However, the rate of conversion of glutamine into glutamate through glutaminase does influence the calculated value of V_{tca}, because the unlabeled carbon flowing into glutamate from glutamine reduces the rate of fractional labeling of glutamate C4. In the conscious human, the C4 labeling of glutamine C4 was shown to be nearly identical to that of glutamate C4, indicating rapid communication of the pools of glutamate and glutamine. Table 6 shows the sensitivities of the TCA cycle rate to the rate of carbon cycling between glutamate and glutamine in the human brain over a span that covers the 95% confidence interval of the

Table 6
Errors Induced in the Value of V_{tca} by Uncertainty in the Rate of Glutamate/Glutamine Carbon Cycling[a]

V_{gln} (μmol min^{-1} g^{-1})	Error (%)
0.05	21.6
0.08	18.0
0.12	14.0
0.16	10.9
0.21	8.3
0.31	3.9
0.41	1.2
0.47	0.0
2.05	−6.4
4.10	−7.3
6.58	−7.6

[a]Glutamate C4 ^{13}C-enrichment time courses were simulated using the expected values of V_{tca} for the human brain (0.73 μmol min^{-1} g^{-1}) with ranges of values for V_{gln}. The simulated time courses were used as input for the model's fitting procedure to determine the error in V_{tca}. The boldfaced value was used for the modeling analysis.

uncertainty distribution of V_{gln}. Over most of the interval, the value of V_{tca} was varied by <10%.

5.10. Summary of Sensitivity Analysis

The use of the time course of glutamate C4 in order to measure TCA cycle activity in the brain provides a robust, noninvasive method. While the analysis contains a variety of assumptions, it has been possible with quantitative modeling to place limits on the effects of the assumptions. The effects of the assumed values were found to be negligible within, and in many cases beyond, limits of uncertainty that have been published for those assumed parameters.

6. METABOLIC COMPARTMENTATION

6.1. Neuronal Activity and Compartmentation

Evidence exists to support a model with two metabolic compartments in the brain (Cooper and Plum, 1987). One compartment contains a large pool of glutamate, and the other contains a small, rapidly metabolized pool. Intracysternal injection of [^{14}C]glutamate and [^{14}C]glutamine into anesthetized monkeys showed that glutamine was more highly labeled than glutamate, which means that a small pool of glutamate must supply label to glutamine in the anesthetized monkey. Cooper *et al.* (1979, 1988) infused $^{13}NH_4^+$ into conscious rats and found that the α-amino nitrogen of glutamine was more rapidly labeled than that of glutamate. Different isotopomer labeling patterns have been observed in the brains of anesthetized rates following infusions of [1,2-^{13}C]acetate (Cerdán *et al.*, 1990). Also in anesthetized rats, brain glutamate and glutamine were labeled, and their C4 fractional enrichments differed from each other after infusions of either [U-^{13}C]β-hydroxybutyrate or [1,2-^{13}C]glucose (Künnecke *et al.*, 1993). [1-^{13}C]Glucose perfusions of guinea pig brain slices caused labeling of glutamate and GABA without labeling of glutamine (Badar-Goffer *et al.*, 1990), while in the presence of depolarizing levels of KCl, low fractions of glutamine were labeled (Badar-Goffer *et al.*, 1992). All of the studies with slow rates of glutamine labeling used brain slices or anesthetized animals.

Two points become significant in the isotopic labeling data presented thus far. One is that glutamate and glutamine can in some circumstances differ in at least part of their substrate choice for the generation of glutamate. The other is that in many preparations, glutamate and glutamine are labeled to different extents, demonstrating a significant lack of mixing of the pools. Were mixing thorough, the different enrichments would not be apparent. Glutamine C4 labeling was higher in depolarized than nondepolarized brain slices. The C4 of glutamine was labeled still more in intact, anesthetized animals, and the final labeling was the same as glutamate C4 labeling in conscious humans, with only a slight lag of glutamine behind glutamate. Therefore, one can infer that a certain degree of normal physiological activity causes mixing of these pools whose precursor glutamate appears to be labeled differently. Such mixing may be necessary for complete labeling of the glutamate pool. Furthermore, a readily available mode of mixing that would be consistent with proximity to normal physiological function would be cycling of glutamate and glutamine between neurons and glia. Neuronal-glial cycling of glutamate and glutamine would be in agreement with enzymatic studies that have localized glutaminase and glutamine synthetase in neurons and glia, respectively (Erecinska and Silver, 1990).

The findings of compartmentation are being extended to GABA as well, with studies of isotopomer analysis (Chapa *et al.*, 1995) and a combination of measurements *in vivo* and *ex vivo* in which GABA C2 was found to be enriched more completely than glutamate C4 (Manor *et al.*, 1996).

6.2. Effects of Glutamate/Glutamine Compartmentation on Measured V_{tca}

The ^{13}C-labeling data measured *in vivo* do not distinguish metabolic compartments, presumably due to the completeness of the mixing of the pools of glutamate and glutamine. However, modeling can be used to determine the potential effects of compartmentation upon the interpretation of the data. The time course of ^{13}C-labeled glutamate probably reflects the large pool overwhelmingly, because the small pool has been estimated to be only 2–11% of the total glutamate concentration (Berl *et al.*, 1961; Cooper *et al.*, 1988), with a turnover time of only ~1 min (Berl *et al.*, 1961). The time course of glutamine C4 lags that of glutamate C4 only slightly, so the data are consistent with three two-compartment possibilities: (1) glutamine formed rapidly from a large pool of precursor glutamate, (2) glutamine synthesized from a small pool of glutamate that is rapidly metabolized, or (3) glutamine formed from both compartments (Fig. 8). The data are consistent with all three hypotheses, but an exchange analysis similar to that of Mason *et al.* (1992b) shows that for the rapid labeling of glutamine, the value of V_{tca} derived from the present data varies negligibly among the three cases. The observed time course of glutamine C4 closely follows that of glutamate C4, so glutamine turns over almost as if it were part of the large pool of glutamate; the [1-^{13}C]glucose labels both pools approximately as if they composed a single pool. A two-compartment analysis of the human NMR data yields a total value of V_{tca} that is the sum of the rates of glutamate C4 labeling by [1-^{13}C]glucose and glutamine C4 labeling by the small glutamate pool. Since the small-pool V_{tca} must be at least as fast as the rate of glutamine C4 labeling by the small pool, the combined pool value of V_{tca} would be a minimum estimate. However, in cases in which glutamine labeling is slow or reduced in the system studied, such as in brain slices or anesthetized animals, compartmentation may play an important role in the analysis of glutamate C4 time course data.

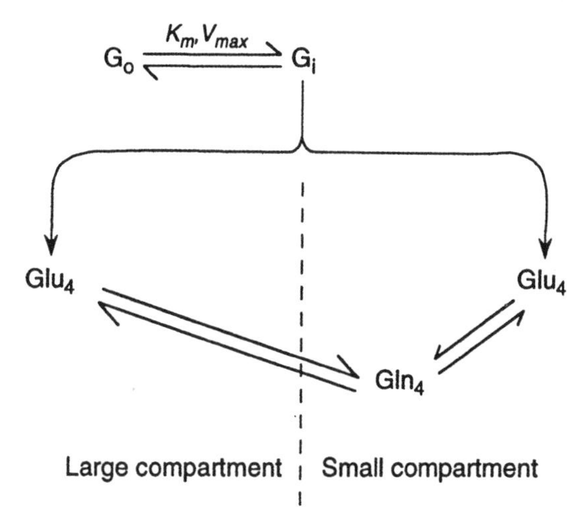

Figure 8. Schematic of a two-compartment model for labeling of glutamate and glutamine. For the case in which labeling of glutamine C4 and of glutamate C4 occurs at nearly the same rate, this scheme is mathematically equivalent to the simpler scheme of Fig. 1.

The primary effect of compartmental analysis is on the estimated value of V_{gln}. Using the assumption of thorough mixing to form the one-compartment model, the value of V_{gln} was found to be 0.47 μmol min^{-1} g^{-1} in four subjects. Assuming no mixing, such that glutamine C4 is labeled solely from a single, rapidly labeled pool of glutamate, a minimum value of V_{gln} in the same subjects was 0.19 μmol min^{-1} g^{-1}. Because some degree of mixing must occur due to neuronal-glial cycling through normal physiologic activity (Erecinska and Silver, 1990; Sibson *et al.*, 1997, 1998), the true value of V_{gln} must lie between 0.19 and 0.47 μmol min^{-1} g^{-1}.

7. FUTURE OF ^{13}C-LABELING STUDIES OF THE BRAIN *IN VIVO*

The future of applications of ^{13}C-labeling studies *in vivo* is affected by several developments that impact the sensitivity, flexibility, and cost of the studies. The effects particularly influence studies of the human brain. Recent improvements have been increases in static magnetic field strength, the introduction of volume coils, and the rapidly decreasing cost of ^{13}C-labeled glucose.

The sensitivity has been increased by the recent application of whole-body NMR imaging systems of 4-T and higher for ^{13}C-labeling studies of the human brain (Pan *et al.*, 1996; Mason *et al.*, 1996b). The high sensitivity of spectroscopic studies at 4 T has permitted the use of volume coils rather than surface coils, thereby introducing the flexibility to measure volumes in locations not limited by the limited spatial coverage of surface coils. The costs of ^{13}C-glucose has fallen severalfold in the past decade, and as an increasing number of sites uses ^{13}C-labeled compounds, the cost is expected to continue to fall.

The improvements are expected to continue, as both animal- and human-size magnets increase in strength. As the costs of ^{13}C-labeled compounds fall, investigators should benefit from doubly labeled [1,6-^{13}C]glucose, which will double the sensitivity of studies that now use [1-^{13}C]glucose. Doubly labeled glucose would improve the feasibility of the use of ^{13}C-labeling studies at field strengths of 1.5 T. Furthermore, glucose labeled at alternative positions, such as C2 and/or C4, will allow the measurement of the activity of pyruvate carboxylase activity as well.

A recent development has been that of rapid spectroscopic imaging (Twieg, 1989; Duyn and Moonen, 1993). Spectroscopic imaging of ^{13}C-labeling would bring NMR advantages normally associated with PET, while retaining the chemical specificity that is a quality of NMR spectroscopy. Indeed, initial spectroscopic images of ^{13}C-labeling have already been made (Inubushi *et al.*, 1993; van Zijl *et al.*, 1993), although with a time resolution that still precludes time course studies.

An additional avenue of anticipated productivity takes advantage of the multiple abilities of NMR, combining methods to form integrated approaches to questions. For example, functional MRI has been combined with both single (Hyder *et al.*, 1996a) and double-volume (Hyder *et al.*, 1996b) measurements of ^{13}C-labeling of glutamate in order to measure the TCA cycle rate in the rat brain *in vivo* during sensory stimulation. Quantitative imaging of tissue type has been combined with measurements of the TCA cycle rate in the human brain in volumes of predominantly grey matter (Mason *et al.*, 1996b).

Many clinical applications are waiting for investigation. While a study of lactate labeling in stroke has already been performed (Rothman *et al.*, 1991), many other disorders of the brain pose interesting metabolic questions. One example is hepatic encephalopathy, shown by ^{13}C and ^{15}N NMR spectroscopy and other methods to have altered metabolism of glutamate and glutamine (Hawkins *et al.*, 1973; Fitzpatrick *et al.*, 1989; Kreis *et al.*, 1992; Kanamori *et al.*, 1993; for a review, see Cooper and Plum, 1987). Recent NMR studies have focused on glutamate/glutamine cycling of carbon, which may be tied to neuronal activity (Martin *et al.*, 1995; Sibson *et al.*, 1996). Disorders in which normal neuronal function and communication is disrupted, such as Alzheimer's disease, multiple sclerosis, and stroke, may provide insight into the role of carbon cycling between these two large cerebral metabolite pools.

Driven by the availability of ^{13}C-labeling data is the development of quantitative modeling for experiment design and analysis, using approaches such as those discussed in this chapter. ^{13}C-labeling studies can be analyzed to obtain metabolic rates using quantitative modeling, as V_{gln}, V_x, and V_{tca} were discussed in this chapter. Models and experimental protocols can be tested for sensitivity to hypotheses or uncertainties in assumed parameters or model characteristics, as was the case for glucose transport, metabolic compartmentation, and many other characteristics discussed in this chapter. The effects of uncertainties in some measured parameters upon other parameters can be tested, as was the case here for the effects of variability in V_x and V_{gln} on the measured value of V_{tca}.

In conclusion, technology is improving the sensitivity for the detection of ^{13}C, compounds labeled with ^{13}C compounds are becoming cheap enough to be used by an increasing number of laboratories for the investigation of brain metabolism in humans and animals, and approaches to the interpretation of labeling data *in vivo* are becoming increasingly quantitative. The future should bring sensitive measurements that yield accurate metabolic rates and precise, quantitative evaluation of hypotheses.

ACKNOWLEDGMENT. The author is supported by grants from the Stanley Foundation and the National Institutes of Health (R01 NS34813, R01 NS37527, P30 MH30929, and P01 HD32573).

REFERENCES

Attwood, P. V., Tipton, P. A., and Cleland, W. W., 1986, *Biochemistry* **25**:8197.

Badar-Goffer, R. S., Bachelard, H. S., and Morris, P. G., 1990, *Biochem. J.* **266**:133.

Badar-Goffer, R. S., Ben-Yoseph, O., Bachelard, H. S., and Morris, P. G., 1992, *Biochem. J.* **282**:225.

Balázs, R., 1970, Carbohydrate Metabolism, in *Handbook of Neurochemistry*, Vol. 3 (A. Lajtha, ed.), Plenum Press, New York, pp. 1–36.

Beckmann, N., Turkalj, I., Seelig, J., and Keller, U., 1991, *Biochemistry* **30**:6362.

Behar, K. L., Petroff, O. A. C., Prichard, J. W., Alger, J. R., and Shulman, R. G., 1986, *Magn. Reson. Med.* **3**:911.

Bendall, M. R., den Hollander, J. A., Arias-Mendoza, F., Rothman, D. L., Behar, K. L., and Shulman, R. G., 1985, *Magn. Reson. Med.* **2**:56.

Berl, S., Takagaki, G., Clarke, D. D., and Waelsch, H., 1962 *J. Biol. Chem.* **237**:2570.

Brainard, J. R., Kyner, E., and Rosenberg, G. A., 1989, *J. Neurochem.* **53**:1285.

Cerdán, S., Künnecke, B., and Seelig, J., 1990, *J. Biol. Chem.* **265**:12916.

Chance, E. M., Seeholzer, S. H., Kobayashi, K., and Williamson, J. R., 1983, *J. Biol. Chem.* **258**:13785.

Chapa, F., Künnecke, B., Calvo, R., Escobar del Rey, F., Morreale de Escobar, G., and Cerdán, S., 1995, *Endocrinology* **136**:296.

Chen, W., and Ackerman, J. J. H., 1989, *Nucl. Magn. Reson. Biomed.* **1**:205.

Chen, W., Novotny, E. J., Boulware, S. D., Rothman, D. L., Mason, G. F., Zhu, Z.-H., Blamire, A., Prichard, J. W., and Shulman, R. G., 1994, *Proc. Soc. Magn. Reson Med., 13th Annual Meeting*, p. 63.

Cohen, S. M., and Shulman, R. G., 1980, *Philos. Trans. R. Soc. London, Ser. B:Biological Sciences* **289**:407.

Cohen, S. M., Shulman, R. G., Williamson, J. R., and McLaughlin, A. C., 1980, *Adv. Exp. Med. Biol.* **132**:419.

Cooper, A. J. L., and Plum, F., 1987, *Physiol. Rev.* **67**:440.

Cooper, A. J. L., Lai, C. K., and Gelbard, A. S., 1988, Ammonia and energy metabolism in normal and hyperammonemic rat brain, in *The Biochemical Pathology of Astrocytes* (M. D. Norenberg, L. Hertz, and A. Schousboe, eds.), Liss, New York, pp. 419–434.

Cooper, A. J. L., McDonald, J. M., Gelbard, A. S., Gledhill, R. F., and Duffy, T. E., 1979, *J. Biol. Chem.* **254**:4982.

Cremer, J. E., Ray, D. E., Gurcharan, S. S., and Cunningham, V. J., 1981, *Brain Res.* **221**:331.

Cremer, J. E., Cunningham, V. J., and Seville, M. P., 1983, *J. Cereb. Blood Flow Metab.* **3**:291.

Davis, D., Artemov, D., Eleff, S., and van Zijl, P. C. M., 1995, *Proc. Soc. Magn. Reson, 3rd Meeting*, p. 524.

DeFronzo, R. A., Tobin, J. E., and Andres, R., 1979, *Am. J. Physiol.* **237**, E214.

Duyn, J. H., and Moonen, C. T. W., 1993, *Magn. Reson. Med.* **30**:409.

Erecinska, M., , and Silver, I. A., 1990, *Prog. Neurobiol.* **35**:245.

Fitzpatrick, S. M., Hetherington, H. P., Behar, K. L., and Shulman, R. G., 1989, *J. Neurochem.* **52**:741.

Fitzpatrick, S. M., Hetherington, H. P., Behar, K. L., and Shulman R. G., 1990, *J. Cereb. Blood Flow Metab.* **10**:170.

Gaitonde, M. K., Evison, E., and Evans, G. M., 1983, *J. Neurochem.* **41**:1252.

Garwood, M., and Merkle, H., 1991, *J. Magn. Reson.* **94**:180.

Gottstein, U., Bernsmeier, A., and Sedlmeyer, I., 1963, *Klin Wochenscht.* **41**:943.

Gruetter, R., Novotny, E. J., Boulware, S. D., Rothman, D. L., Mason, G. F., Shulman, G.I., Shulman, R. G., and Tamborlane, W. V., 1992, *Proc. Natl. Acad. Sci. U.S.A.* **89**:1109.

Gruetter, R., Novotny, E. J., Boulware, S. D., Mason, G. F., Rothman, D. L., Shulman, G. I., Prichard, J. W., and Shulman, R. G., 1994, *J. Neurochem.* **63**:1377.

Gruetter, R., Adriany, G., Merkle, H., and Andersen, P. M., 1996, *Proc. Int. Soc. Magn. Reson. Med., 4th Meeting*, p. 380.

Hanstock, C. C., Rothman, D. L., Prichard, J. W., Jue, T., and Shulman, R. G., 1988, *Proc. Natl. Acad. Sci. U.S.A.* **85**:1821.

Hawkins, R. A., and Mans, A. M., 1983, Intermediary metabolism of carbohydrates and other fuels, in *Handbook of Neurochemistry*, Vol. 3 (A. Lajtha, ed.), Plenum Press, New York, pp. 259–294.
Hawkins, R. A., Miller, A. L., Nielsen, R. C., and Veech, R. L., 1973, *Biochem. J.* **134**:1001.
Hawkins, R. A., Mans, A. M., Davis, D. W., Hibbard, L. S., and Lu, D. M., 1983, *J. Neurochem.* **40**:1013.
Hawkins, R. A., Mans, A. M., Davis, D. W., Viña, J. R., and Hibbard, L. S., 1985, *Am. J. Physiol.* **248**:C170.
Heiss, W.-D., Pawlik, ., Herholz, K., Wagner, R., Göldner, H., and Weinhard, K., 1984, *J. Cereb. Blood Flow Metab.* **4**:212.
den Hollander, J. A., Ugurbil, K., and Shulman, R. G., 1986, *Biochemistry* **25**:212.
Hyder, F. H., Chase, J. R., Behar, K. L., Mason, G. F., Siddeek, M., Rothman, D. L., and Shulman, R. G., 1996a, *Proc. Natl. Acad. Sci. U.S.A.* **93**:7612.
Hyder, F. H., Rangarajan, A., Behar, K. L., Duncan, J., Rothman, D. L., and Shulman, R. G., 1996*b* *Proc. Int. Soc. Magn. Reson. Med., 4th Meeting*, p. 102.
Inubushi, T., Morikawa, S., Kito, K., Kimura, R., and Handa, J., 1993, *Proc. Int. Soc. Magn. Reson. Med., 12th Meeting*, p. 1508.
Juhlin-Dannfelt, A., 1977, *Scand. J. Clin. Lab. Invest.* **37**:443.
Kanamori, K., Parivar, F., and Ross, B. D., 1993, *NMR Biomed.* **6**:21.
Katz, J., and Rognstad, R., 1976, *Curr. Top. Cell. Regul.* **10**:237.
Knudsen, G. M., Paulson, O. B., and Hertz, M. M., 1991, *J. Cereb. Blood Flow Metab.* **11**:581.
Kreis, R., Ross, B. D., Farrow, N. A., and Ackerman, Z., 1992, *Radiology* **182**:19.
Künnecke, B., Cerdán, S., and Seelig, J., 1993, *NMR Biomed.* **6**:264.
Lajtha, A., and Waelsch, H., 1961, *J. Neurochem.* **7**:186.
Lebrun-Grandié, P., Baron, J.-C., Soussaline, F., Loch'h, C., Sastre, J., and Bousser, M.-G., 1983, *Arch. Neurol.* **40**:230.
Lowry, O. H., and Passonneau, J. V., 1964, *J. Biol. Chem.* **239**:31.
Malloy, C. R., Sherry, A. D., and Jeffrey, F. M. H., 1990, *Am. J. Physiol.* **259**:H987.
Manor, D., Rothman, D. L., Mason, G. F., Hyder, F., Petroff, O. A. C., and Behar, K. L., 1996, *Neurochem. Res.* **21**:1031.
Martin, M. A., Mason, G. F., Behar, K. L., and Shulman, R. G., 1995, *Proc. Soc. Magn. Reson., 3rd Meeting*, p. 1783.
Mason, G. F., Behar, K. L., Rothman, D. L., and Shulman, R. G., 1992a, *J. Cereb. Blood Flow Metab.* **12**:448.
Mason, G. F., Rothman, D. L., Behar, K. L., and Shulman, R. G., 1992b, *J. Cereb. Blood Flow Metab.* **12**:434.
Mason, G. F., Gruetter, R., Rothman, D. L., Behar, K. L., Shulman, R. G., and Novotny, E. J., 1995, *J. Cereb. Blood Flow Metab.* **15**:12.
Mason, G. F., Behar, K. L., and Lai, J. C. K., 1996a, *Metab. Brain Disease* **11**:283.
Mason, G. F., Pan, J. W., Chu, W. J., Zhang, Y., Khazaeli, M. B., Williams, R., Newcomer, B. D., Orr, R., Conger, K., Pohost, G. M., and Hetherington, H. P., 1996b, *Proc. Int. Soc. Magn. Reson. Med., 4th Meeting*, p. 407.
Melzer, E., and Schmidt, H.-L., 1987, *J. Biol. Chem.* **262**:8159.
Novotny, E. J., Ogino, T., Rothman, D. L., Petroff, O. A. C., Prichard, J. W., and Shulman, R. G., 1990, *Magn. Reson. Med.* **16**:431.
Novotny, E. J., Gruetter, R., Rothman, D. L., Boulware, S., Tamborlane, W. V., and Shulman, R. G., 1993, *Proc. Soc. Magn. Reson. Med., 12th Annual Meeting*, p. 324.
Ordidge, R. J., Connelly, A., and Lohman, A. B., 1986, *J. Magn. Reson.* **66**:283.
Otsuka, T., Wei, L. Bereczki, D., Acuff, V., Patlak, C., and Fenstermacher, J., 1991, *Am. J. Physiol.* **261**:R265.
Pan, J. W., Mason, G. F., Pohost, G. M., and Hetherington, H. P., 1996, *Magn. Reson. Med.* (in press).
Pardridge, W. M., and Oldendorf, W. H., 1977, *J. Neurochem.* **28**:5.

Perry, T. L., Young, V. W., Bergeron, C., Hansen, S., and Jones, K., 1987, *Ann. Neurol.* **21**:331–336.

Petroff, O. A. C., Burlina, A. P., Black, J., and Prichard, J. W., 1991, *Neurochem. Res.* **16**:1245.

Portais, J. C., Pianet, I., Allard, M., Merle, M., Raffard, G., Kien, P., Biran, M., Labouesse, J., Caille, J. M., and Canioni, P., 1991, *Biochimie* **73**:93.

Reibstein, D., den Hollander, J. A., Pilkis, S. J., and Shulman, R. G, 1986, *Biochemistry* **25**:219.

Rothman, D. L., 1987, *Application of Multipulse 1H and 13C NMR for measuring in vivo rates of metabolism*, Ph.D. dissertation, Yale University.

Rothman, D. L., Behar, K. L., Hetherington, H. P., den Hollander, J. A., Bendall, M. R., Petroff, O. A. C., and Shulman, R. G., 1985, *Proc. Natl. Acad. Sci. U.S.A.* **82**:1633.

Rothman, D.L., Howseman, Graham, G. D., Petroff, O. A. C., Lantos, G., Fayad, P. B., Brass, L. M., Shulman, G. I., Shulman, R. G., and Prichard, J. W. 1991 *Magn. Reson. Med.* **21**:302.

Rothman, D. L., Novotny, E. J., Shulman, G. I., Howseman, A. M., Petroff, O. A. C., Mason, G., Nixon, T., Hanstock, C. C., Prichard, J. W., and Shulman, R. G., 1992, *Proc. Natl. Acad. Sci. U.S.A.* **89**:9603.

Shank, R. P., Leo, G., and Zielke, H. R., 1993, *J. Neurochem.* **61**:315.

Shulman, R. G., Brown, T. R., Ugurbil, K., Ogawa, S., Cohen, S. M., and den Hollander, J. A., 1979, *Science* **205**:160.

Sibson, N. R., Dhankar, A., Mason, G. F., Behar, K. L., Rothman, D. L., and Shulman, R. G., 1996 *Proc. Int. Soc. Magn. Reson. Med., 4th Meeting*, p. 99.

Sibson, N. R., Dhankhar, A., Mason, G. F., Behar, K. L., Rothman, D. L., and Shulman, R. G., 1997, *Proc. Natl. Acad. Sci. U.S.A.* **94**:2699.

Sibson, N. R., Dhankhar, A., Mason, G. F., Behar, K. L., Rothman, D. L., and Shulman, R. G., 1998, *Proc. Natl. Acad. Sci. U.S.A.* **95**:316.

Sillerud, L. O., and Shulman, R. G., 1983, *Biochemistry* **22**:1087.

Sillerud, L. O., Alger, J. R., and Shulman, R. G., 1981, *J. Magn. Reson.* **45**:142.

Sonnewald, U., Gribbestad, I. S., Westergaard, N., Krane, J., Unsgård, G., Petersen, S. B., and Schousboe, A., 1991, *Neurosci. Lett.* **128**:235.

Sonnewald, U., Westergaard, N., Hassel, B., Müller, T. B., Unsgård, G., Fonnum, F., Hertz, L., Schousboe, A., and Petersen, S. B., 1993, *Dev. Neurosci.* **15**:351.

Tipton, P. A., and Cleland, W. W., 1988, *Biochemistry* **27**:4325.

Twieg, D. B., 1989, *Magn. Reson. Med.* **12**:64.

van Zijl, P. C. M., Chesnick, A. S., DesPres, D., Moonen, C. T. W., Ruiz-Cabello, J., and van Gelderen, P., 1993, *Magn. Reson. Med.* **30**:544.

van Zijl, P. C. M., Davis, D., Eleff, S. M., Moonen, C. T. W., Parker, R., and Strong, J., 1995, *Proc. Soc. Magn. Reson, 3rd Meeting* p. 271.

Vaughan, J. T., Hetherington, H. P., Out, J. O., Pan, J. W., and Pohost, G. M., 1994, *Magn. Reson. Med.* **32**:206.

Westergaard, N., Sonnewald, U., Unsgård, G., Peng, L., Hertz, L., and Schousboe, A., 1994, *J. Neurochem.* **62**:1727.

7

In Vivo ^{13}C NMR Spectroscopy

A Unique Approach in the Dynamic Analysis of Tricarboxylic Acid Cycle Flux and Substrate Selection

Pierre-Marie Luc Robitaille

1. INTRODUCTION

More than 50 years have past since Hans Krebs (Krebs and Johnson, 1937; Krebs ,1970) first identified the reactions which constitute the tricarboxylic acid (TCA) cycle (see Fig. 1). During this period of time, scientists have added greatly to our understanding of this intricate series of reactions. In addition to providing us with a detailed mechanistic and structural analysis of the enzymes directly associated with the TCA cycle, they have sought to determine how it interacts with associated reactions and pathways. These associated reactions are linked to the flow of substrates and oxidizable fuels into and out of this central pathway. Scientists are also working to understand the mechanism by which substrates are selected for oxidation by the cycle. In addition, they are investigating the means by which the

---

Pierre-Marie Luc Robitaille • Departments of Medical Biochemistry and Radiology, The Ohio State University, Columbus, Ohio 43210.

Biological Magnetic Resonance, Volume 15: In Vivo Carbon-13 NMR, edited by L. J. Berliner and P.-M. L. Robitaille. Kluwer Academic / Plenum Publishers, New York, 1998.

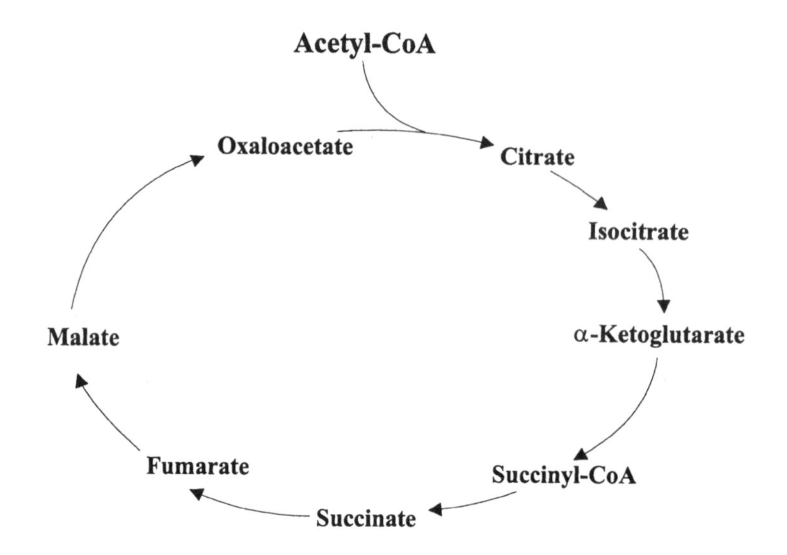

Figure 1. Schematic representation of the tricarboxylic acid cycle.

rate of this pathway is regulated, not only at the enzymatic level in the test tube, but also in the *in vivo* organ.

At the present time, it is not possible to provide a unified description of the intermediary metabolism which surrounds the TCA cycle. As such, only the most primitive knowledge exists as to how the TCA cycle interacts with other biochemical pathways. This is because scientists still do not have any real understanding of the manner in which the key pathways of intermediary metabolism act in a concerted effort to adapt to physiological and pathophysiological pressures and, in so doing, sustain life. There are several reasons for this lack of progress in the study of unified intermediary biochemistry.

The primary reason is undoubtedly our reliance on simplified models. Experimentalists are often dictating a priori the outcome of their studies (perhaps unknowingly) through the presentation of unusual experimental settings, far removed from the natural state, involving unique biochemical and physiological challenges (buffer, substrate, workstates). Clearly, the use of cuvettes and isolated tissues removes the reactions of interest from their natural environment. Consequently, we are deprived from studying the full complement of physiological and neurohumaral control mechanisms which can impact a given pathway. In addition, the transport of substrates, especially lipid substrates, into organs involves particularly complex interactions between carrier proteins and substrates (Ballard *et al.*, 1960; Fournier, 1987; Gordon and Cherkes, 1957; Shipp *et al.*, 1961; Stam *et al.*, 1987). These complex carrier trains may well become compromised in the perfused setting.

The second reason is perhaps our lack of adequate scientific methods. It is not always apparent to the biochemist that we remain, in a very real sense, technique limited. While this may seem like a poorly considered statement, it is a reality that scientists have very few tools enabling them to study many metabolites or reactions simultaneously under *in vivo* conditions. Indeed, it seems at times that we have made very little progress since the days of artrial-venous difference experiments (Bing, 1955).

It is important, however, to note that only NMR spectroscopy can be used to assay biochemical reactions and reaction rates within a given organ in its native state. No other technique can assay biochemical reactions in the whole organ *in vivo* setting! This is a truly remarkable aspect of magnetic resonance methods and herein lies the power of these approaches. Through the use of the chemical shift and of varying nuclei (^{1}H, ^{2}H, ^{3}H, ^{13}C, ^{31}P, ^{14}N, ^{15}N, and ^{17}O) the NMR spectroscopist has at his/her disposal a unique means of elucidating biochemical pathways and of probing the interplay among these reactions. Now that many spectroscopic techniques have been validated in simpler settings, spectroscopists are beginning to apply these methods to the study of biochemical reactions under *in vivo* conditions in both animals and humans. As research groups move away from the simple application of techniques to the desire to understand real biochemical problems, it is certain that NMR spectroscopy holds in store for us all a new world of biochemical discovery.

The third reason is that biochemists cannot easily resample a given biochemical system. In fact, many of the most powerful biochemical techniques are destructive in nature. In this regard, only a single probe of the system is permitted. Magnetic resonance, however, is noninvasive, permitting the repeated sampling of the same tissue under a wide variety of possible external conditions and stimuli. This once again speaks to the strength of these approaches.

It is also important that we consider the shear complexity of the problem. It is always easy to identify biochemical building blocks. The subtlety of their interactions, however, is quite another matter. The study of intermediary metabolism in a sense is entering a new period in its development. Now that we understand the structures and enzymatic steps, we are beginning to have a heightened awareness of the importance of compartmentalization. We are also beginning to address issues on a dynamic time scale. To understand the interplay between biochemical pathways, we will be forced to investigate changes dynamically.

In this work, a summary is presented of our crude attempts to increase the understanding of intermediary metabolism under the *in vivo* setting. In doing so, a commentary is first presented on the interaction between major pathways. Post-steady-state methods are then reviewed. This work is revisited in the context of the importance of these methods in linking the TCA cycle to the reactions of the malate–aspartate shuttle. In addition, we also summarize in this chapter the elegant mathematics of the post-steady-state approach. In the last section of the chapter, we briefly discuss the use of dynamic NMR methods in the analysis of substrate selection under *in vivo* conditions. We summarize our findings with acetate, pyruvate, and lactate and utilize these results to highlight our imperfect understanding of intermediary metabolism in the context of substrate selection, oxidation, and the regulation of pool sizes.

2. THE INTERACTION BETWEEN MAJOR BIOCHEMICAL PATHWAYS AND REACTIONS

The interaction between major biochemical pathways and reaction sites is a fascinating aspect of intermediary metabolism. In a concerted effort to maintain life, all cellular reactions and pathways must be fine-tuned to one another, both to avoid the unnecessary expenditure of energy and to avoid the buildup of unnecessary amounts of substrates and products. For instance, when the cell is operating under conditions of maximal oxygen consumption, it is likely that many intracellular reactions are operating near their maximal rates or $V_{max.}$ If the oxygen consumption, however, is permitted to rapidly drop to basal levels (which may only be a small fraction of V_{max}), it is evident that not all reactions will be able to suddenly move to the new steady-state levels. For instance, the TCA cycle may adjust more rapidly, while glycolysis may require some time to clear itself of the large amounts of substrates now flowing between its enzymes. The same principle would be true in the case of upward jumps in workstates. Oxidizable substrates would be required in this case which can immediately act to prime the TCA cycle, while glycolysis in turn could be permitted a longer lag time to reach the new steady state. Accordingly, in the case of upward jumps in workstates, it appears that the cell requires certain key priming substrates, while during downward shifts in workstates, the cell requires certain key dumps for flowing substrates. Further insight into this problem can be gained by examining the principles of oxygen delivery and the presence of myoglobin in the myocardial tissue.

2.1. The Myoglobin System

It is well understood that in the myocardium, oxygen is first transferred from the hemoglobin molecule to myoglobin prior to its final release to serve as an oxidizing agent in the electron transport chain. The flow of oxygen through this series of reactions can be represented as follows: ($Hb + O_2 \rightarrow HbO_2$); ($HbO_2 + Mb \rightarrow MbO_2 + Hb$); ($MbO_2 \rightarrow Mb + O_2$). If one considers only the interaction with myoglobin and neglects the hemoglobin–oxygen interaction, one can write: ($Mb + O_2 \rightleftharpoons MbO_2$). In this last reaction, MbO_2 is a dead end product and it can only react in a reversible manner to give back the initial reactants.

During steady-state operation, the simple delivery of oxygen from HbO_2 would be sufficient to meet the demands of the working myocardium. However, when the heart is required to make a rapid and significant jump in workstates, MbO_2 is present to immediately provide the required oxygen. For this purpose, the myoglobin system is an excellent example of a biochemical capacitor. It acts to store a substrate in such a way that it can provide for a sudden increase in oxygen demand. Conversely, when the myocardium jumps back down to basal workstates, the MbO_2 system could easily be replenished to prepare for the next significant increase in workloads. Note also that it is present at the juncture of two very important biochemical processes, namely, oxygen delivery and utilization.

2.2. The Lactate and Alanine Systems

In the study of intermediary metabolism, one of the most interesting junctures lies at the interface of glycolysis and the TCA cycle. Once pyruvate is produced as the aerobic end product of glycolysis, it is left to four possible fates: (1) it can enter the TCA cycle for oxidation, either directly through the synthesis of acetyl-CoA, or indirectly through carboxylation; (2) it can be carboxylated and transamitated to form cytoplasmic aspartate; (3) it can be transaminated to alanine through the action of alanine aminotransferase; and (4) it can be reduced to lactate. The lactate, alanine, and aspartate systems are particularly important because they can also represent true biochemical capacitors.

It has been taught in general biochemistry texts that the cell is making lactate, since this process enables the recovery of NAD^+ and thereby the continued operation of glycolysis under ischemic conditions. This hypothesis, however, may not be well considered. This is because such a system cannot be justified based on natural selection principles. Specifically, evolutionary pressures must exists such that a distinct advantage can be conferred by a selected pathway or reaction. For the lactate system to be created and enable a better myocardial response to ischemia, two requirements must hold true. The first is that the myocardium must be confronted with ischemic episodes. The second, that the ability to overcome this ischemic episode must allow the individual to survive, reproduce, and thereby transfer to his/her offspring this evolutionary advantage. However, while it is true that genetics is an important component of the predisposition to ischemic heart disease, the incidence of extensive heart disease in modern society is more often the result of poor diet, sedimentary lifestyles, and smoking. It is well recognized, for instance, that scientists have tremendous difficulty finding good animal models of ischemic heart disease. That is precisely because other mammals are well-conditioned, do not smoke, and have diets with lower caloric intakes. As such, other mammals are not faced with ischemic heart disease, yet they nonetheless have the ability to produce lactate. This clearly points to an alternative role for the lactic acid system. Interestingly, the increasing incidence of both ischemic heart disease and congestive heart failure in the population is also a reflection of increased longevity. Brought on by modern medical techniques, this longevity, however, is unlikely to be associated with a significant increase in reproductive advantages. Indeed, for most of the human race children are conceived relatively early in life and usually long before any bouts with ischemic heart disease. Moreover, it is also true that the ability to maintain glycolysis in the absence of oxygen through the formation of lactate presents no real advantage. This is because the myocardium requires such extensive production of energy that the left ventricle will fibrillate soon after the beginning of a significant ischemic event. In this respect, the production of lactate in itself cannot sufficiently address the needs of the distressed ischemic myocardium. Once again, this provides another argument against this role for lactic acid. The evolutionary advantage of a lactate system in the myocardium which is based on the ability to more effectively deal with ischemic episodes is simply not present.

Therefore, it is proposed that the real role of lactic acid is to act as a compensatory mechanism for differential rates between glycolysis and the TCA cycle in association with normal changes in workload. If the flow of substrate through glycolysis exceeds the maximal rate at which the electron transport chain and the TCA cycle can operate, then lactate will be produced. This does not necessarily imply that the myofibrils are ischemic. This may be the case, for instance, in an immediate drop in workload by the myocardium. Alternatively, this may also occur during a sudden change in substrate selection by the TCA cycle in such a manner that pyruvate is no longer required. If the TCA cycle slows down faster than glycolysis, it is clear that lactate would temporarily be produced until a new steady state is achieved. Conversely, during an increase in workstates, the lactic acid pool could be used to immediately prime the TCA cycle, until such time that glycolytic rates are increased and a new steady state is achieved. In this process, the production of lactic acid would play a very important role in the myocardium since, much like the myoglobin system, it could permit the heart to undergo jumps in workstates. In addition, the presence of the lactate system would also enable the TCA cycle to make an immediate change in substrates away from glycolytic end products. These may well be the reasons why this reaction is so ubiquitous in the animal kingdom. The ability to provide a rapid jump in workstates and a switch in substrates does present a strong evolutionary advantage, during flight for instance. This evolutionary advantage would exist prior to the reproductive age, and that is why this system, much like the myoglobin system, would encounter strong positive selection.

In looking at the disposal of products from glycolysis in the mammalian heart, we come to realize that there are four possible disposal mechanisms for pyruvate as stated above. In the production of acetyl-CoA, however, metabolism is disposing only of pyruvate itself, while the NADH is shuttled through the malate–aspartate shuttle to the electron transport chain. As a result, these two pathways are being fed from the production of acetyl-CoA. During the production of lactate, both pyruvate and NADH are dumped to form lactate. Note in this case that both glycolytic end products are utilized. When lactate is produced, however, this enables both the TCA cycle and the electron transport chain to derive their substrates from sources other than glycolysis.

Interestingly, alanine provides a reservoir for pyruvate, NADH, and NH_4^+. In doing so, alanine serves as a link between four important pathways, namely, glycolysis, the TCA cycle, the electron transport chain, and the urea cycle. In a sense, this amino acid can be used when the relative rates between glycolysis and the TCA cycle need to be adjusted. It should be noted, however, that in this instance alanine is also making recourse to the urea cycle. It is important to consider that the junction between glycolysis and the TCA cycle is being served in a sense by at least two separate capacitors. It may be that lactate acid, for instance, has a greater role in a downward change in TCA cycle flux and this may well explain why scientists have erroneously attributed its production to an ischemic response. Conversely, alanine appears to be a good priming substrate for the TCA cycle, and this may explain why the myocardium has such variable amounts of this substrate. Both alanine and pyruvate, nevertheless, have the ability to help adjust differences between the rates of glycolysis and the TCA cycle during both increases and decreases in workloads. It is worth noting that the TCA cycle is also linked to both malate–aspartate shuttle and to the urea cycle through its α-ketoacids (namely, α-ketoglutarate and oxaloacetate) and their transamination products (glutamate and aspartate). As such, glutamate and aspartate not only provide links to the TCA cycle through the malate–aspartate shuttle but, like fumarate, they also are important direct and indirect participants in the urea cycle (see Fig. 2). In this context, this glutamate and aspartate may be important in adjusting relative rates between these two pathways.

2.3. The Creatine Kinase System

The creatine kinase system is ubiquitous in mammals and indeed various phosphagens systems (phophocreatine, phosphoarginine, phosphoguanidotaurine, etc.) are found in much of the animal kingdom. Nonetheless, the role of these phosphagen systems is incompletely understood (Wallimann *et al.*, 1992) In the mammalian heart, scientists have hypothesized that the creatine kinase system is acting as an energy shuttle between the mitochondria and the myocin-ATPase. On the other hand, it has been proposed that the creatine kinase system helps to regulate the levels of ADP and keep these levels down. In this manner, ADP is prevented from inactivating cellular ATPases and is also permitted to participate in the control of respiration. Creatine kinase, through its phosphorylation of ADP, also acts to prevent the loss of adenine nucleotides outside the cell. It has also been proposed that the creatine kinase system is present to provide an “energy store” in the myocardium. This stored energy would be invoked during ischemia, for instance. Wallimann *et al.* (1992) have proposed that the creatine kinase system in the cell

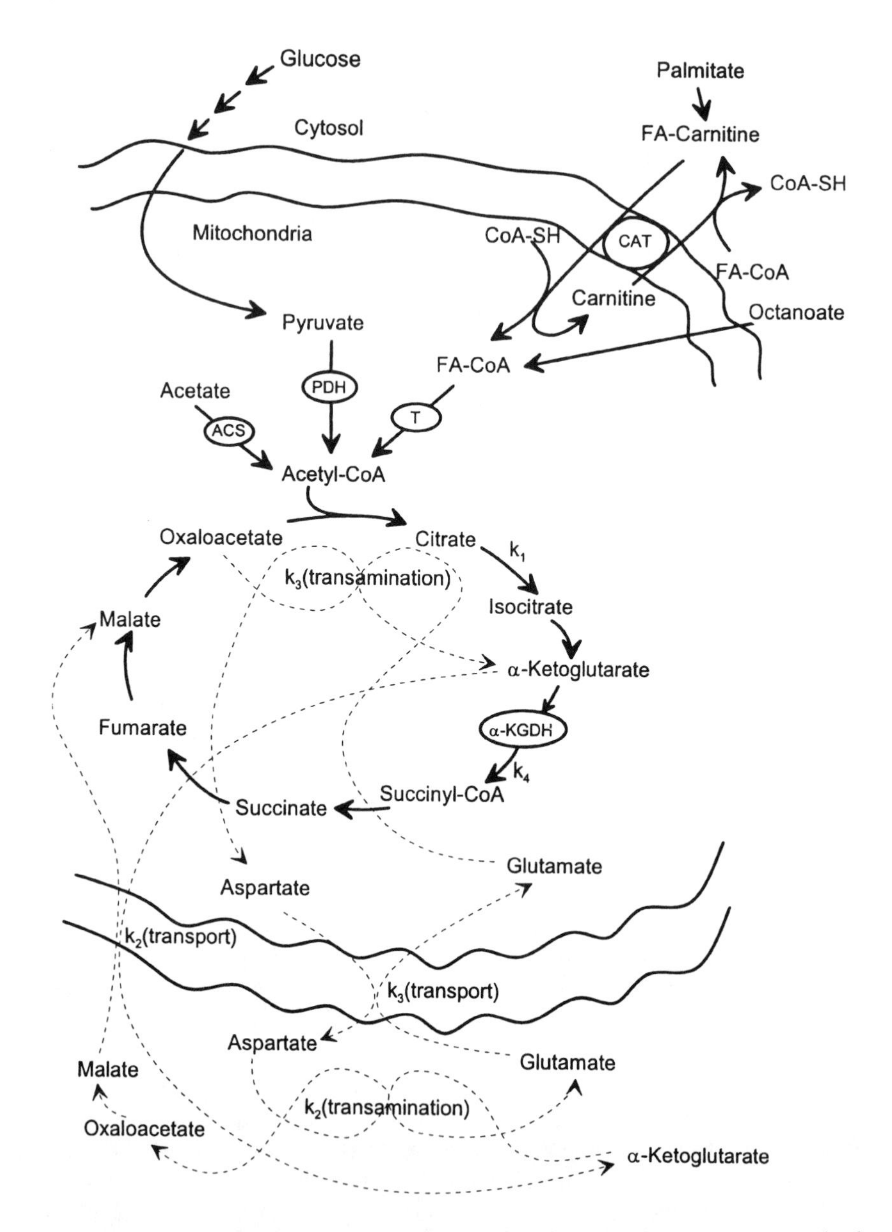

Figure 2. Schematic representation of the tricarboxylic acid cycle, the malate–aspartate shuttle, β-oxidation, glycolysis, anaplerotic and associated pathways. Labeled enzyme systems correspond to carnitine acyl transferase (CAT), thiolase (T), pyruvate dehydrogenase (PDH), acetyl Co-A synthetase (ACS), and α-ketoglutarate dehydrogenase (α-KGDH). Rate constants for transamination and transport steps of the malate–aspartate shuttle and TCA cycle flux correspond to previously applied conventions (Robitaille *et al.*, 1993b). Reproduced from Robitaille (1997).

acts as an energy circuit which is critical to the distribution of energy throughout the cell. Much of these concepts have been the subject of an excellent review (Wallimann *et al.*, 1992).

In the context of the previous discussion, however, the creatine kinase system may simply be viewed as a biological capacitor which enables rapid upward or downward changes in workload, by providing readily available substrate (in the case of the high energy bond) upon demand, and a readily available dump as well. This could easily explain why the creatine kinase system would be found at major sites of energy production (the mitochondria) and utilization (the myosin ATPase). It is perhaps for this reason that neonates have incompletely developed creatine kinase systems (precisely since their myocardium is working at a relatively fixed and elevated workload). In addition, this may help explain why creatine levels often drop in failure. This may simply be a consequence of the fact that the failing heart is no longer able to undergo a rapid increase in workload and therefore no longer has any use for this system. This may also explain why creatine phosphate levels are very low in organs which do not undergo pronounced rapid changes in work (the liver and smooth muscle system, for instance).

3. THE HISTORY OF CELLULAR ^{13}C NMR

As Hans Krebs has insightfully pointed out: "Those ignorant of the historical development of science are not likely ever to understand fully the nature of science and scientific research" (Krebs, 1970). As such, it is important to reflect on the history of ^{13}C NMR spectroscopy. While it is a tremendous task to recall the numerous studies which have molded modern ^{13}C NMR, it is perhaps educational to examine some of the seminal works in this field. Thus, the power of ^{13}C NMR for the study of intermediary metabolism was initially revealed to us by studies of the microorganism *S. cerevisiae* (Dickerson *et al.*, 1983; Den Hollander *et al.*, 1986), the isolated liver (Cohen *et al.*, 1979; Cohen *et al.*, 1981; Cohen *et al.*, 1983), and the isolated perfused heart (Bailey *et al.*, 1981; Sherry *et al.*, 1985; Lavanchy *et al.*, 1984; Hoekenga *et al.*, 1988). The laboratories of Professors Robert Shulman and George Radda were instrumental in many of these studies. These groups also realized that ^{13}C NMR could be applied to study intermediary metabolism and monitor metabolic flux with a new level of sophistication. Furthermore, Shulman and Radda noted the importance of labeling patterns within glutamic acid and other metabolites.

The next advance in the study of the TCA cycle with ^{13}C NMR methods occurred when Chance provided an elegant mathematical analysis of isotope labeling within this series of reactions (Chance *et al.*, 1983). In his work, Chance clearly recognized the usefulness of isotopomer analysis and the dynamic nature of the changes in resonance intensity. While these two findings had been recognized by others, this work provided an important transition in ^{13}C NMR since it acted to launch two distinct approaches, namely, isotopomer analysis and dynamic methods. As a result, the flow of metabolites through the major pathways of intermediary metabolism can now be monitored through the use of the stable ^{13}C isotope (Bailey *et al.*, 1981; Sherry *et al.*, 1985; Chance *et al.*, 1983; Neurohr *et al.*, 1983; London, 1988; Malloy *et al.*, 1988; Malloy *et al.*, 1990). Indeed, one of the key features of metabolic studies with NMR is the presence of metabolic isotopomers (Sherry *et al.*, 1985; Malloy *et al.*, 1988; Malloy *et al.*, 1990). Isotopomers differ from one another only in the nature of their ^{13}C enrichment. Therefore, isotopomer analysis has been applied to the study of substrate selection. Importantly, much of the progress in ^{13}C NMR can be attributed to the efforts of Malloy and Sherry whose contributions in this area have had a profound impact on the field (Sherry *et al.*, 1985; Malloy *et al.*, 1988; Malloy *et al.*, 1990). Isotopomers have also been analyzed with mass spectroscopy (Katz *et al.*, 1993). This provides an important complementary approach to ^{13}C NMR methods since many more intermediates can be detected with these methods, and since in this case isotopomer patterns can be analyzed to yield true flux measurements using mass spectroscopy (Katz *et al.*, 1993). Consequently, while studies with mass spectroscopy require tissue digestion, they are able to provide much more specific information about intermediary metabolism (Katz *et al.*, 1993).

Since ^{13}C NMR linewidths are broadened under *in vivo* conditions, isotopomer methods, which depend on high spectral resolution, cannot be simply applied in these situations. Thus, *in vivo* ^{13}C NMR studies must rely on dynamic methods; these include both pre- (Chance *et al.*, 1983; Mason *et al.*, 1992; Weiss *et al.*, 1992; Lewandowski and Hulbert, 1991; Lewandowski, 1992a; Lewandowski, 1992b; Robitaille *et al.*, 1993a) and post-steady-state (Robitaille *et al.*, 1993b) analysis. These methods extend several aspects of Chance's work (Chance *et al.*, 1993) and thereby enable the dynamic analysis of substrate selection and TCA cycle fluxes in the isolated myocardium. For instance, by applying dynamic ^{13}C NMR methods (Weiss *et al.*, 1992; Rath *et al.*, 1994; Lewandowski and Hulbert, 1991; Lewandowski, 1992a; Lewandowski, 1992b), it has been shown that the heart has unchanging fractional enrichments and glutamate pool sizes under greatly varying workloads. Dynamic methods also hold the advantage of yielding information on the TCA cycle which was independent of intermediate pool size and fractional enrichment (Rath *et al.*, 1994; Robitaille *et al.*, 1993b; Lewandowski and Hulbert, 1991; Lewandowski, 1992a; Lewandowski, 1992b).

4. POST-STEADY-STATE ANALYSIS

Several years ago, this laboratory advanced the post-steady-state analysis of the TCA cycle (Robitaille *et al.*, 1993b). The key feature of post-steady-state analysis lies in the inherent simplicity of the method, since it transforms the analysis of the TCA cycle from a cyclic problem, which cannot be solved explicitly, to a linear problem, for which a closed-form mathematical solution could be derived (Fig. 3). This is a key distinction between pre-steady-state and post-steady-state dynamic methods. Thus, in pre-steady-state methods, the experimentalist is examining the uptake of label into the glutamate pool. In this case, the label can be viewed as first entering the glutamate pool at the C4 position and completing a turn of the cycle to label either the C2 or C3 resonances of glutamate. Only following a complete turn through the TCA cycle is sampling completed. This is because the α-ketoglutarate dehydrogenase reaction, a key rate-limiting step of the TCA cycle, cannot be sampled simply by examining the pre-steady-state labeling of the C4 resonance of glutamate. In fact, this reaction is only sampled by examining the C2 or C3 labeling of glutamate. That is why some pre-steady-state methods have recourse to the ratio C2/C4 or C3/C4. By having recourse to this ratio, these methods ensure that the α-ketoglutarate reaction is indeed sampled. However, when one examines the washout of label from the C4 carbon of glutamic acid, it is clear that one is immediately sampling both the malate–aspartate shuttle activity (Robitaille *et al.*, 1993b) and the α-ketoglutarate dehydrogenase reaction. In performing ^{13}C NMR measurements of the TCA cycle, it should always be remembered that: . . . "the rate of label washout depends not only on the flux through α-ketoglutarate

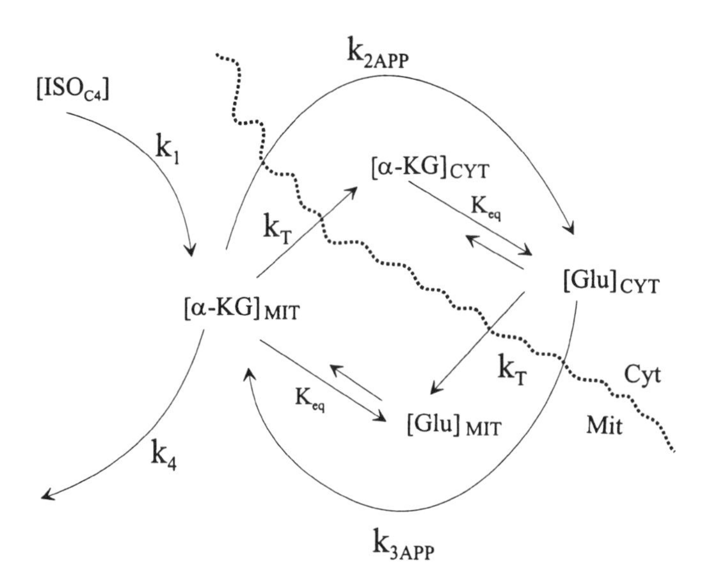

Figure 3. Isolation diagram of the TCA cycle as it relates to post-steady-state analysis. In this figure all relevant metabolite pools are indicated (cytosolic vs. mitochondrial α-ketoglutarate and glutamate) along with the relevant equilibrium constants, K_{eq}, and rate constants.

dehydrogenase, but most importantly on the activity of the malate–aspartate shuttle as determined by the forward and reverse fluxes through the transaminases and by the rate of transport of glutamate and α-ketoglutarate across the mitochondrial membrane" (Robitaille *et al.*, 1993b). This statement also applies to the label incorporation into the glutamate pool with pre-steady-state methods. In this chapter, however, focus will not be placed on the malate–aspartate shuttle (Safer, 1975; LaNoue and Schoolwerth, 1979; LaNoue *et al.*, 1974; LaNoue and Tischler, 1974; Schoolwerth *et al.*, 1982; Safer and Williamson, 1973), particularly since this topic has been so heavily addressed recently in the literature by this group and others and since it is also eloquently addressed in other chapters in this volume. Rather, we will focus on those aspects of ^{13}C NMR methodology which are truly unique to *in vivo* measurements.

To the novice in ^{13}C NMR methods, it may be somewhat confusing that pre-steady-state methods cannot be easily applied to monitor TCA cycle flux in the myocardium under *in vivo* conditions. This is because the *in vivo* myocardium maintains full latitude in substrate selection and therefore may not select the substrate which is being presented for the measurement. Thus, the heart may select a substrate for oxidation at low workloads which is specifically excluded at high work. This would prevent the analysis over the full range of workloads. This complication, however, is surmounted with post-steady-state analysis (Robitaille *et al.*, 1993b) since the labeling in this setting is always performed at low work. In this regard, one of the key incentives for advancing post-steady-state methods was that this technique had the advantage of being largely independent of substrate selection.

4.1. The Theory of Post-Steady-State Analysis

Figure 3 provides a schematic representation of the TCA cycle, as it relates to post-steady-state analysis. In this experiment (Robitaille *et al.*, 1993b), k_{2APP} and k_{3APP} are apparent rate constants which act to link the mitochondrial α-ketoglutarate pool ($[\alpha\text{-KG}]_{MIT}$) to the cytosolic glutamate pool ($[\text{GLU}]_{CYT}$). These two pools were referred to as B_o and C_o respectively in the initial presentation of the post-steady-state experiment (Robitaille *et al.*, 1993b). They are related as follows:

$$\frac{k_{2APP}}{k_{3APP}} = \frac{[\text{GLU}]_{CYT}}{[\alpha\text{-KG}]_{MIT}} \tag{1}$$

In the post-steady-state experiment, the experimentalist is monitoring the washout of the glutamate C4 label (see Fig. 3). During these studies the glutamate pool is initially labeled to steady-state enrichment levels with a ^{13}C-enriched substrate. For this task [2-^{13}C] sodium acetate is usually selected, since this avoids kinetic complications with substrate entry into the cell (such as would occur when glucose is utilized). Once a steady state in labeling is achieved, the post-steady-state experiment can be initiated by switching to unenriched substrate. When the switch to unenriched substrate is made, the level of glutamate enrichment at the C4 carbon will begin to drop, reflecting the entry of unlabeled substrate into the TCA cycle and the subsequent washout of labeled C4 carbons (see Fig. 4). The washout time constant for the glutamate C4 carbon in post-steady-state washout data (τ) can be related to kinetic parameters as follows (Robitaille *et al.*, 1993b):

$$\tau = \frac{1}{k_4}\left(\frac{k_{2APP}}{k_{3APP}} + 1\right) + \frac{1}{k_{3APP}} \tag{2}$$

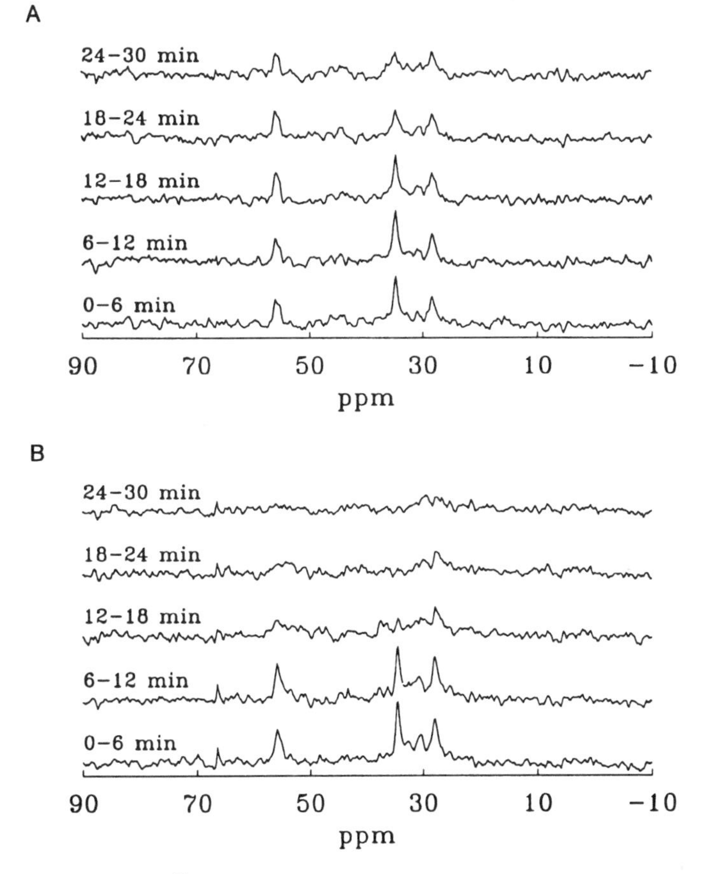

Figure 4. Post-steady-state ^{13}C NMR washout spectra obtained from a single animal using 8 mM acetate infusion at basal rate pressure products. Note the fall in the glutamate C4 resonance as the post-steady-state experiment is initiated through the infusion of unlabeled acetate (see text).

Note that in Eq. (2), the k_4 term corresponds to the rate constant for α-ketoglutarate dehydrogenase. In addition, in Eqs. (1), and (2), k_{2APP} and k_{3APP} correspond to the apparent rate constants for transamination and transport. These apparent rate constants are key components of the malate–aspartate shuttle. In this fashion, the post-steady-state treatment explicitly linked the effect of the malate–aspartate shuttle to measurements of TCA cycle flux. Importantly, in advancing this new method, it was also stated that it should be possible to monitor the effects of changing malate–aspartate shuttle activity on labeling kinetics within the glutamate pool (Robitaille *et al.*, 1993b).

When Eq. (2) was derived, no rate pressure product dependence was introduced and no approximations were made other than (1) requiring that the post-steady-state washout data could be fitted to a simple exponential, and (2) having the condition that the $[\text{GLU}]_{\text{CYT}}$ pool is much larger than the $[\alpha\text{-KG}]_{\text{MIT}}$ pool (Robitaille *et al.*, 1993b).

However, Eq. (2) can also be utilized to examine the rate pressure product (workload or oxygen consumption) dependence of the TCA cycle. Thus Eq. (2) can be re-expressed with rate pressure product dependence:

$$\tau = \frac{A}{\text{RPP}} + B \tag{3}$$

In this case, $A = (k_{2APP}/k_{3APP} + 1)/\alpha$ and $B = 1/k_{3APP}$, and a direct relationship is assumed between k_4 and rate pressure product ($k_4 = \alpha\,\text{RPP}$). It must be noted that a plot of τ vs. RPP permits us to determine both of the constants A and B. The latter is the asymptotic limit as RPP is increased. At the simplest level, the apparent rate constants k_{2APP} and k_{3APP} can be assumed to remain constant as a function of rate pressure product. However, given adequate signal to noise and sufficient data points, it should be possible to assume any dependence of these constants with rate pressure products. In this case, the experimentalist would simply make the appropriate substitution in Eq. (2). In practice, however, we have not found a need for a more complex dependence for the apparent rate constants.

Now that an exact expression for the rate of label washout from the C4 carbon of glutamate has been obtained, post-steady-state kinetics can be utilized to provide an exact expression for TCA cycle flux. Using a first-order kinetics model, the rate of the TCA cycle can be expressed as:

$$r_{\text{TCA}} = k_4\,[\alpha - \text{KG}]_{\text{MIT}} \tag{4}$$

By merging Eqs. (1), (2), and (4), we immediately obtain that:

$$r_{\text{TCA}} = \frac{k_4 k_{2APP} + k_4^2}{\tau k_4 k_{2APP} - k_{2APP}}\,[\text{GLU}]_{\text{CYT}} \tag{5}$$

As such, we now provide an exact closed-form expression for TCA cycle flux which is derived mathematically and which can be solved without recourse to numerical solutions. This exact expression can be further simplified if $k_{2APP} >> k_4$, in which case:

$$r_{TCA} \approx \frac{k_4 k_{2APP}}{\tau k_4 k_{2APP} - k_{2APP}} [GLU]_{CYT} \tag{6}$$

In addition, this solution becomes remarkably simple if $\tau\, k_4 >> 1$:

$$r_{TCA} \approx \frac{[GLU]_{CYT}}{\tau} \tag{7}$$

Equation (7) implies that TCA cycle flux can be approximated at a given workload by estimating the size of the cytosolic glutamate pool and dividing this value by the time constant for label washout from the C4 carbon of glutamic acid. In this case, one is making the assumption that $k_{2APP} >> k_4$ and $\tau\, k_4 >> 1$. Note, however, that no such assumptions were made when deriving Eqs. (2), (3), and (5).

It can also be noted that the flux through the TCA cycle is directly related to the oxygen consumption as follows:

$$r_{TCA} = (\text{F:O})(\text{MVO}) \tag{8}$$

In this equation, F:O is the ratio of TCA cycle flux to oxygen consumption, while MVO corresponds to the oxygen atom consumption. The behavior of MVO as a function of rate pressure product for the canine myocardium is given approximately by $\text{MVO} \approx \text{MVO}_0\,(\text{RPP}) \approx [(2.30 \pm 0.08) \times 10^{-3}](\text{RPP})$ (Brown *et al.*, 1988). In this case, the MVO_0 value, in units of μmol/gdw mmHg, was established by finding the best linear fit for the MVO_2 vs. RPP data presented by Brown *et al.* (1988).

Equation (8) also makes recourse to the F:O ratio. However, an understanding of F:O in the heart is a complex problem since it is linked to the nature of the oxidized substrate (Bing, 1955; Opie, 1991; Kobayashi and Neely, 1979; Randle *et al.*, 1970). Thus, the F:O in an acetate perfused heart should be 0.25, while when the perfusion is accomplished with butyrate the value should be about 0.2 (Bing, 1955; Opie, 1991; Kobayashi and Neely, 1979; Randle *et al.*, 1970). It has also been determined that the minimum F:O value for a glucose perfused heart should be 0.167 (Malloy *et al.*, 1996).

Post-steady-state methods can also be used to obtain absolute values for k_{2APP}, k_{3APP}, α, $[\alpha\text{-KG}]_{MIT}$, and $[GLU]_{CYT}$ from Eqs. (1), (2), (4), and (8). As previously stated, these four independent equations were obtained from an understanding of (1) post-steady-state washout kinetics (Robitaille *et al.*, 1993b), (2) the value of F:O (Bing, 1955; Opie, 1991; Kobayashi and Neely, 1979; Randle *et al.*, 1970; Malloy *et al.*, 1996), and (3) the myocardial oxygen consumptions in the canine heart (Brown *et al.*, 1988). Knowledge of the behavior of τ over a wide range of workloads as given by Eq. (2), however, provides both the behavior of k_4 and the ratio of k_{2APP}/k_{3APP} as a function of increasing oxygen consumption. In this case, the following relationships also hold:

$$[\alpha\text{-KG}]_{MIT} = (A)(F{:}O)(MVO)_0 - [GLU]_{CYT} \tag{9}$$

$$k_{2APP} = \frac{[GLU]_{CYT}(k_{3APP})}{(A)(F{:}O)(MVO)_0 - [GLU]_{CYT}} \tag{10}$$

$$\alpha = \frac{(F{:}O)\,(MVO)_0}{(A)(F{:}O)(MVO)_0 - [GLU]_{CYT}} \tag{11}$$

From these equations we can begin to probe the dependence of post-steady-state kinetic parameters and pools sizes on factors such as substrate selection and oxygen consumption. This represents a further advantage of post-steady-state mathematics, since the simple form of the post-steady-state equations now enables us to link kinetic parameters to intracellular and intramitochondrial pool sizes for the first time.

5. *IN VIVO* ^{13}C NMR ANALYSIS OF SUBSTRATE SELECTION

It is always essential to realize that there is an important difference between substrate selection and TCA cycle flux measurements. In this regard, substrate selection refers to the substrate that the TCA cycle is actually utilizing to obtain its reduction equivalents. Conversely, when we speak of TCA cycle flux, we are referring directly to the rate at which the cycle is operating. While substrate selection and TCA cycle flux may be inherently related, they are thus clearly not the same entity. When ^{13}C NMR-based methods are utilized to study substrate selection, the experimentalist is often probing the ability of one substrate (usually labeled and exogenous) to compete for oxidation in the TCA cycle in the presence of all other substrates (usually unlabeled and endogenous).

Recall from above that pre-steady-state methods cannot be easily applied to monitor TCA cycle flux in the *in vivo* myocardium. This is because these methods have a strong dependence on the nature of the oxidized substrate as a function of workload. Interestingly, it is precisely this aspect of pre-steady-state methods that makes them so important in examining changes in substrate selection under *in vivo* conditions. Therefore, in acquiring data on substrate selection based on the pre-steady-state protocol (Robitaille *et al.*, 1993a), a set of control spectra is typically acquired for approximately 40 minutes with the infusion of unenriched substrate. When these experiments are performed in the canine myocardium, the label is typically infused through an angiocatheter placed into the proximal left anterior descending coronary artery of the heart. This is typically accomplished under basal conditions with a cardiac rate pressure product of ≤10,000 mmHg/min. The use of a 40-minute equilibration period helps ensure that the myocardium has reached metabolic steady-state. Once equilibration is completed, the experimentalist then changes the infused substrate to its ^{13}C-enriched form. Acquisition is then resumed for an additional 40 minutes. The presence of enrichment at low rate pressure products is recorded once the label has reached steady-state levels of enrichment. Once a high quality steady-state spectrum is obtained, the workload of the heart is raised, typically using a dopamine drip and the substrate changed to its unlabeled form. This permits the label to fully wash out of the myocardium. At this stage, no labeling can be observed. Following an additional 40 minutes of equilibration, the infused substrate will be changed to its enriched form. However, in this case, the extent of infusion of enriched substrate is increased in a manner which corresponds to the relative increase in workloads. Acquisition is once again resumed for 40 minutes and the extent of glutamate, aspartate, or other pool labeling is compared with results obtained under basal workload conditions. In this manner, one can analyze in a single heart the differences in metabolite labeling which occur simply as a function of increasing cardiac workloads. These changes, in turn, can be linked either to changing metabolic pool sizes or to actual changes in substrate selection. It is important to note that once the *in vivo* experiments are completed that a section of the myocardium is typically excised and subjected to fractional enrichment measurements. These measurements are based on isotopomer analysis and are performed in extracts. Such fractional enrichment measurements are a critical step in confirming the *in vivo* results.

Using pre-steady-state analysis we have been able to examine the selection of acetate by the *in vivo* myocardium as a function of increased work (Robitaille *et al.*, 1993a). Surprisingly, we have found that the *in vivo* myocardium acts to exclude acetate oxidation when workloads are increased (Fig. 5). Thus we observed that the myocardium usually selected in favor of the oxidation of acetate at low workloads. This was confirmed by extensive labeling of the glutamate pool under these conditions. However, when workloads were increased to very high levels, acetate oxidation was specifically excluded, and little if any labeling of the glutamate pool occurred as evidenced by the absence of signal in the NMR spectra (Robitaille *et al.*, 1993a). This was a most remarkable finding (see Fig. 5). When examining extract data we were able to conclude that while canine hearts can have up to 90% fractional enrichment of glutamate at a rate pressure product of 10,000 or less, they will have absolutely no enrichment above 25,000 (Robitaille *et al.*, 1993a). As a result over less than a threefold change in rate pressure products, the fractional enrichment drops to zero, not to 30%. Yet, this happened despite the fact that the acetate infusion rate had been tripled. Theoretically, therefore, we should have observed little or no change in fractional enrichments (Robitaille *et al.*, 1993a). We

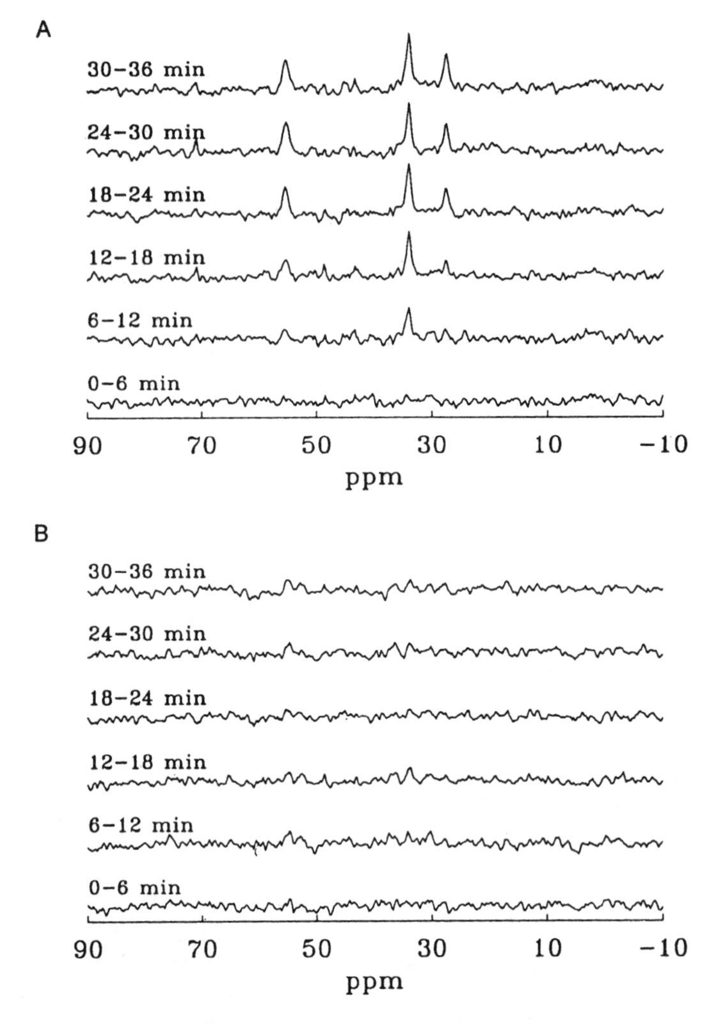

Figure 5. Pre-steady-state ^{13}C NMR difference spectra acquired from the *in vivo* canine myocardium at rate pressure products of 10,800 mmHg/min (A) and 24,400 mmHg/min (B) during infusion of 4 mM [2-^{13}C]sodium acetate. Resonances correspond to the C2, C4, and C3 carbons of glutamate (55.5, 34.2, and 27.6 ppm respectively). Note the absence of glutamate pool labeling at the higher RPP (B). Reproduced from Robitaille *et al.* (1993a).

have now performed extensive extract analysis of glutamate pool sizes as a function of increasing rate pressure products and have found that this pool remains relatively constant as a function of workload. Therefore, these results cannot be explained by a drop in glutamate pool size (Rath *et al.*, 1994). Thus, over a nearly eightfold change in RPP (5,000–40,000) there was about a 25% change in the mean glutamate pool. Studies in the perfused heart also revealed that glutamate pools do not change with workload (Lewandowski, 1992a,b). It is also worth noting that in some instances the infusion of acetate was insufficient to label the glutamate pool even under basal conditions. This was indeed unexpected and brings into question many conclusions reached on the sole basis of acetate exclusion. This, of course, has implications with respect to viability results obtained with PET. It is also interesting that, when studying acetate, the use of this substrate failed to result in any *in vivo* labeling of the aspartate pool either at low workloads or at intermediate rate pressure products. However, at low rate pressure products, in most cases, acetate was able to label the glutamate pool fully at C2, C3, and C4. This implies, by extension, that oxaloacetate, malate, and fumarate are also being labeled. Since aspartate is a transamination product of oxaloacetate, it is truly surprising that aspartate labeling *in vivo* was not observed when acetate was utilized as a substrate. This is noteworthy also since aspartate is strongly labeled by pyruvate and lactate at intermediate rate pressure products, as will be discussed below.

The fact that the left ventricle excludes sodium acetate from oxidation has also been confirmed in the swine (Angelos *et al.*, 1996). In one animal, for example (Angelos *et al.*, 1996), the glutamate pool was strongly enriched by the infusion of 8 mM acetate at a rate pressure product of about 9,000. When we attempted to repeat the experiment in the same animal at a rate pressure product of 14,000, however (while infusing 20 mM acetate), there was almost complete absence of labeling. In this experiment, however, we are examining only a 50% increase in RPP in the presence of a 2.5-fold increase in acetate infusion. This was a truly remarkable example of the strong selection against acetate.

We have also performed extensive studies of substrate selection as a function of workload with both lactate and pyruvate. The behavior of these two substrates was rather similar in that they were typically able to label the glutamate, aspartate, and alanine pools at basal workloads (Rath *et al.*, 1997). However, in these studies, the extent of this labeling was highly variable under low workloads (Fig. 6). The extent of this labeling nonetheless could usually be increased with the addition of dichloroacetic acid, a well-known indirect activator of pyruvate dehydrogenase, through its action on pyruvate dehydrogenase kinase. Once again, it is important to note that in some animals it was not possible to label the glutamate pool with pyruvic acid even at the lowest workloads. In some cases, this occurred in spite of the presence of DCA (Rath *et al.*, 1997). Nonetheless, as a general rule, the addition of DCA to the infusate resulted in increased labeling of the glutamate pool. This is a clear indication that PDH was at least partially inactive in these animals and that DCA was successful, through the action of pyruvate dehydrogenase kinase, in increasing the activity of this key enzyme. In this work, it was also found that pyruvate could not label the glutamate pool at the most elevated workloads, in a manner which is quite reminiscent of acetate. The same conclusion was also reached

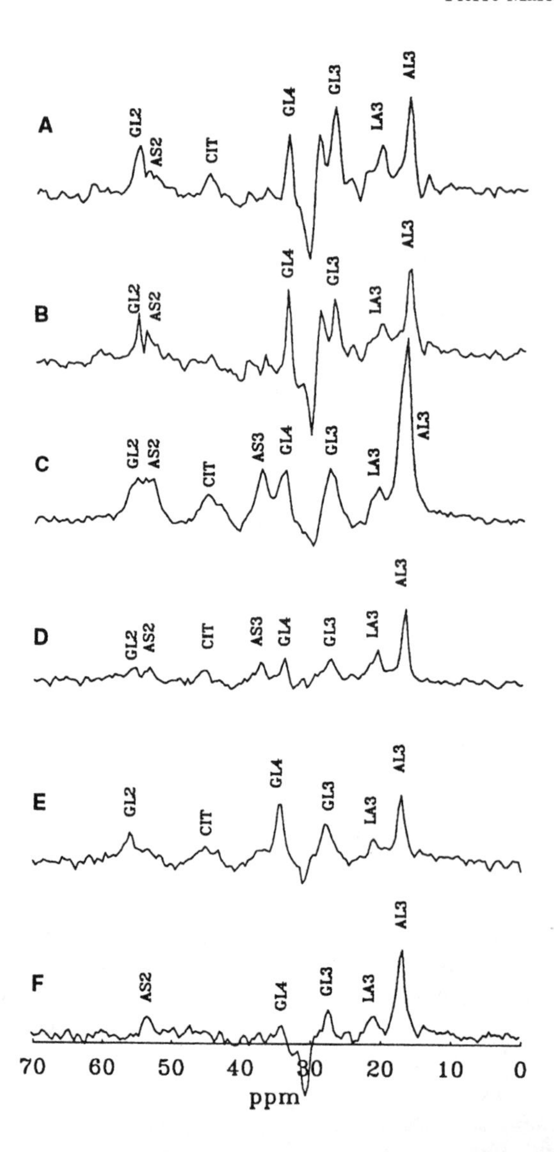

Figure 6. Pre-steady-state ^{13}C NMR difference spectra acquired from the *in vivo* canine myocardium at low (5,000–12,000 mmHg/min) rate pressure products (A–F) during infusion of [3-^{13}C]sodium pyruvate in the absence of pyruvate dehydrogenase (PDH) activation with dichloroacetate (DCA). With the exception of spectrum (F), which was obtained with 4 mM pyruvate infusion, all other spectra were obtained at 8 mM infusion. Nonetheless, it is clear that sufficient pyruvate was provided in (F) by examining the alanine resonance. Resonances correspond to the C2, C4, and C3 carbons of glutamate (GL2, GL4, and GL3/PY3, respectively), the C2 and C3 carbons of aspartate (AS2 and AS3), the C2 and C4 carbons of citrate (CIT), the C3 carbons of pyruvate, lactate, and alanine (GL3/PY3, LA3, and AL3). Note the highly variable enrichments of these metabolites. Reproduced from Rath *et al.* (1997).

for lactate. It should be mentioned that the selection against these substrates at very high rate pressure products could not be removed even in the presence of DCA.

6. CONCLUSIONS

While it is surely fascinating that NMR, for the first time, is permitting us to examine substrate selection in the same organ in real time and as a function of various physiological interventions, we are actually only beginning to probe this exciting new branch of ^{13}C NMR. Much remains to be answered. For instance, one will recall that acetate, without exception, was unable to label the aspartate pool in the *in vivo* spectra. Yet, acetate is clearly oxidized and able to label glutamate at basal workloads. So, if aspartate is truly in equilibrium with mitochondrial oxaloacetate, why was this labeling not observed? Yet, aspartate was usually labeled when pyruvate or lactate was utilized as a substrate. Furthermore, when the heart was taken to intermediate rate pressure products with pyruvate and aspartate, the glutamate pool enrichment decreased while the aspartate resonance intensity increased (see Fig. 7). This enrichment of the aspartate pool is unlikely to be coming from label which has entered the TCA cycle. If this was the case, the glutamate enrichment in fact should not have dropped.

Still, since the glutamate pool is not labeled with acetate at high workloads (see Fig. 8), it could be argued that the exchange between the glutamate pool and the ketoglutarate pool is reduced as rate pressure products are increased. We have evidence from post-steady-state kinetics, however, that this is not the case. Namely, we are able to observe rapid label washout from the glutamate pool in all cases. It is of significant consequence that if the washout of the glutamate pool is still observable at elevated rate pressure products, then the inflow of label into this pool should not be kinetically limited. Otherwise, the intensity of the glutamate pool would drop with increasing rate pressure products. This was not observed.

Perhaps the labeling of the aspartate pool is actually telling us something about the interface of major pathways. It appears that when rate pressure products are increased, the myocardium is making a switch away from pyruvate for oxidation in the TCA cycle. At the instant that the switch is being made, the heart must find a dump for this pyruvate. It is likely that one possible dump involves carboxylation of pyruvate to ultimately result in the synthesis of cytosolic oxaloacetate which is then transaminated to aspartate. Through this mechanism, the cytosolic aspartate pool would be increased. However, it is difficult to reconcile this mechanism with the isometric labeling of the aspartate C2 and C3 resonances which was observed under *in vivo* conditions. There is still a great deal that ^{13}C NMR methods could contribute to our understanding of substrate selection with ^{13}C NMR methods. Perhaps the key to much of this lies in the application of simultaneous ^{13}C NMR and ^{14}C radiotracer methods. Nonetheless, it is clear that the application of ^{13}C

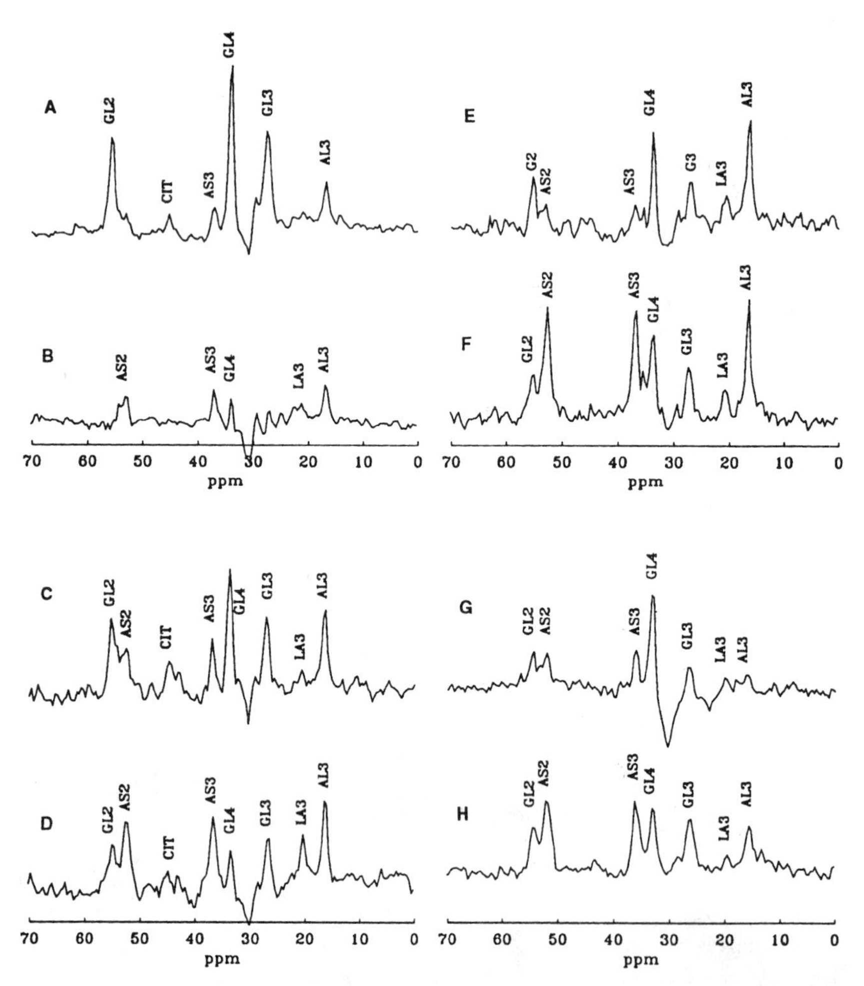

Figure 7. Pre-steady-state ^{13}C NMR difference spectra acquired from the *in vivo* canine myocardium in four animals (A, B; C, D; E, F; G, H) at low (A, C, E, G) and intermediate (B, D, F, H) rate pressure products. Spectra were acquired with infusion of 8 mM [3-^{13}C]sodium pyruvate and 5 mM DCA at low rate pressure products. In contrast, 8 mM (B) or 20 mM [3-^{13}C]sodium pyruvate (D, F, H) and 5 mM (B), 10 mM (D), or 15 mM DCA (F, H) was infused at intermediate workloads. Note the decrease in glutamate resonance intensities in response to the increased work, and the accompanying increase in aspartate resonance intensities. Importantly, the alanine resonance intensity was constant in three of these four cases, and in one case (G, H) it increased with increasing rate pressure products. It is also interesting to note the change in an unenriched endogenous metabolite at 30.8 ppm observed in (G) and (H). This decreasing endogenous resonance probably corresponds to a triglyceride $-CH_2-$ resonance (Bailey *et al.*, 1981). Reproduced from Rath *et al.* (1997).

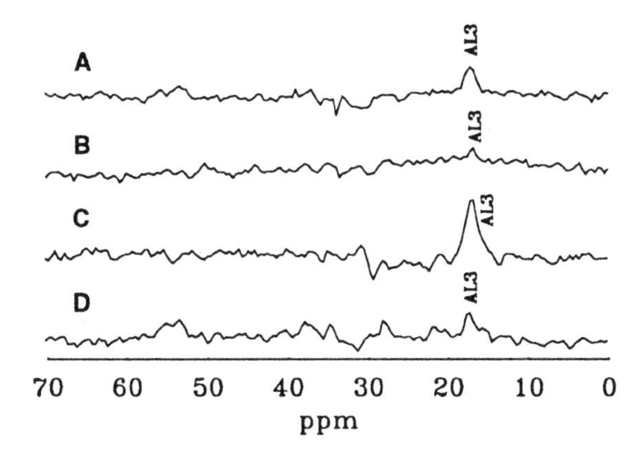

Figure 8. Pre-steady-state ^{13}C NMR difference spectra acquired from the *in vivo* canine myocardium at elevated rate pressure products (18,000–32000 mmHg/min) during infusion of 8 mM [3-^{13}C]sodium pyruvate. All of the cases displayed pyruvate uptake and oxidation at low rate pressure products. Note the absence of glutamate pool labeling, despite clear alanine enrichment. Very slight aspartate enrichment is also noted (D). Reproduced from Rath *et al.* (1997).

NMR methods in the *in vivo* myocardium has brought with it not only powerful new techniques (Robitaille *et al.*, 1993a,b; Rath *et al.*, 1997), but perhaps, more interestingly, fascinating new questions.

ACKNOWLEDGMENTS. This work was supported in part by NIH grant #HL-45120.

I would like to take this opportunity to thank the many members of the Division of Magnetic Resonance at The Ohio State University who have made this work possible and recognize the contribution of Mr. Ryan Augé in the preparation of Fig. 2. This work is dedicated to the memory of Jacqueline Alice Roy.

REFERENCES

Angelos, M. G., Little, C. M., Torres, C. A. A., Rath, D. P., Zhu, H., Tong, X., Jiang, Z., and Robitaille, P. M. L., 1996, *Acad. Emerg. Med.* **3**:A3.

Bailey, I. A., Gadian, D. G., Matthews, P. M., Radda, G. K., and Seeley, P. J., 1981, *FEBS Lett.* **123**:315.

Ballard, F. B., Danforth, W. H., Naegle, S., and Bing, R. J., 1960, *J. Clin. Invest.* **39**:717.

Bing, R. J., 1955, *Harvey Lect.* **50**:27.

Brown, M. A., Myears, D. W., and Bergman, S. R., 1988, *J. Am. Coll. Cardiol.* **12**:1054.

Chance, E. M., Seeholzer, S. H., Kobayashi, K., and Williamson, J. R., 1983, *J. Biol. Chem.* 258:13785.

Cohen, S. M., 1983, *J. Biol. Chem.* **258**:14291.

Cohen, S. M., Shulman, R. G., and McLaughlin, A. C., 1979, *Proc. Natl. Acad. Sci.* U.S.A. **76**:4808.

Cohen, S. M., Rognstad, R., Shulman, R. G., and Katz, J., 1981. *J. Biol. Chem.* **256**:3428.

Den Hollander, J. A., Ugurbil, K., Brown, T. R., Bednar, M., Redfield, C., and Shulman, R. G., 1986, *Biochemistry* **25**:203.
Dickerson, J. R., Dawes, I. W., Boyd, A. S. F., Baxter, R. L., 1983, *Proc. Natl. Acad. Sci. U.S.A.* **80**:5847.
Fournier, N. C., 1987, *Basic Res. Cardiol.* **82**:11.
Gordon, R. S., and Cherkes, A., 1957, *J. Clin. Invest.* **36**:810.
Hoekenga, D. E., Brainard, J. R., and Hutson, J. Y., 1988, *Circ. Res.* **62**:1065.
Katz, J., Wals, P., and Lee, W. N. P., 1993, *J. Biol. Chem.* **268**:25509.
Kobayashi, K., and Neely, J. R., 1979, *Circ. Res.* **44**:166.
Krebs, H. A., 1970, *Perspect. Biol. Med.* **14**:154.
Krebs, H. A., and Johnson, W. A., 1937, *Enzymologia* **4**:148.
LaNoue, K. F., and Tischler, M. E., 1974, *J. Biol. Chem.* **249**:7522.
LaNoue, K. F., and Schoolwerth, A. C., 1979, *Annu. Rev. Biochem.* **48**:871.
LaNoue, K. F., Bryla, J., and Bassett, D. J. P., 1974, *J. Biol. Chem.* **249**:7514.
Lavanchy, N., Martin, J., and Rossi, A., 1984, *FEBS Lett.* **178**:34.
Lewandowski, E. D.,1992a, *Circ. Res.* **70**:576.
Lewandowski, E. D., 1992b, *Biochemistry* **31**:8916.
Lewandowski, E. D., and Hulbert, C., 1991, *Magn. Reson. Med.* **19**:186.
London, R. E., 1988, *Prog. NMR Spectrosc.* **20**:337.
Malloy, C. R., Sherry, A. D., and Jeffrey, F. M. H., 1988, *J. Biol. Chem.* **263**:6964.
Malloy, C. R., Sherry, A. D., and Jeffrey, F. M. H., 1990, *Am. J. Physiol.* **259**:H987.
Malloy, C. R., Jones, J. G., Jeffrey, F. M., Jessen, M. E. and Sherry, A. D., 1996, *MAGMA* **4**:35.
Mason, G. F., Rothman, D. L., Behar, K. L., and Shulman, R. G., 1992, *J. Cereb. Blood Flow Metab.* **12**:434.
Neurohr, K. J., Barrett, E. J., and Shulman, R. G., 1983, *Proc. Natl. Acad. Sci. U.S.A.* **80**:1603.
Opie, L. H., 1991, *The Heart: Physiology and Metabolism*, Raven Press, New York, pp. 208–246.
Randle, P. J., England P. J., and Denton, R. M., 1970, *Biochem. J.* **117**:677.
Rath, D. P., Zhang, H., Jiang, Z., Hamlin, R. L., and Robitaille, P. -M. L., 1994, *Proc. Soc. Magn. Reson.* **2**:1231.
Rath, D. P., Zhu, H., Tong, X., Jiang, Z., Hamlin, R. L., and Robitaille, P. M. L., 1997, *Magn. Reson. Med.* **38**:in press.
Robitaille, P. -M. L., 1997, *Current and Future Applications of Magnetic Resonance in Cardiovascular Disease*, Futura Publ.:in press.
Robitaille, P. -M. L., Rath, D. P., Abduljalil, A. M., O'Donnell, J. M., Jiang, Z., Zhang, H., and Hamlin, R. L., 1993a, *J. Biol. Chem.* **268**:26296.
Robitaille, P. -M. L., Rath, D. P., Skinner, T. E., Abduljalil, A. M., and Hamlin, R. L., 1993b, *Magn. Reson. Med.* **30**:262.
Safer, B., 1975, *Circ. Res.* **37**:527.
Safer, B., and Williamson, J. R., 1973, *J. Biol. Chem.* **248**:2570.
Schoolwerth, A. C., LaNoue, K. F., and Hoover, W. J., 1982, *J. Biol. Chem.* **258**:1735.
Sherry, A. D., Nunnally, R. L., and Peshock, R. M., 1985, *J. Biol. Chem.* **260**:9272.
Shipp, J. C., Opie, L. H., and Challoner, D. C., 1961, *Nature* **189**:1018.
Stam, H., Schroonderwoerd, K. and Hulsmann, W. C., 1987, *Basic Res. Cardiol.* **82**:19.
Wallimann, T., Wyss, M., Brdiczka, D., Nicolay, K., and Eppenberger, H. M., 1992, *Biochem. J.* **281**:21.
Weiss, R. G., Gloth, S. T., Kalil-Filho, R., Chacko, V. P., Stern, M. D., and Gerstenblith, G., 1992, *Circ. Res.* **70**:392.

Contents of Previous Volumes

VOLUME 1

VOLUME 6

VOLUME 7

Index